SAP PRESS e-books

Print or e-book, Kindle or iPad, workplace or airplane: Choose where and how to read your SAP PRESS books! You can now get all our titles as e-books, too:

- By download and online access
- For all popular devices
- And, of course, DRM-free

Convinced? Then go to www.sap-press.com and get your e-book today.

Object-Oriented Programming with ABAP® Objects

SAP PRESS

SAP PRESS is a joint initiative of SAP and Rheinwerk Publishing. The know-how offered by SAP specialists combined with the expertise of Rheinwerk Publishing offers the reader expert books in the field. SAP PRESS features first-hand information and expert advice, and provides useful skills for professional decision-making.

SAP PRESS offers a variety of books on technical and business-related topics for the SAP user. For further information, please visit our website: *www.sap-press.com*.

Jeffrey Boggess, Colby Hemond, James Wood, Joseph Rupert

Object-Oriented Programming with ABAP® Objects

Rheinwerk
Publishing

Editor Megan Fuerst
Copyeditor Rachel Gibson
Cover Design Graham Geary
Photo Credit iStockphoto: 1316648821/© matejmo
Layout Design Vera Brauner
Production Hannah Lane
Typesetting III-satz, Germany
Printed and bound in the United States of America, on paper from sustainable sources

ISBN 978-1-4932-2714-3
3rd edition 2026

© 2026 by:
Rheinwerk Publishing, Inc.
2 Heritage Drive, Suite 305
Quincy, MA 02171
USA
info@rheinwerk-publishing.com
+1.781.228.5070

Represented in the E.U. by:
Rheinwerk Verlag GmbH
Rheinwerkallee 4
53227 Bonn
Germany
service@rheinwerk-verlag.de
+49 (0) 228 42150-0

Library of Congress Cataloging-in-Publication Control Number: 2025045969

Contents at a Glance

Contents

4   Object Initialization and Cleanup

5   Inheritance and Composition

6   Polymorphism

7   Component-Based Design Concepts

8   Error Handling with Exception Classes   239

9   Unit Tests with ABAP Unit

12   ABAP Cloud

13   Best Practices and Design Patterns

Appendices 427

Preface

It's been more than seventeen years since the first edition of this book was published, and it's amazing how much SAP technology adoption we've seen during that timeframe. Indeed, in the ten years that have passed since we wrote the second edition of this book, we saw the introduction of both SAP S/4HANA and SAP Business Technology Platform (SAP BTP). Most recently, the launch of the SAP BTP ABAP environment has allowed developers to take their ABAP talents to the cloud.

While these are exciting developments, the core concepts of object-oriented programming (OOP) haven't really changed. What *has* changed is the widespread use of ABAP Objects across an ever-expanding SAP landscape. Knowledge of ABAP Objects used to be seen as a "nice-to-have," but now it's an essential part of every ABAP developer's toolbox. So, once again, we felt the time was right to take a fresh look at OOP with ABAP Objects.

Our goal with this revised and expanded edition is the same as the first edition: to teach you how to *think* about writing ABAP-based software from an object-oriented perspective. While it takes a fair amount of time and effort to embrace this way of thinking, we believe you'll find the investment worthwhile from both a personal and professional standpoint.

Target Group and Prerequisites

This book is for ABAP developers who have basic experience with writing ABAP programs. While it presents most concepts with background information for developers who are new to OOP, this book is not a comprehensive introduction to ABAP. If you haven't worked with ABAP before, we recommend that you take a training course or read a beginner-friendly book such as *ABAP: An Introduction* (SAP PRESS, 2020) before proceeding. Aside from that, no prior OOP experience is expected.

The majority of the tools and features described in this book are supported on almost any application server (AS) ABAP system. Since the object-oriented extensions to ABAP (i.e., the *Objects* part of ABAP Objects) were made available in SAP R/3 4.6C in 1999, it's rare to find a system that doesn't support object-oriented ABAP these days. Still, we'll point out release-specific features and additions where appropriate.

Structure of the Book

As we put together this edition, we aimed to strike a balance between theoretical concepts and practical examples demonstrating the use of ABAP Objects in real-world scenarios. This approach led us to organize the book as follows:

- First, we introduce you to core OOP concepts and the ABAP Objects functionality and syntax that corresponds with those concepts.

- Second, we build on this knowledge by presenting case studies that demonstrate how you can use ABAP Objects in practical development scenarios.

At the end of most chapters, we include brief tutorials on the *Unified Modeling Language* (UML). These tutorials show you how to describe your OOP designs with a graphical notation that's commonly used throughout the software industry.

Despite this logical approach, we let the code do most of the talking throughout the book, so that you don't feel like you're reading a college textbook on OOP. Let's take a closer look at what we'll be covering:

- **Chapter 1: Introduction to Object-Oriented Programming**
  In this chapter, we set the stage for the rest of book by introducing you to object-oriented programming and its basic concepts.

- **Chapter 2: Working with Objects**
  In Chapter 2, you'll start getting your hands dirty with classes in ABAP Objects. We explore ABAP Objects syntax for defining classes and methods. Along the way, we introduce the core development tools you'll use to develop classes throughout the book: the classic ABAP Workbench and the newer ABAP development tools.

- **Chapter 3: Encapsulation and Implementation Hiding**
  Though inheritance usually steals the spotlight, we could argue that encapsulation is the most important aspect of OOP. In this chapter, we review encapsulation concepts by comparing the object-oriented development approach with the procedural approach. We also introduce the important topic of *component visibilities* in this chapter.

- **Chapter 4: Object Initialization and Cleanup**
  In Chapter 4, we explore the lifecycle of objects, from the point they're created to the point they're removed by the garbage collector service. As you make your way through the chapter, you'll learn about constructor methods and other ways of influencing the object lifecycle.

- **Chapter 5: Inheritance and Composition**
  This chapter looks at two common ways of achieving code reuse in OOP: *inheritance* and *composition*. You'll discover how these techniques, when coupled with the encapsulation techniques introduced in Chapter 3, allow you to extend your code libraries in ways you never thought possible.

- **Chapter 6: Polymorphism**
  In this chapter, you'll learn how to use inheritance relationships to build solutions that rely on interchangeable parts. We also introduce *interfaces*, another key OOP concept.

- **Chapter 7: Component-Based Design Concepts**
  While the previous chapters looked at classes under a microscope, this chapter

broadens its focus by showing you how to use ABAP packages to organize class libraries into coarse-grained development components.

- **Chapter 8: Error Handling with Exception Classes**
  This chapter explains how to deal with exception situations in ABAP using the class-based exception handling concept.

- **Chapter 9: Unit Tests with ABAP Unit**
  Chapter 9 shows you how to develop automated unit tests using the ABAP Unit test framework. This expanded chapter also introduces modern ABAP features, including test seams and test doubles for SAP HANA and core data services (CDS). These automated testing tools help you ensure that your classes deliver on the functionality specified in their application programming interface (API) contracts.

- **Chapter 10: Business Object Development with BOPF**
  Chapter 10 introduces SAP's *Business Object Processing Framework* (BOPF). In this chapter, you'll see basic object-oriented principles applied on a macro scale toward the development of reusable business objects. Internally, there are lots of object-oriented concepts on display—both within the business objects themselves and in the generic API used to interface with the business objects.

- **Chapter 11: ABAP RESTful Application Programming Model**
  In Chapter 11, we explore SAP's ABAP RESTful application programming model. This model blends the declarative elements of the BOPF with ABAP Objects to create a new and effective paradigm for API development with OData.

- **Chapter 12: ABAP Cloud**
  In Chapter 12, we boldly go where few ABAP developers have gone before by exploring ABAP development in the cloud using the SAP BTP ABAP environment. Here, you'll find a fascinating look into the future of ABAP development.

- **Chapter 13: Best Practices and Design Patterns**
  In this final chapter, we'll take stock of what you learned and review some best practices and common design patterns that you can use to take your OOP skills to the next level.

Conventions

This book contains code examples that demonstrate concepts such as syntax and functionality. To distinguish these examples and improve code readability, we use a monospaced font like the one used by many integrated development environments (IDEs):

```
CLASS lcl_test DEFINITION.
  PUBLIC SECTION.

    ...
ENDCLASS.
```

As new syntax is introduced, we'll highlight the syntax using a bold listing font as demonstrated in the previous excerpt with the highlighted PUBLIC SECTION statement. The use of the ellipsis indicates that irrelevant portions of the code were omitted for brevity's sake.

Source Code and Examples

As noted in the previous section, this book includes code examples that demonstrate the use of ABAP Objects in different settings. Though many of these examples are small and self-contained, we have included the more involved examples in a source code bundle that you can download from the book's webpage (*www.sap-press.com/6093*). Most of the code is text that you can easily paste into your preferred ABAP editor tool and test out for yourself. For more complex development objects, we provide detailed instructions for installation and testing.

You can work with these examples in any ABAP system by simply creating the relevant objects in your local development package (i.e., the $TMP package). If you don't have access to an ABAP system, you can use *SAP Cloud Appliance Library* (*https://cal.sap.com/*) to set up a trial instance on your preferred cloud provider (e.g., Microsoft Azure, Amazon Web Services [AWS], or Google Cloud Platform).

Acknowledgments

Putting a book like this together is no simple feat, and it requires help from many important individuals behind the scenes. We were blessed to have the support of Megan Fuerst and the entire Rheinwerk Publishing team, who guided us through every step of the writing process. We can't thank you enough for your support—we simply couldn't have done it without you.

—The Authors

I would also like to thank another set of unsung heroes in this process: my wife Andrea and my kids Andersen, Paige, Parker, and Xander. They've been amazingly patient with me as I've burned the midnight oil putting this book together. I would call them my muse, but alas, I don't think they would appreciate me drawing such nerdy inspiration from them. If only ABAP were cooler in the minds of children...

—James Wood

I would like to thank my wonderful and supportive family: my wife Jennifer and my lovely daughters Amelia and Lillian.

—Joseph Rupert

I would like to thank my family for their patience and support: my wife Emily and my daughter Harper.

—*Colby Hemond*

I would like to thank my wife Nicole for her unwavering support and patience putting this together. It's so nice having a copyeditor in house to help bring my ideas to life.

—*Jeffrey Boggess*

Chapter 1

Introduction to Object-Oriented Programming

This chapter provides an overview of the basic concepts and philosophy behind object-oriented programming (OOP). The concepts we'll introduce in this chapter lay the foundation for the remainder of the book.

OOP is a software design approach that models solutions using *objects* representing real-world items or entities from your problem domain. This shift in design makes programs feel more *intuitive*, so they're easier to understand, maintain, and improve. In this chapter, we'll walk through the basics of OOP to help you design and build effective OOP solutions. These concepts are key for most modern OOP languages, including C++, Java, and ABAP Objects. We'll also introduce the *Unified Modeling Language* (UML), a modeling language that's used extensively for *object-oriented analysis and design* (OOAD).

1.1 The Need for a Better Abstraction

In the field of software engineering, few topics incite more controversy than OOP. Loyalists defend the merits of OOP with near-religious fervor, while detractors often roll their eyes at the very thought of it. If you're reading this book, perhaps you find yourself somewhere in the middle of this seemingly endless debate. And if that's the case, the most pressing questions on your mind may include the following:

- Why should I bother learning OOP?
- Is OOP really better than procedural programming or other methodologies?
- What makes OOP so special?

In the sections that follow (and throughout this book), we'll attempt to answer these questions by demonstrating how OOP sets itself apart from other programming methodologies and provides a better, more intuitive form of abstraction.

1.1.1 The Evolution of Programming Languages

The effectiveness of any language—spoken or otherwise—comes down to how well it can express complex thoughts and ideas. If you look at programming languages through this lens, you can see a clear evolution from low-level assembly languages to higher-level procedural languages like C, COBOL, or even classic ABAP. As languages evolve, they become easier to read and write. Of course, this begs the question: What exactly makes a programming language more expressive?

There are countless opinions on this topic (hence the sheer number of programming languages we have today), but most approaches share a common goal: to improve the quality of the abstractions that developers have to work with. Better abstractions make it easier to focus on solving problems instead of getting bogged down in technical details, which boosts productivity.

To put this concept into perspective, let's consider C—one of the most widely used and influential programming languages of all time. Back in the days of assembly programming, developers' days were spent twiddling bits and bytes, manipulating CPU registers, and dealing with a host of low-level details. C changed the game by giving developers basic tools like:

- Variables with meaningful names and intuitive data types (e.g., integers and strings).
- Conditional statements that made code look and feel like spoken language (e.g., "if this, then that").
- Callable functions that let developers break big problems into smaller, more manageable pieces (or *modules*).

These innovations shifted the focus away from technical minutiae and toward solving real problems, making developers' lives significantly easier.

1.1.2   Moving Toward Objects

By the 1960s, software researchers noticed a lingering issue. While early procedural languages improved readability and logic, they still forced developers to think in terms of the computer's structure rather than the problem they were solving. This gap—what we'll call *semantic dissonance*—meant that code didn't often resemble the real-world problems it was meant to address.

This disconnect put tremendous pressure on developers to translate requirements into code accurately. And if something got lost in translation or a mistake was made? Well, let's just say it usually resulted in a few long and stressful days of debugging.

Recognizing this, several influential language designers took a collective step back and began contemplating what the most ideal type of abstraction would be. Think of it this way: If you could choose *any* kind of element to model your program designs, what would you ask for? Would you stop at abstract data types (ADTs) like structures, functions, and subroutines? What if instead someone offered you a series of magical *objects* that look and behave like the entities you interact with in the real world?

While the latter approach may sound too good to be true, it turns out that conjuring up such objects in programs is achievable if you think about your program designs just a little bit differently. We'll explore this thought process in depth beginning in Section 1.2, but before we segue into more practical matters, let's take a moment to understand *why* OOP matters. At the end of the day, OOP is all about bridging the gap between whatever problem domain you're working with and the solution space where your program code lives and operates. The goal is to model your code in such a way that it resembles (or *simulates*) the problem space you're working in (e.g., purchasing and accounting). In his book, *Thinking in C++, Vol. 1: Introduction to Standard C++, Second Edition* (Prentice Hall, 2000), Bruce Eckel summarizes the benefits of this approach: "Casting the solution in the same terms as the problem is tremendously beneficial because you don't need a lot of intermediate models to get from a description of the problem to the description of the solution."

Once you begin to see the world through object-oriented glasses, you open yourself up to all kinds of new and exciting possibilities. For example, with objects, it's easier to achieve reuse because you're dealing with self-contained entities that have defined responsibilities as opposed to a scattering of data structures and subroutines. This will all become clearer as you progress through the book, but for now, we simply ask that you open your mind to the possibility of a system overrun by lots of tiny little objects.

1.2   Classes and Objects

Students learning pure object-oriented languages like Java are often taught that everything is an object. While this is not necessarily the case in a hybrid language like ABAP Objects (where it's still possible to use procedural and event-based programming constructs), it's still a good way to start thinking about how to design programs using an object-oriented approach. Of course, it helps if you know what an object is. In this section, we'll attempt to unravel the mysteries surrounding objects while also considering a closely related concept in OOP: *classes*.

1.2.1   What Are Objects?

From a technical perspective, an object is a special kind of variable that has distinct characteristics and behaviors. The characteristics (or *attributes*) of an object are used to describe the *state* of an object. For example, a Car object might have attributes to capture information such as the color of the car, its make and model, or even its current driving speed. Behaviors (or *methods*) represent the *actions* performed by an object. In our Car example, there might be methods that can be used to drive, turn, and stop the car, for instance.

With these two concepts in mind, our initial definition of an object would read something like this: "An object is a variable that combines data and behaviors together in a self-contained package." However, if we were to stop there, our definition would be rather limiting. In his book, *Design Patterns Explained: A New Perspective on Object-Oriented Design, Second Edition* (Addison-Wesley, 2004), Alan Shalloway emphasizes that objects make it possible to "...package data and functionality together by *concept*, so you can represent an appropriate problem-space idea rather than being forced to use the idioms of the underlying machine." This distinction, though subtle, is important in getting you where you really want to go with OOP: to create autonomous entities with defined roles and responsibilities that can think and act for themselves.

1.2.2 Introducing Classes

Now that you have a sense of what objects *are*, you might be wondering how objects are *defined* in the first place. Unlike other variable types that you might be accustomed to working with (e.g., integers or strings), this process requires a fair amount of thought. Since an object, by definition, can literally refer to *anything* (i.e., a person, place, thing, or idea), you first need to figure out the *types* of objects you need to model your problem domain. For example, if you're building a financial accounting solution, you would probably need objects to represent items like accounts and ledgers.

Sometimes this analysis process is intuitive; other times, not so much. In these latter cases, you must collect your thoughts using an ordered and methodical process. A common approach for initiating this process is to go through requirements documents and underline all the nouns that are used to describe various aspects of the problem domain. Then, from there, you can further organize the objects by examining their roles and responsibilities as well as their relationships to other objects. Early OOP researchers observed that the nature of this analysis process bore many similarities to the classification techniques used by biologists to identify, categorize, and understand the relationships between plants and animals. Consequently, the term *class* was used to describe (or *classify*) these abstract data types, and over the years the name has stuck.

In practical terms, you can think of a class as being rather like a specialized type declaration. This is to say that a class declaration defines the *type* of an object. This typing concept should feel intuitive for ABAP developers who are accustomed to working with structures and internal tables. For example, in Listing 1.1, you can see how we've defined a structure variable called LS_PERSON in terms of a custom type we've declared called TY_PERSON. This custom type declaration tells the ABAP compiler what the LS_PERSON structure will look like at runtime (i.e., what component fields the structure will have).

```
TYPES: BEGIN OF ty_person,
         first_name TYPE string,
         last_name TYPE string,
       END OF ty_person.

DATA ls_person TYPE ty_person.
```

Listing 1.1 Declaring a Custom Structure Data Type

With class type declarations, you're essentially trying to accomplish the same thing; The only difference is that you're declaring a class of objects as opposed to a structure or internal table. For example, in Listing 1.2, you can see how we've defined a custom class type called LCL_PERSON. We'll unpack this syntax further in Chapter 2, but for now, notice the similarities between this class type declaration and the TY_PERSON type declaration from Listing 1.1. This is not by accident, since classes are, in many respects, just another form of ADTs.

```
CLASS lcl_person DEFINITION.
  PRIVATE SECTION.
    DATA mv_first_name TYPE string.
    DATA mv_last_name TYPE string.
ENDCLASS.

DATA lo_person TYPE REF TO lcl_person.
```

Listing 1.2 Declaring a Custom Class Type

When you look at class type declarations in this light, you can begin to appreciate the relationship between classes and objects. Conceptually speaking, it's appropriate to think of classes like templates (or *blueprints*) that an OOP runtime environment can use to figure out how to create object instances at runtime. Therefore, the relationship between an object and a class is normally described as *an object is an instance of a class* in OOP lingo.

This relationship is illustrated from a runtime perspective in Figure 1.1. Here, you can see how an arbitrary number of instances of the LCL_PERSON class are created (or *instantiated*) at runtime. Each created object instance is unique and independent in its own right (i.e., it has its own memory space). For example, notice how each object instance contains its own values for the FIRST_NAME and LAST_NAME attributes. If you were to change the FIRST_NAME attribute for the first object instance, the other object instances would not be affected because they're independent entities. Indeed, the only thing that these objects really have in common at runtime is their shared definition class.

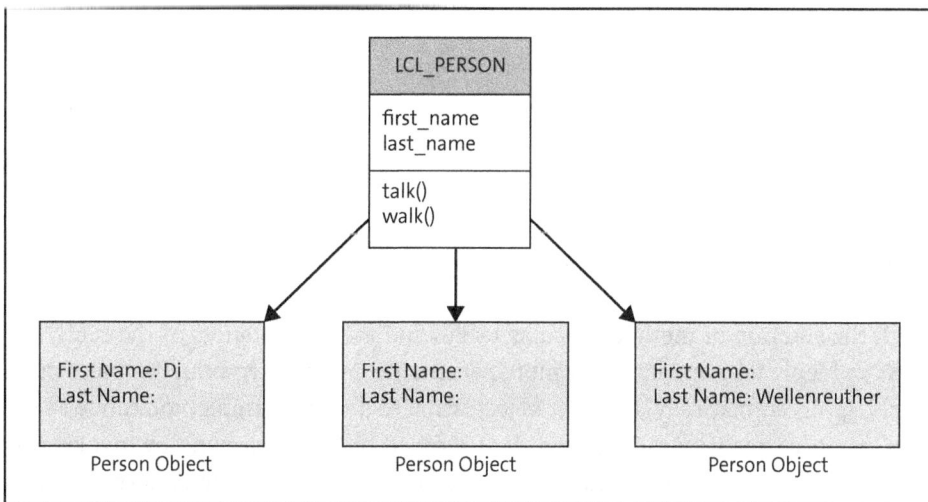

Figure 1.1 Understanding the Relationship Between Classes and Objects

If all this seems confusing, don't worry; we'll delve into some hands-on examples in Chapter 2 that should make the relationship between classes and objects much clearer.

1.2.3 Defining a Class Interface

In the previous section, we highlighted some of the similarities between class types and other familiar ADTs such as structures or internal tables. While this analogy is useful in understanding how objects are defined conceptually, it starts to break down when you approach the specification of a class's methods. Methods, which are conceptually like subroutines (*forms* in ABAP parlance) or functions in procedural programming, define the interaction points that make it possible to *communicate* with object instances at runtime. In more formal terms, methods make up a class's *interface*.

To illustrate the idea of a class's interface, consider the revised LCL_PERSON class type declaration shown in Listing 1.3.

```
CLASS lcl_person DEFINITION.
  PUBLIC SECTION.
    METHODS:
      talk,
      walk.

  PRIVATE SECTION.
    DATA mv_first_name TYPE string.
    DATA mv_last_name TYPE string.
ENDCLASS.

CLASS lcl_person IMPLEMENTATION.
  METHOD talk.
    WRITE: / 'Hello'.
  ENDMETHOD.

  METHOD walk.
    ...
  ENDMETHOD.
ENDCLASS.
```

Listing 1.3 Declaring a Class with Methods

With the addition of methods such as talk() and walk(), instances of the LCL_PERSON class suddenly take on quite a bit more personality. In effect, you can use these methods to (programmatically) tell your objects what to do. For example, you can tell a person instance to talk by calling the talk() method or to walk around by invoking the walk() method. Such behaviors transform objects from inanimate data structures to living, breathing entities that are self-aware and, to a certain extent, autonomous. Plus, with their attribute data in context, objects inherently know what they are, what their current state is, and what sort of operations they can perform.

1.3   Establishing Boundaries

In OOP, objects aren't meant to exist in isolation—they're designed to work together, much like a team that collaborates to solve real-world problems. Think of your program as a simulation running within a virtual problem domain, with each class or object taking on specific roles and responsibilities. Instead of tasking a few overburdened "god objects" to oversee and handle everything, you break the solution into smaller, more focused classes, each handling a specific piece of the puzzle. These specialized classes have *high cohesion* because all their operations are closely related and work together intuitively.

While having this level of collaboration among objects in a program design is a good thing, it does require you to define some boundaries up front. In this section, you'll learn how these boundaries are established within classes and why they're important.

1.3.1   Encapsulation and Implementation Hiding

Before we look at the mechanics of boundary definitions within classes, let's briefly take a step back and think about why boundaries are needed in the first place. After all, this sort of thing is rarely (if ever) a concern in other programming paradigms (notably, procedural programming).

To understand why boundaries matter, let's look at a real-world example from manufacturing: the *smartphone*. A smartphone, much like a software object, has attributes (like its size and memory capacity) and behaviors (making calls, sending emails, etc.). It also has an *interface* that you can interact with, supporting gestures like typing on a virtual keyboard or tapping an icon to open an app. Despite its ability to handle all kinds of complex tasks, a smartphone is surprisingly intuitive and easy to use. How can something so intricate feel so simple? The answer lies in a principle called *encapsulation*.

Encapsulation is all about hiding sensitive parts of an object inside a "capsule," shielding them from the outside world, as shown in Figure 1.2. The primary goal here is to keep unnecessary details out of sight. It's not that these hidden components aren't important—they're just not something the user needs to deal with to make the object work. The less the user has to understand about how a product works internally, the easier it is for them to use. In software design, encapsulation plays the same role: It shortens the learning curve by keeping complex details tucked away, so developers and users can focus on what really matters.

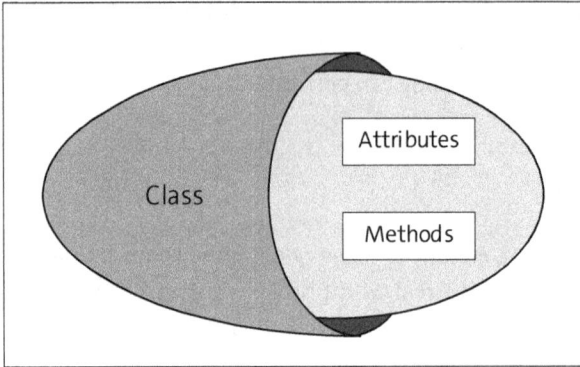

Figure 1.2 Encapsulation of Data and Behavior Inside a Class

Apart from decluttering an object's interface, there are several other important benefits to be gained from applying encapsulation techniques to your class/object designs:

- **Targeted change management**
  From an implementation standpoint, encapsulation helps protect the more volatile, sensitive parts of a system by keeping them hidden. This approach makes it easier to adapt to changing requirements over time, as it allows you to pinpoint and isolate specific pieces of functionality without disrupting the entire system.

- **Elimination of dependencies**
  Simplifying the interface promotes *loose coupling* between components, eliminating unnecessary connections or dependencies between them. This streamlined design not only makes objects easier to use but also enhances their reusability in different contexts.

- **Seamless user experience**
  Encapsulation allows you to update or modify the implementation behind the scenes without affecting users. Returning to the smartphone analogy, imagine a manufacturer swaps out the existing processor for a smaller, faster chip in a new model. For the consumer, this change is seamless—they continue using the phone the same way as before, except now it's faster and more responsive. Similarly, you can achieve the same flexibility with objects by updating the implementation code behind the class interface. For example, you might optimize a database lookup by using a stored procedure in SAP HANA instead of a standard SQL SELECT statement. For the consumer, nothing changes—as long as you deliver the expected data, the process remains seamless from their perspective.

We'll cover these concepts in much more detail in Chapter 3. For now, though, the main takeaway is that it makes sense to draw clear boundaries between the various collaborating objects in your object-oriented designs.

Of course, to enforce such boundaries, you're going to need some support from ABAP and the ABAP runtime environment. Enter *visibility sections*.

1.3.2   Understanding Visibility Sections

To clearly define a class's interface, you need a way to distinguish between elements you want to share publicly and those you want to keep private. Most object-oriented languages address this by organizing elements into three distinct visibility categories:

- **Public section**
  Components (and by component, we mean attributes and methods) that you define in this section can be accessed from *any* context. These components make up the class's external interface.

- **Private section**
  Components defined in this section are completely shut off and hidden from external consumers. Consumers cannot access them in any way. The sole purpose of these components is to facilitate the internal workings of the class (e.g., helper methods and internal data).

- **Protected section**
  Components defined in this section are visible only to the superclass and its subclasses. To the outside world, it's as if these components are defined in the private section of the class. To subclasses, it's a different story. You'll see this visibility section come into play when we discuss inheritance in Chapter 5.

Once components are organized into their respective visibility sections, you can rest assured that the ABAP runtime environment will make sure that these boundaries are enforced no matter what. In Chapter 3, we'll look at ways of exploiting this functionality to improve the robustness of your designs.

1.4   Reuse

One of the biggest advantages of using an object-oriented approach to program design is the potential for promoting widespread code reuse. While the promise of huge productivity gains can be exciting, it's important to stay grounded; developing truly reusable classes takes time and experience. In the following sections, we'll explore some fundamental techniques for reusing classes. We'll look at this concept in more detail in Chapter 5 and Chapter 6.

1.4.1   Composition

The easiest way to reuse a class is to simply create an instance of it and start calling its methods. Such instances can be created in isolation or as attributes of new classes that you may decide to build. This latter usage type is often referred to as *composition*, where new classes are *composed* from existing classes. These classes are *aggregates*, using existing classes as building blocks to construct arbitrarily complex assemblies. Designs based on composition are easy to understand and highly flexible. Since member objects can be hidden just like any other attribute, it's easy to change the way you utilize these objects at design time and at runtime.

1.4.2   Inheritance

Another way to reuse a class is through *inheritance*. The concept of inheritance is an extension of the classification metaphor used to describe the nature of classes and their relationships. Here, you define *specialization* relationships between families of related classes. These relationships begin to reveal themselves as an object-oriented design matures.

The idea of inheritance is best explained by an example. Imagine that you're working on an object-oriented design for a banking system. Initially, you come up with a series of classes, including one to represent an individual bank account. After studying the requirements further, you discover that there are certain peculiarities unique to checking and savings accounts. At this point, you're faced with a dilemma. You could copy the code you put together for the account into new checking and savings account classes. However, this seems wasteful, since it would introduce a lot of redundant code.

Another option would be to use *inheritance* to describe this specialization relationship. In this scenario, you still create new checking and savings account classes, but you create them as *subclasses* derived from the original account class (the parent class or superclass). The checking and savings account subclasses *inherit* the attributes, behaviors, and type of the account superclass (see Figure 1.3). Now, you can apply the relevant changes (or specializations) to each of the subclasses independently without having to reinvent the wheel.

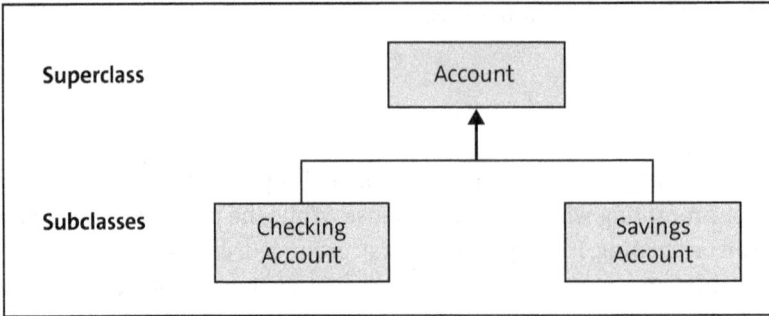

Figure 1.3 Understanding Inheritance Relationships

It's important to understand that inheritance describes a *relationship*; it's not just a fancy term for copying and pasting one piece of code into another. Initially, a subclass looks like a clone of the superclass. However, over time, a subclass can be extended to add new attributes and methods as needed. Additionally, subsequent changes to the superclass are automatically applied to the subclass (although you'll see exceptions to this rule in Chapter 5). In general, it's possible to create class hierarchies with arbitrarily deep inheritance relationships.

The connection between a subclass and its parent is often described using the *is-a* relationship concept. In the example in Figure 1.3, a checking account *is an* account.

The is-a relationship is a simple way of saying that the subclass and superclass share the same type. As you may recall from our discussion in Section 1.2.3, a class's type describes how you can communicate with objects of that class. Since objects of a superclass and its subclass share the same type, you can interact with both in exactly the same way. *Polymorphism* exploits this capability, allowing for code reuse across multiple dimensions.

1.4.3   Polymorphism

The definition of an inheritance relationship implies that a subclass inherits both the *type* and the *implementation* of its superclass. At the subclass level, however, it's possible to *redefine* a method's implementation to further specialize certain behaviors. Redefining a method does not change the interface of the method (i.e., the way it's called) in any way. Rather, it simply redefines the behavior of the method to suit the needs of the subclass.

To the ABAP runtime environment, however, all these class-specific object references look the same. In its eyes, the requirements for calling a particular method on a subclass instance are no different than the requirements to call the same method on the superclass. After all, the method names are the same, and they have the same number of parameters.

As a programmer, you can take advantage of this relationship to make your designs more flexible. To put this phenomenon into perspective, let's consider an example. Figure 1.4 depicts an Employee class hierarchy that can be used to model the types of employees managed within a human resources (HR) system. In this scenario, the Employee superclass is used to describe the basic characteristics and behaviors of all types of employees.

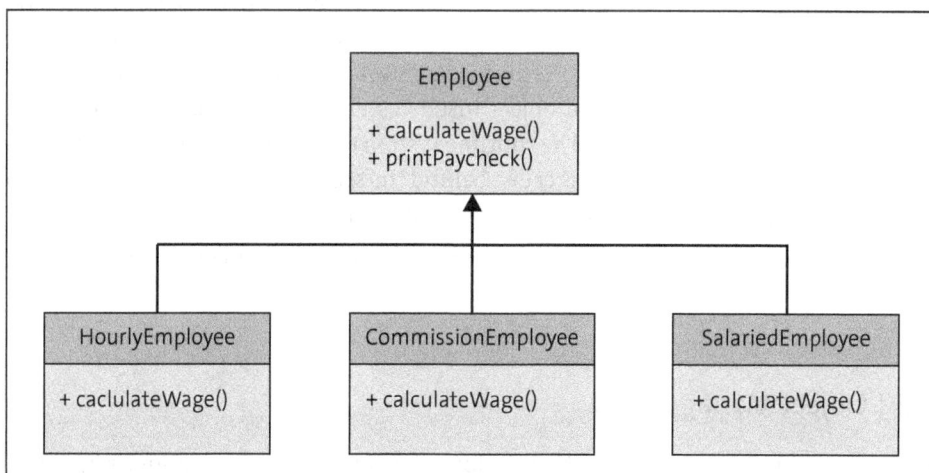

Figure 1.4 An Employee Class Hierarchy

The three specialized subclasses (HourlyEmployee, CommissionEmployee, and SalariedEmployee) are extensions of the Employee superclass used to represent employees paid by the hour, employees working on commission, and salaried employees, respectively.

For the purposes of our example, let's assume that the calculateWage() method has been redefined in each of the subclasses to properly calculate the employee's wage based on the actual employee type.

Now, imagine that the company wants to use this Employee class hierarchy to enhance their accounts payable system to automate the creation of monthly paychecks. Listing 1.4 shows an example of the pseudocode for such an enhancement.

```
Get Employees
For Each Employee
    Call calculateWage() to calculate the employee's wages
    Call printPaycheck() to print the employee's paycheck
End For
```

Listing 1.4 Pseudocode for Generic Payroll Run Algorithm

If you look closely at the logic in Listing 1.4, you can see that there's no real distinction between employee types in the main loop that drives the payment run. This isn't simply a casual omission of detail in the pseudocode. Rather, we're purposefully utilizing the is-a relationship between Employee class types to build a generic algorithm that can process payments for *any* kind of Employee type. In formal terms, we're exploiting a core object-oriented language feature called *polymorphism*.

The term polymorphism can be translated from its Greek origins as *many forms*. In the example in Listing 1.4, each subclass represents a different form (or type) of Employee. However, since the subclasses take part in an inheritance relationship with the Employee superclass, each subclass *is* an Employee. Consequently, since both the superclass and subclass share the same public interface, any method that can be called on the superclass can also be called on the subclass. Our fictitious accounts payable system is taking advantage of this feature by defining its interface to work with generic Employee instances. At runtime, the actual object instances being processed could be of type Employee or any of its subclasses. The runtime system takes care of making sure that the proper method implementation is called behind the scenes. This is another example of how an object is smart enough to know how to do its job.

In Chapter 6, we'll look at ways of utilizing polymorphism to build more intelligent and flexible designs.

1.5 Object Management

At this point, we've covered the three major pillars of OOP: *encapsulation*, *inheritance*, and *polymorphism*. However, before we wrap up this introductory chapter, we'll briefly touch on another topic of similar importance: *object management*.

As you learned in Section 1.2, objects are specialized variables that are defined by class type. The class type defines a blueprint that provides the runtime environment with the information it needs to build object instances. Of course, such object instances aren't created automatically. After all, how would the ABAP runtime environment know how many object instances you might need at runtime? Because of the abstract nature of these types, it's up to you as a developer to explicitly tell the runtime environment when and where you want object instances to be created.

Somewhere in between the point when you request that an object instance be created and the point where you actually get our hands on the allocated object reference, the runtime environment provides you with a mechanism for initializing the object. Here, you can define specialized callback methods called *constructors* within your classes that the runtime environment will invoke as instances of a class are being allocated. The constructor's job is to make sure that the object is initialized in a consistent state before it's used. That way, it's ready to perform its requisite tasks when called upon.

We'll investigate the details of object management from a development perspective in Chapter 4. In that chapter, you'll also learn a few tricks for improving performance and influencing the object creation process.

1.6   UML Tutorial: Class Diagram Basics

As you've seen, object-oriented software development places a considerable amount of emphasis on design. Before you start coding, it's imperative that you have all your ducks lined up in a row. In particular, you must know what types of objects you'll need as well as how those objects will interact with one another at runtime.

Object-Oriented Analysis and Design (OOAD) is a software development methodology used to analyze system requirements and formulate a system design from an object-oriented perspective. OOAD practitioners often use graphical modelling techniques to communicate their designs more effectively.

The *Unified Modeling Language* (UML) contains a set of graphical notations for building diagrams that depict various aspects of the system model. UML is used extensively throughout the software development industry, so it's important that you understand how to use UML diagrams to express and interpret object-oriented designs.

Throughout this book, we'll examine the usage types of various UML diagrams. Our discussions will be based on version 2.0 of the UML standard. We'll begin our introduction to UML by first looking at the *class diagram*. For now, we'll keep it simple and reinforce the concepts covered in this introductory chapter. In Chapter 5 and Chapter 6, we'll cover more advanced features of class diagrams.

Note

The UML standard is maintained by the *Object Management Group* (OMG). For more information on the OMG, check out their website at *http://www.uml.org*.

1.6.1 What Are Class Diagrams?

Class diagrams are used to illustrate the static architecture of an object-oriented system. They depict the various classes used in the system, as well as their relationships. Figure 1.5 shows a simple class diagram that describes a scaled-down model of a sales order system used to process orders for an e-commerce website. Here, you can observe the basic class types that will make up the sales order system (i.e., the rectangular boxes), their elements, and their relationships to other classes. As simple as this may seem, it's surprising how much information you can glean just by visualizing the key players in an object-oriented design.

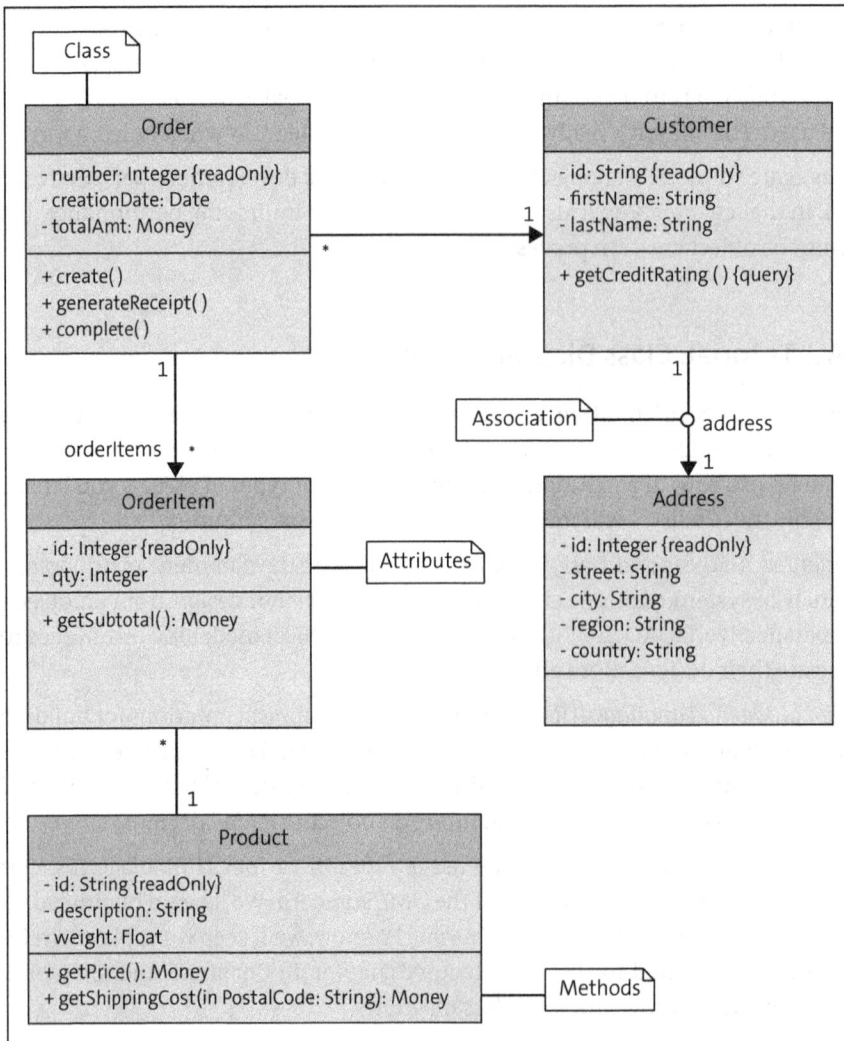

Figure 1.5 An Example UML Class Diagram

In the upcoming sections, we'll explore specific elements of the UML class diagram in more detail.

1.6.2   Classes

The diagram in Figure 1.5 contains five classes: Order, OrderItem, Product, Customer, and Address. As you can see, classes are represented in class diagrams as rectangular boxes partitioned into three sections (as shown in Figure 1.6):

- The top (shaded) section contains the class name, as well as some other optional modifiers that we'll cover in Chapter 5 and Chapter 6.
- The optional middle section contains some of the more prominent attributes defined by the class.
- The bottom section contains relevant operations (or methods) of the class.

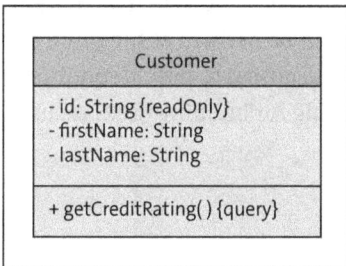

Figure 1.6 Understanding UML Class Notation

As you read through this description, did you notice how we used the terms *some* and *optional*? This is because the UML class diagram notation does not require you to specify *every* element defined within a class. Instead, the goal is to include the key elements that are needed to convey the point of the class. Too much detail is actually a bad thing since it clutters up the diagram and makes it difficult to visualize the system at a glance.

The point is to not get too carried away with the details, as this can complicate the model to the point that the diagram is unreadable. Developers who are new to UML sometimes fall into this trap, worrying that there isn't enough information in their class diagram to start writing code. If you find yourself in this position, remember that UML provides a multitude of diagrams that can be used to express the various aspects of your design, and class diagrams only tell one part of the story.

1.6.3   Attributes

Attributes can be specified on the class diagram using the syntax shown in Listing 1.5.

```
visibility name: type-expression = initial-value
                {property-string}
```

Listing 1.5 Attribute Notation for a UML Class Diagram

Technically speaking, you're only required to provide the name when specifying an attribute in a class diagram. However, you can use the other syntax elements shown in Listing 1.5 to provide some additional information about the attribute:

- The visibility part of an attribute definition describes the accessibility of the attribute from an external perspective. Possible values for visibility include + for public attributes, - for private attributes, and # for protected attributes.

- The type-expression describes the attribute's type. UML defines standard types such as *integer* or *string*, but you can also specify custom types here as well. The type-expresssion expression can also be used to express the cardinality of an attribute (e.g. for an internal table), and the initial value of the attribute (if one is assigned).

The property-string expression is an optional element that you can use to describe certain additional properties for an attribute. For example, in the OrderItem class from Figure 1.5, the id attribute has the readOnly property assigned to indicate that an item's ID number never changes. Values for these properties can be defined at the discretion of the person designing the class diagram. The primary purpose is to provide additional details that are helpful to the developer who is responsible for building the class using a specific programming language like ABAP Objects.

1.6.4   Operations

Operations (i.e., methods) can be expressed using the syntax shown in Listing 1.6.

```
visibility name(parameter-list) : return-type
                               {property-string}
```

Listing 1.6 Operation Notation in a UML Class Diagram

For brevity's sake, developers will often just specify the name of an operation when creating a class diagram. The remaining optional syntactical elements from Listing 1.6 are typically used strategically to emphasize a certain aspect of the operation:

- The visibility of an operation defines its accessibility. Possible values include + for public operations, - for private operations, and # for protected operations.

- The parameter-list in parentheses specifies a comma-separated list of parameters for the operation. Each parameter is of the form kind name : type = default-value. Here, kind signifies the type of parameter. Valid values include in for inbound parameters passed by value, out for outbound parameters passed by value, and inout for inbound parameters passed by reference. The name token symbolizes the parameter name. Each parameter can have an optional type associated with it using the type token. This can be a generic type or a type specific to a particular programming language. Finally, you can specify an initial value for the parameter using the default-value expression.

- The return-type element specifies the data type of values returned by functional operations.

- The optional property-string indicates certain properties assigned to an operation. An example of this would be the {query} property string assigned to the getCreditRating() operation of class Customer. Such operations are *read-only* operations that do not alter the state of the object. Applying these property strings can give hints to aid the developer in implementing the class in a particular programming language.

An example of the syntax described in Listing 1.6 is given in Listing 1.7. This example declares a public operation called getShippingCost() that receives a single inbound parameter called postalCode (of type String). The operation returns a value of type Money to represent the derived shipping cost.

```
+ getShippingCost(in postalCode: String) : Money
```

Listing 1.7 An Example of an Operation Definition

1.6.5   Associations

The lines drawn between classes in a class diagram represent a type of *association*. You can think of an association as another way to specify an attribute for a class.

For example, the directed line drawn between the Customer and Address classes in Figure 1.5 describes an attribute of type Address for class Customer. The arrow in the association between classes Customer and Address indicates that instances of class Address can be reached through an attribute defined in class Customer. If the association line had contained arrows pointing in both directions, then the association would have been *bidirectional*. In this case, an attribute of type Customer would also have been defined for class Address, making it possible to navigate between attributes in both directions.

The numbers affixed to each endpoint represent the cardinality of the association from the perspective of the nearby class (see Table 1.1). For example, in Figure 1.5, the association between classes Order and OrderItem denotes a *one-to-many* relationship between an order and its items. In this case, an order can contain zero or more items, and any given item can exist for exactly one order.

Cardinality	Description
0..1	Zero or one instances of a class
1	Exactly one instance of a class
*	Zero or more instances of a class
m..n	A range of instances with upper/lower bounds (e.g., 2..4)

Table 1.1 UML Cardinality Notation

You might be wondering why you'd bother creating an association when you could just use a simple attribute instead. There's no strict rule for choosing one over the other, but a good guideline is to use an association when employing composition to reuse a class within another class. This approach makes the composition relationship more explicit and makes it easier to adjust the diagram as you refine your class model.

1.6.6   Notes

You can add comments to your UML diagrams using notes, which are represented by a dog-eared sticky note element (see Figure 1.7). You can use notes in any kind of UML diagram to include comments related to a particular element (linked via a dashed line) or to the diagram as a whole. Notes are often used to help clarify a certain requirement that's too difficult to express using standard UML notation.

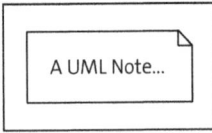

Figure 1.7 UML Note Notation

1.7   Summary

In this chapter, we explored how classes serve as a blueprint for creating object instances. By combining attributes and methods, classes model real-world phenomena in a software context. Visibility sections play a key role by enforcing rules and constraints, controlling access to the attributes and methods within a class. We also touched on the foundational reuse capabilities that classes offer.

This chapter covered a lot of ground very quickly. If you're finding yourself a little lost, don't worry—we'll have much more to say about each of these topics in the upcoming chapters. This begins in Chapter 2, where we'll start to unpack the ABAP Objects syntax used to define and interact with classes and objects.

Chapter 2
Working with Objects

This chapter introduces you to some basic ABAP Objects syntax and the relevant development tools that you'll need to start building object-oriented programs in ABAP.

Object-oriented programming (OOP), like many abstract concepts, is best learned by example. Now that we've gotten the basic definitions out of the way in Chapter 1, we're ready to turn our attention toward more practical matters and look at some basic syntax and sample code using ABAP Objects.

Because the primary unit of development for object-oriented programs is the class, we'll spend quite a bit of time in this chapter exploring the syntax used to define classes and their internal components. Once you're up to speed with basic syntax rules, we'll take a look at the tools used to define and maintain classes.

2.1 Defining Classes

Classes in ABAP Objects are declared using the CLASS statement block. This statement block is a wrapper of sorts, grouping all relevant class component declarations into two distinct sections:

- **Declaration section**
  This section specifies all the components defined within the class including attributes, methods, and events.
- **Implementation section**
  This section provides implementations (i.e., the source code) for the methods defined within the declaration section.

In the following sections, we'll unpack the syntax used to build out these sections and fully specify class types. For the purposes of this introductory section, our focus will be on defining *local classes* (i.e., classes that are defined within ABAP report programs and function group includes). However, in Section 2.4, you'll learn that this same syntax applies to the definition of global class types as well. The primary difference in the case of global classes is that you have a form-based editor in the Class Builder tool that spares you from manually typing out some of the declaration syntax.

2.1.1   Creating a Class

To define a new class type, you must declare it within a CLASS...DEFINITION...ENDCLASS statement block as shown in Listing 2.1. This statement block makes up the aforementioned declaration section of the ABAP class definition. As we noted earlier, this section is used to declare the primary components that make up a class, such as attributes and methods.

```
CLASS {class_name} DEFINITION [class_options].
  PUBLIC SECTION.
    [components]
  PROTECTED SECTION.
    [components]
  PRIVATE SECTION.
    [components]
ENDCLASS.
```

Listing 2.1 ABAP Class Declaration Section Syntax

If you look closely at Listing 2.1, you can see that the components of a class definition are organized into three distinct *visibility sections*: the PUBLIC SECTION, the PROTECTED SECTION, and the PRIVATE SECTION. Each of these visibility sections is optional, so it's up to you as a developer to determine which components go where—a subject that we'll consider at length in Chapter 3.

Besides the definition of the components that makeup the class's interface, the next most important task in defining a class is coming up with a good and meaningful name for it. As trivial as it may sound, this task is often harder than it looks. Part of the challenge stems from the fact that ABAP only gives you 30 characters to work with. You must come up with a meaningful name that fits within the confines of the syntax shown in Listing 2.2.

```
[{Namespace}|{Prefix}]CL_{Meaningful_Name}
```

Listing 2.2 ABAP Class Naming Convention

Listing 2.3 shows how this class naming syntax is applied to the various class types that may exist within the ABAP Repository. You'll see many more examples of this naming convention at work as you progress through the book.

```
LCL_LOCAL_CLASS          "Local Customer Class
ZCL_GLOBAL_CLASS         "Global Customer Class
CL_ABAP_MATCHER          "SAP-Standard Class (no namespace)
/BOWDK/CL_STRING_UTILS   "3rd-Party Class w/Namespace Prefix
```

Listing 2.3 Class Naming Examples

2.1.2 Component Declarations

As you've seen, the structure and makeup of a class is determined by its component definitions. Therefore, in this section, you'll learn about the different component types that you can define within a class. Before you get started, though, you first need to understand how components are grouped from a scoping perspective. Within a class declaration, there are two different types of components:

- **Instance components**
  Instance components, as the name suggests, are components that define the state and behavior of individual object instances. For example, an `Employee` class might have an instance attribute called `id` that uniquely identifies an employee within a company. Each instance of the `Employee` class maintains its own copy of the `id` attribute, which has a distinct value. Instance methods operate on these instance attributes to manipulate the object's state and perform instance-specific tasks.

- **Class components**
  Class components, on the other hand, are defined at the class level, meaning class components are shared across *all* object instances. Such components can come in handy in situations where you want to share data or expose utility functions on a wider scale. For example, in the `Employee` class scenario, you might use a class attribute called `next_id` to keep track of the next available employee ID number. This value could be used as a primitive number range object to assign the `id` instance attribute for newly created `Employee` objects.

In practice, most of the classes you define will contain few class components. After all, it's hard to establish identity at the object level if all the data and/or functionality resides in global class components. However, you'll see that class components come in handy in certain situations, such as dealing with complex object creation scenarios or finding a home for utility functions.

Static Components

Class components are sometimes referred to as *static components* since they are statically defined and maintained at the class level. This is especially the case in other object-oriented languages such as Java or C#.

Internal Namespaces

Regardless of where you decide to define your components, it's important to keep in mind that all component names within an ABAP Objects class belong to the same internal namespace. For example, it's not possible to define an attribute and a method using the same name—even if they belong to different visibility sections. In the sections that follow, you'll learn that the adoption of good naming conventions makes it easy to avoid such naming collisions.

Attributes

As you learned in Chapter 1, attributes are essentially variables defined internally within a class or object. Attributes are defined in the same way that variables are defined in other ABAP programming modules. The primary difference in the case of classes is that you're contending with different contexts.

To put these contexts into perspective, consider the LCL_CUSTOMER sample class in Listing 2.4. Within this class definition, we've defined three different types of attributes:

- **Instance attributes**
  To define the properties that are unique to a particular customer instance, we've created several instance attributes such as mv_id, mv_customer_type, mv_name, and ms_ address. As you can see in Listing 2.4, these instance attributes are defined using the familiar DATA keyword. Here, you can choose from any valid ABAP data type including structure types, table types, reference types, or even other class types.

- **Class attributes**
  The sv_next_id attribute is an example of a class attribute. As you can see, the only real difference syntax-wise between class attributes and instance attributes is the use of the CLASS-DATA keyword in lieu of the typical DATA keyword.

- **Constants**
  In the PUBLIC SECTION of our customer class, we've also defined several constants to represent the different customer types modeled in our class: CO_PERSON_TYPE for individuals, CO_ORG_TYPE for organizations, and CO_GROUP_TYPE for customer groups. These constants are defined just like any other constant using the CONSTANTS keyword. However, in the case of class constants, what we're really talking about is a specialized case of a class/static attribute (one that can't be modified at runtime).

```
CLASS lcl_customer DEFINITION.
    PUBLIC SECTION.
        CONSTANTS: CO_PERSON_TYPE TYPE c VALUE '1',
                   CO_ORG_TYPE    TYPE c VALUE '2',
                   CO_GROUP_TYPE  TYPE c VALUE '3'.
    PRIVATE SECTION.
        DATA: mv_id TYPE i,
              mv_customer_type TYPE c,
              mv_name TYPE string,
              ms_address TYPE adrc.
        CLASS-DATA: sv_next_id TYPE i.
ENDCLASS.
```

Listing 2.4 Declaring Attributes Within a Class

Though the ABAP compiler will generally allow you to define attributes with whatever name you prefer, we strongly recommend that you adopt a naming convention that makes it easier to identify the scope of a given attribute. Table 2.1 describes the naming convention that will be used within this book.

Attribute Type	Naming Convention	Description
Instance attributes	M{Type}_{Meaningful_ Name} Examples: mv_id ms_address mt_contacts	Here, the M implies that you're defining a *member variable*. The {Type} designator helps you more easily determine whether you're dealing with elementary variables (V), structures (S), internal tables (T), and so on. Aside from these scoping details, the rest of the instance attribute name is freeform and should be defined in such a way that it conveys meaning.
Class (static) attributes	S{Type}_{Meaningful_ Name} Examples: sv_next_id	This convention is almost identical to instance attributes. However instead of the M for member variable, static attributes are prefixed with an S to imply that the attribute belongs to the static context.
Constants	CO_{MEANINGFUL_NAME}	Constants are typically defined in all caps using the CO_ prefix.

Table 2.1 Naming Convention for Defining Attributes

Methods

Methods are defined using either the METHODS statement for instance methods or the CLASS-METHODS statement for class methods. The syntax for both statement types is shown in the syntax diagram in Listing 2.5. Here, you can see that a method definition consists of a method name, an optional parameter list, and an optional set of exceptions that might occur. For this introductory section, we'll focus on the first two parts of a method definition. We'll circle back and cover exceptions in Chapter 8.

```
{CLASS-}METHODS {method_name}
    [IMPORTING parameters]
    [EXPORTING parameters]
    [CHANGING  parameters]
    [RETURNING VALUE(parameter)]
    [{RAISING}|{EXCEPTIONS}...].
```

Listing 2.5 Method Definition Syntax

As you can see in Listing 2.5, the first thing you specify in a method definition is the method's name. Since methods define the behavior of classes, it's important that you come up with meaningful names that intuitively describe the method's purpose. Normally, it makes sense to prefix a method name with a strong action verb that describes the type of operation being performed. The sample class in Listing 2.6 provides some examples of this convention.

```
CLASS lcl_date DEFINITION.
  PUBLIC SECTION.
    METHODS:
      add IMPORTING iv_days TYPE i,
      subtract IMPORTING iv_days TYPE i,
      get_day_of_week RETURNING VALUE(rv_day) TYPE string,
      ...
ENDCLASS.
```

Listing 2.6 Defining Meaningful Names for Methods

After you come up with meaningful names for your methods, your next objective is to determine what sort of parameters (if any) the methods will need to perform their tasks. If you look at the syntax diagram from Listing 2.5, you can see that there are four different types of parameters that can be defined within a method's parameter list. Table 2.2 describes each of these parameter types in detail.

Parameter Type	Description
Importing	Importing parameters define the input parameters for a method. The values of an importing parameter cannot be modified inside the method implementation.
Exporting	Exporting parameters represent the output parameters for a method.
Changing	Changing parameters are input/output parameters that allow you to update or modify data within a method.
Returning	Returning parameters are used to define *functional methods*. You'll learn more about this parameter type when we look at functional methods in Section 2.2.7.

Table 2.2 Parameter Types for Method Definitions

To distinguish between the various parameter types within a method definition, method parameters are normally prefixed according to the convention described in Table 2.3. Here, the {Type} designator is once again used to differentiate between elementary data types (V), structure types (S), table types (T), and so on.

Parameter Type	Naming Convention
Importing	`I{Type}_{Parameter_Name}`
Exporting	`E{Type}_{Parameter_Name}`
Changing	`C{Type}_{Parameter_Name}`
Returning	`R{Type}_{Parameter_Name}`

Table 2.3 Method Parameter Naming Conventions

Regardless of the parameter's type, the syntax for declaring a parameter p1 is shown in the syntax diagram in Listing 2.7. As you can see, this syntax provides you with several configuration options for defining a parameter:

- The optional VALUE addition allows you to specify that a parameter will be passed by *value* instead of by reference. For more details on this concept, check out the upcoming text box.

- You can use the TYPE addition to specify the parameter's data type. The addition is used in this context in the exact same way it's used to define normal variables or form parameters.

- You can use the OPTIONAL addition to mark a parameter as *optional*. Such parameters can be omitted during method calls on the consumer side.

- You can use the DEFAULT addition to specify a default value for a given parameter (which makes the parameter optional from a consumer perspective). The caller of the method can override this value as needed.

```
{ p1 | VALUE(p1)} TYPE type [OPTIONAL | {DEFAULT def1}]
```

Listing 2.7 Formal Parameter Declaration Syntax

Pass-by-Value versus Pass-by-Reference

At runtime, whenever a method that contains parameters is invoked, the calling program will pass parameters by matching up *actual parameters* (e.g., local variables in the calling program and literal values) in the method call with the *formal parameters* declared in the method signature (see Figure 2.1). Here, parameters are passed in one of two ways: *by reference* (default behavior) or *by value*.

Pass-by-value semantics is enabled via the aforementioned VALUE addition. Performance-wise, pass-by-value implies that a *copy* of an actual parameter is created and passed to the method for consumption. As a result, changes made to value parameters inside the method only affect the copy; the contents of the variable used as the actual parameter are not disturbed in any way. This behavior is illustrated at the top of Figure 2.1 with the mapping of parameter a. Here, whenever the method is invoked, a copy of variable x is made and assigned to parameter a. As you might expect, this kind of operation can become rather expensive when you start dealing with large data objects.

Reference parameters, on the other hand, contain a reference (or *pointer*) to the actual parameter used in the method call. Therefore, changes made to reference parameters *are* reflected in the calling program. In Figure 2.1, this is illustrated in the mapping of parameter b. Here, if you were to change the value of parameter b inside the method, the change would be reflected in variable y in the calling program.

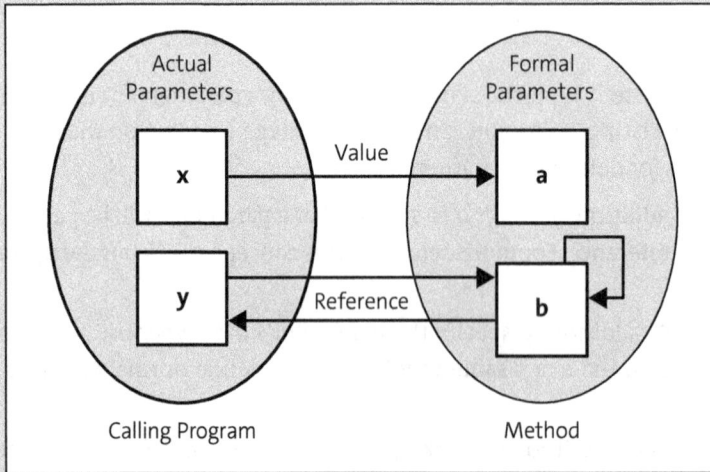

Figure 2.1 Mapping Actual Parameters to Formal Parameters

Since this behavior can potentially cause dangerous side effects, ABAP allows you to lock down reference parameters for editing inside methods by defining them as importing parameters. For example, if you define parameter b as an importing parameter, the compiler would complain if you try to modify its contents within the method body. In effect, importing parameters allow you to attain all the performance benefits of reference passing without the negative effects.

Collectively, a method's name and parameter list make up the method's *signature*. From the perspective of class consumers, method signatures determine the exact requirements for calling a particular method: which parameters to pass, the data types of the parameters being exchanged, and so on. As a method designer, it's important that you get these details right so that your methods are intuitive and easy to use. To that end, here are some design points to consider when defining method signatures:

- In general, keep the number of parameters being passed to or from methods to the bare minimum. You should assume that an object already has most of the information it needs (via its instance attributes) to perform a particular task, so you should only require a handful of parameters when defining a method.

- Define methods to perform one task. Avoid defining methods such as copyDataAndWashCat().

- When performing generic operations where specific data types don't matter, incorporate the use of generic ABAP types so that the methods can be (re)used in a variety of contexts. For a list of available generic types, search for "generic ABAP types" in the ABAP Keyword Documentation.

Events

Besides the more common attributes and methods that you see in most object-oriented languages, ABAP Objects also allows you to define events that model certain types of occurrences within an object's lifecycle. Once again, you can distinguish between instance events that occur within a specific object instance and class events that are defined at the class level.

Listing 2.8 contains the basic syntax used to define instance events and class events. The parameters defined for an event are used to pass additional information about the event to interested event handler methods. Since this is a one-way data exchange, you're only allowed to define exporting parameters in an event definition. The syntax is pretty much identical to the syntax used to define exporting parameters in methods. The only difference in this case is that event parameters must be passed by value. Aside from the formally defined exporting parameters in an event definition, the system also supplies an implicit parameter called sender that contains a reference to the sending object (i.e., the object that raised the event).

```
EVENTS evt [EXPORTING parameters].
CLASS-EVENTS evt [EXPORTING parameters].
```

Listing 2.8 Event Declaration Syntax

Types

You can define custom data types within a class using the ABAP TYPES statement. These types are defined at the class level and are therefore not specific to any object instance. You can use these custom types to define local variables within methods, method parameter types, and so on. It's also possible to declare the use of global type pools defined within the ABAP Dictionary using the TYPE-POOLS statement.

The definition of class LCL_PERSON in Listing 2.9 demonstrates how types can be declared and used in a class definition. Here, we've defined a custom structure type called TY_NAME that's being used to define the person's ms_name attribute. The use of the TY_ prefix in this case is by convention: Class-defined types are normally defined using the naming convention TY_{Type_Name}.

```
CLASS lcl_person DEFINITION.
  PRIVATE SECTION.
    TYPES: BEGIN OF ty_name,
             first_name TYPE char40,
             middle_initial TYPE char1,
```

```
            last_name TYPE char40,
          END OF ty_name.
    TYPE-POOLS: szadr.       "Business Address Services

    DATA: ms_name     TYPE ty_name,
          ms_address TYPE szadr_addr1_complete.
ENDCLASS.
```

Listing 2.9 Defining and Working with Class-Level Types

If you look closely at Listing 2.9, you can also see how type groups from the ABAP Dictionary are declared using the TYPE-POOLS statement. In this case, the class has declared the use of the SZADR type group from the *Business Address Services* package. Once this declaration is in place, you can use types such as the SZADR_ADDR1_COMPLETE type in attribute definitions and method signatures.

2.1.3   Implementing Methods

Any time you define methods within the declaration section of a class, you need to provide implementations for them in the implementation section. Such implementations are provided using METHOD...ENDMETHOD statement blocks that are nested inside of a CLASS...IMPLEMENTATION...ENDCLASS statement block, as shown in Listing 2.10.

```
CLASS lcl_date DEFINITION.
  ...
ENDCLASS.

CLASS lcl_date IMPLEMENTATION.
  METHOD add.
    mv_date = mv_date + iv_days.
  ENDMETHOD.

  METHOD subtract.
    mv_date = mv_date - iv_days.
  ENDMETHOD.

  METHOD get_day_of_week.
    "Implementation goes here..
  ENDMETHOD.
ENDCLASS.
```

Listing 2.10 Providing Implementations for Methods

As you can see in Listing 2.10, method implementations allow you to jump right into the code. There's no need to provide any further details about the method context, since you've already defined its signature in the declaration section. Within the method processing block, you can implement the behavior of the class using regular ABAP statements in much the same way that you would implement subroutines and function modules from the procedural world. You'll see many examples of this in the sections that follow.

Syntax Restrictions

If you're coming to ABAP Objects from a procedural background, we should point out that there are a handful of ABAP language constructs that have been rendered obsolete/deprecated from within the object-oriented context. These changes came about as part of a language cleanup effort when SAP first introduced object-oriented extensions to ABAP. SAP saw an opportunity to do some internal housekeeping and ensure that deprecated language elements didn't make their way into new ABAP Objects classes.

For the most part, developers following current best practices shouldn't encounter these statements, as their use is generally frowned upon in any context. Still, if you're not sure which statements have become deprecated over the years, don't worry; the compiler will tell you if you've used one.

Before we wrap up our discussion on method implementations, let's take a moment to discuss *variable scoping rules* in an object-oriented context. Unlike procedural contexts, where the context is pretty cut-and-dry between global variables and local variables, method implementations get their hands on variables at several different scoping levels:

- Class attributes that behave like global variables
- Local variables whose scope is limited to the method that defines them
- Instance variables that sit somewhere in the middle, defining the state of a given object instance

With these additional options in play, you should be careful when qualifying variables so that their usage is clear. This makes the code more readable and prevents you from accidentally *hiding* instance or class attributes behind method-local variables with the same name. As you can expect, hiding instance or class attributes within a method can have some nasty side effects. Fortunately, if you stick to the naming conventions outlined in Section 2.1.2, this shouldn't ever be a concern.

2.2 Working with Objects

Now that you have a feel for how classes are defined in ABAP Objects, let's take a look at how these classes can be utilized from a consumer standpoint. In the sections that follow, you'll learn to define object reference variables, create new object instances, and put them to work in ABAP programs.

2.2.1   Object References

Before you begin creating new object instances, you first need to define variables to hold onto these objects so that you can address them within your programs. For reasons that will be explained in Chapter 4, the ABAP runtime environment does not allow direct access to an object in your programs. Instead, you're given indirect access to allocated objects via a special kind of variable called an *object reference variable*.

Listing 2.11 demonstrates the syntax used to define an object reference variable. Notice the use of the TYPE REF TO addition to indicate that lo_date is a reference variable. When reading this statement, keep in mind that the lo_date is an object reference variable that can point to objects of (class) type LCL_DATE.

```
DATA lo_date TYPE REF TO lcl_date.
```

Listing 2.11 Syntax to Define an Object Reference Variable

You can use this type of syntax to define object reference variables as instance attributes, local variables within method implementations, local variables within form routines, or even global variables.

2.2.2   Creating Objects

Once you define the appropriate object reference variables, you can begin creating object instances using the CREATE OBJECT statement shown in Listing 2.12. Functionally, this statement is processed behind the scenes as follows:

1. First, the ABAP runtime environment dynamically creates a new object of type LCL_ DATE.
2. Then, after the object instance is created, control is handed off to a special method called a *constructor* that provides you with the ability to initialize the object instance before it's used. You'll learn more about constructor methods in Chapter 4.
3. Finally, once the object instance is initialized, the ABAP runtime environment fills in the lo_date variable with a reference that points to the newly created object.

```
DATA lo_date TYPE REF TO lcl_date.
CREATE OBJECT lo_date.
```

Listing 2.12 Instantiating an Object at Runtime

From a syntax perspective, that's all there is to instantiating objects. Anytime you want a new object reference, you simply use the CREATE OBJECT statement to allocate one on the fly. Of course, if you're not careful in maintaining your object reference variables, these objects can become orphaned. With that in mind, the next section focuses on the important topic of object reference assignments.

2.2.3 Object Reference Assignments

Since object reference variables are basically a special kind of variable, they can be used in assignment statements using the familiar equals (=) operator. Of course, when assigning object reference variables, it's important to remember *what* you're assigning. To put this concept into perspective, consider the assignment scenario in Listing 2.13.

```
DATA lo_date1 TYPE REF TO lcl_date.
DATA lo_date2 TYPE REF TO lcl_date.

CREATE OBJECT lo_date1.
CREATE OBJECT lo_date2.

lo_date1 = lo_date2.
```

Listing 2.13 Understanding Object Reference Assignments

Within the code excerpt in Listing 2.13, we have two object reference variables called lo_date1 and lo_date2 that point to newly created LCL_DATE objects. Prior to the assignment statement at the bottom of the code excerpt, the variable assignments resemble what is shown in Figure 2.2. Notice that the objects themselves are not stored within the object reference variables. Instead, the object reference variables contain an address for where the object exists in memory.

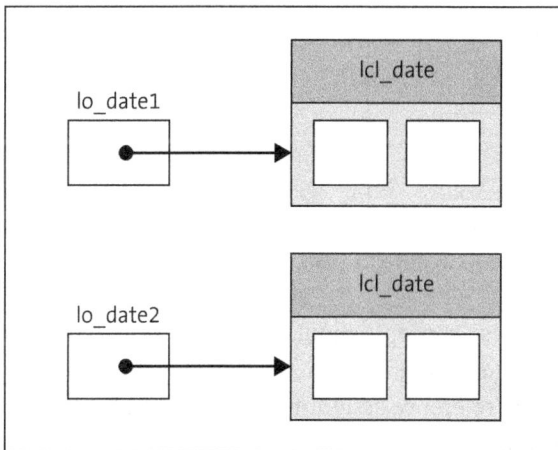

Figure 2.2 Understanding Object-Reference Assignments (Part 1)

The diagram in Figure 2.3 illustrates what things look like *after* the object reference assignment is performed at the bottom of Listing 2.13. Here, you can see that the assignment statement has copied the address of the LCL_DATE object instance pointed to by the lo_date2 object reference into lo_date1. Now, both lo_date1 and lo_date2 point to the same object instance (i.e., the instance at the bottom of Figure 2.3).

If you look closely at the before and after memory snapshots in Figure 2.2 and Figure 2.3, you can draw several important conclusions about object reference assignments:

1. First, it should be clear that object reference assignments only copy the *addresses* of objects, and not the objects themselves. This implies that object reference assignments are relatively inexpensive from a performance standpoint.

2. Second, any time you have two or more object reference variables that point to the same object instance, changes made to the object via one object reference variable will be reflected in the other object reference variables. This should come as no surprise, since the object reference variables all point to the same object instance.

3. Finally, if an object instance is no longer pointed to by any live object reference variables, the object instance becomes *orphaned* and is no longer accessible from a programming context. In Chapter 4, you'll see how a special service of the ABAP runtime environment called the *garbage collector* cleans up these orphaned objects to recoup unused memory.

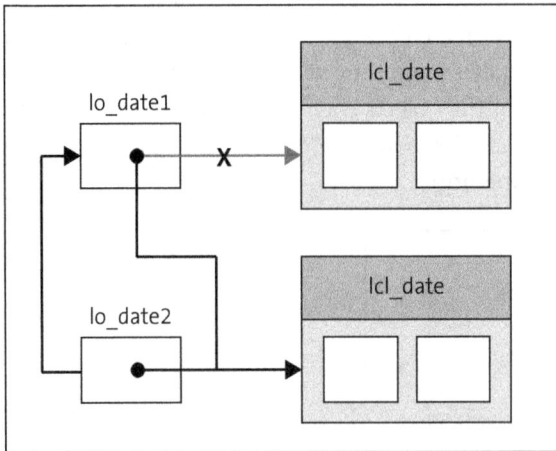

Figure 2.3 Understanding Object-Reference Assignments (Part 2)

With time and a little bit of practice, these concepts should become second nature. In the meantime, though, we recommend taking a methodical approach to creating object instances and performing object reference assignments. For example, consider the code excerpt in Listing 2.14. The intent was to create 10 date objects but, since there's only one object reference variable, the first 9 date objects are created and then subsequently orphaned.

```
DATA lo_date TYPE REF TO lcl_date.
DO 10 TIMES.
  CREATE OBJECT lo_date.
ENDDO.
```

Listing 2.14 An Invalid Idiom for Creating a Collection of Objects

Listing 2.15 corrects the error from Listing 2.14 by introducing an internal table of object references. Now, each new date object that's created is stored in a separate object reference variable within the table. As obvious as this may seem, these are the types of issues that can occur if you aren't careful with your object handling.

```
DATA lt_dates TYPE STANDARD TABLE OF REF TO lcl_date.
FIELD-SYMBOLS <lo_date> LIKE LINE OF lt_dates.

DO 10 TIMES.
  APPEND INITIAL LINE TO lt_dates ASSIGNING <lo_date>.
  CREATE OBJECT <lo_date>.
ENDDO.
```

Listing 2.15 Defining a Collection of Objects

Thinking (Object)ively

For many ABAP developers, the notion of reference variables is a foreign concept. If you find yourself getting tripped up by all this indirection, perhaps an analogy will help. Consider the relationship between a remote control and a TV set.

Remote controls are small, lightweight devices that make it easy for you to control a TV set. As long as you have your remote control, you can turn on the TV, change the channel, and control the volume as desired. However, if you were to lose the remote, then you'd no longer be able to access the TV (at least, not without getting off the couch). To guard against such occurrences, you could buy a universal remote as a backup. That way, you could program the universal remote to point to the TV's remote frequency. Once the universal remote is programmed, you could control the TV with either remote, since they both effectively point to the same TV.

Relating this back to our object reference discussion, object reference variables are rather like remote controls. As long as an object reference variable points to a particular object instance, you can use the object reference to control the object it points to. However, if you reassign the object reference or clear it out using the ABAP CLEAR statement, then you can no longer use it to access the object instance. This doesn't mean the object is deleted, the same way that your TV would still exist even if you lost your remote control. What it does mean is that you may no longer be able to access the object if you don't have another object reference variable on hand that happens to point to that object. This is the OOP equivalent of losing all of your remotes in the couch cushions.

The moral of the story is to treat object references with care. Make sure that you're really done with an object before clearing its object reference variables.

2.2.4 Accessing Instance Components

As you learned in the previous sections, object reference variables provide a handle for addressing object instances. Using this handle, you can access the instance components of an object by building compound expressions using the object component selector operator (->), as shown in Listing 2.16. Here, you can see that the object component selector allows you to specify which instance component you want to access within a given object instance.

```
oref->attribute
oref->method()
CALL METHOD oref->method( )
```

Listing 2.16 Working with the Object Component Selector (Part 1)

What's the Proper Syntax for Calling a Method?

As you can see in Listing 2.16, there are two different ways to call methods. These days, the direct oref->method() option is the preferred option, as it more closely resembles syntax used in other object-oriented languages. The CALL METHOD statement is still valid, of course, but you should avoid it as a rule. In Section 2.2.7 and Section 2.2.8, you'll understand why it's a good idea to get into the habit of calling methods directly.

To demonstrate the use of the object component selector operator, let's take a look at an example. Imagine that you're modeling a 2D graphics system and want to create an object to represent points in the Cartesian coordinate system. If it's been a while since your last high school geometry class, a Cartesian coordinate system (or plane) is a two-dimensional grid that contains a horizontal x-axis and vertical y-axis (see Figure 2.4). To plot points on the graph, all you have to do is specify an x-coordinate and a y-coordinate. This is demonstrated in Figure 2.4 where we've plotted a point at (1,2).

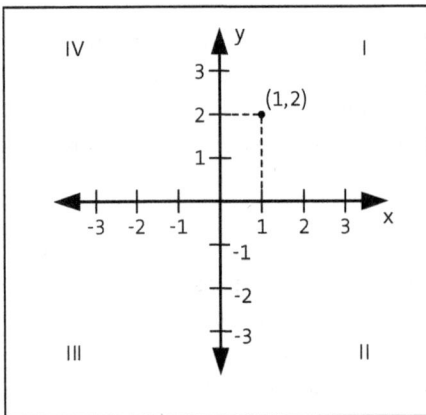

Figure 2.4 Modeling a Point Object in the Cartesian Coordinate System

To model our point object, we'll create a new class called LCL_POINT, as shown in Listing 2.17. This class contains three instance components: two instance attributes called mv_x and mv_y to represent the x and y coordinates, respectively, and an instance method called get_distance() that can be used to calculate the Euclidian distance between the current point and some other point within the plane.

```
CLASS lcl_point DEFINITION.
  PUBLIC SECTION.
    DATA: mv_x TYPE p DECIMALS 2   "X-Coordinate
          mv_y TYPE p DECIMALS 2.  "Y-Coordinate

    METHODS get_distance IMPORTING io_point2
                           TYPE REF TO lcl_point
                           RETURNING VALUE(rv_distance) TYPE f.
ENDCLASS.

CLASS lcl_point IMPLEMENTATION.
  METHOD get_distance.
    DATA: lv_dx TYPE f,       "Diff. X
          lv_dy TYPE f.       "Diff. Y

    "Calculate the Euclidean distance between the points:
    lv_dx = io_point2->mv_x - me->mv_x.
    lv_dy = io_point2->mv_y - me->mv_y.

    rv_distance =
        sqrt( ( lv_dx * lv_dx ) + ( lv_dy * lv_dy ) ).
  ENDMETHOD.
ENDCLASS.
```

Listing 2.17 Working with the Object Component Selector (Part 2)

If you look closely at the implementation of the get_distance() method, you can see that the object component selector is used to access the instance attributes of two different objects: the io_point2 object passed to the method and the current point object. In the latter case, we're referring to the current point object's instance attributes using the me self-reference variable.

Where Does the me Self-Reference Variable Come From?

The me self-reference variable is a special instance attribute implicitly defined by the ABAP runtime environment whenever an object instance is created. As its name implies, the me reference variable points back to its containing object. If you've worked with other object-oriented languages such as Java, you can think of the me reference variable as being equivalent to the this self-reference variable.

From a usage perspective, the me self-reference variable can be used just like any other object reference variable. For example, in Listing 2.17 we used me to access the mv_x and mv_y instance attributes of the LCL_POINT class. Technically speaking, we didn't have to use me to access these attributes. Instead, we could have simply referenced the attributes directly and the system would have quietly resolved the reference behind the scenes. The advantage of qualifying these references directly is that we make our intentions clear to the reader.

Another place where the me reference variable is used is in situations where you want to pass the current object instance as a parameter to another method. In this case, me provides you with a convenient mechanism for accessing the current object directly within a method implementation.

The code excerpt in Listing 2.18 demonstrates how to use the object component selector to access attributes and methods from outside of a class. Here, we're using the selector to:

- Initialize the instance attributes of a pair of point objects (lo_point_a and lo_point_b, respectively).
- Invoke the get_distance() method to calculate the distance between the two points.

```
DATA: lo_point_a  TYPE REF TO lcl_point,
      lo_point_b  TYPE REF TO lcl_point,
      lv_distance TYPE f.

"Instantiate both of the point objects:
CREATE OBJECT lo_point_a.
lo_point_a->mv_x = 1.
lo_point_a->mv_y = 1.

CREATE OBJECT lo_point_b.
lo_point_b->mv_x = 3.
lo_point_b->mv_y = 3.

"Calculate the distance & display the results:
lv_distance = lo_point_a->get_distance( lo_point_b ).
WRITE: 'Distance between point a and point b is: ',
       lv_distance.
```

Listing 2.18 Working with the Object Component Selector (Part 3)

2.2.5   Accessing Class Components

To demonstrate how to access class components, Listing 2.19 shows how we can enhance the LCL_POINT class we developed in the previous section to incorporate class components. Here, we'll introduce a new class method called create_from_polar() that can be used to create point objects using polar coordinates. To drive the conversion routine, we've also created a constant called CO_PI to represent the value of pi.

```
CLASS lcl_point DEFINITION.
  PUBLIC SECTION.
    CONSTANTS CO_PI TYPE f VALUE '3.14159265'.

    CLASS-METHODS:
      create_from_polar IMPORTING iv_r TYPE f iv_theta TYPE p
                        RETURNING VALUE(ro_point) TYPE REF TO lcl_point.
    ...
ENDCLASS.

CLASS lcl_point IMPLEMENTATION.
  METHOD create_from_polar.
    "Convert the angle measure to radians:
    DATA lv_theta_rad TYPE f.
    lv_theta_rad = ( iv_theta * CO_PI ) / 180.

    "Create a new point object and calculate the
    "X & Y coordinates:
    CREATE OBJECT ro_point.

    ro_point->mv_x = iv_r * cos( lv_theta_rad ).
    ro_point->mv_y = iv_r * sin( lv_theta_rad ).
  ENDMETHOD.
  ...
ENDCLASS.
```

Listing 2.19 Defining Class Components

Since our new create_from_polar() method is defined at the class level, we don't require an object reference to access it. Instead, we can access it via the static/class context using the class component selector operator (=>), as shown in Listing 2.20. Here, you can see how we're also accessing the CO_PI constant using the same kind of syntax: {class_name}=>{class_component}.

```
DATA lo_point TYPE REF TO lcl_point.
DATA lv_message TYPE string.

lo_point = lcl_point=>create_from_polar( iv_r = '3.6' iv_theta = '56.31' ).
lv_message =
  |Coordinates: ({ lo_point->mv_x }, { lo_point->mv_y })|.
WRITE: / lv_message.
lv_message = |PI is { lcl_point=>CO_PI }|.
WRITE: / lv_message.
```

Listing 2.20 Working with the Class Component Selector

If you look back at Listing 2.19, you'll notice that we didn't qualify the use of the `CO_PI` constant within the `create_from_polar()` method. Within the class itself, such qualifications are optional, since the class context is implicitly known. Whether or not you choose to formally qualify such references is strictly a matter of preference.

2.2.6   Working with Events

For developers coming into ABAP Objects with a background in other object-oriented languages such as Java or C#, the concept of events as first-class citizens of class definitions may seem a bit foreign. However, once you understand how events work, it's easy to see how they relate to common object synchronization patterns employed in those environments (e.g., the observer pattern).

From a conceptual perspective, events are a special kind of component you can use to model important milestones that might occur during an object's lifecycle. Such milestones could be unique to a particular object instance (instance events) or to the class itself (class events). In either case, whenever a particular milestone is reached, you can highlight the occurrence by *raising* an event. You can have interested parties (i.e., other objects) listen for these events by registering them as *event handlers*. This allows the ABAP runtime environment to automatically notify objects of the event.

This exchange is illustrated by the diagram in Figure 2.5. Here, we have a class called `LCL_PUBLISHER` that defines an instance event called `MESSAGE_ADDED`.

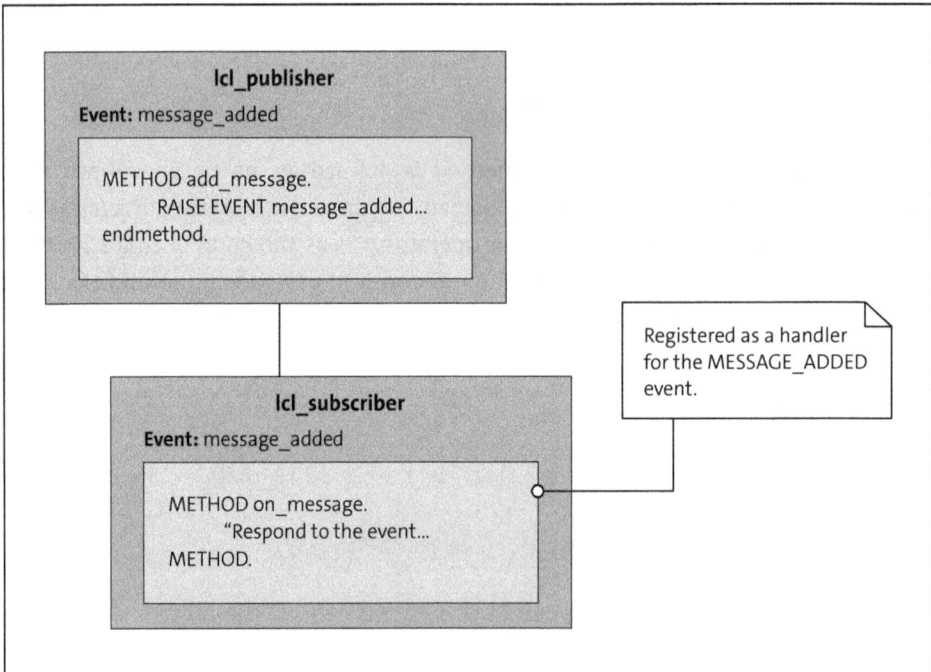

Figure 2.5 Understanding the Event Process Flow

This event is triggered whenever the publisher receives a new message via its add_
message() instance method. On the other end of the exchange, we have another class
called LCL_SUBSCRIBER that is registered as a *listener* for the MESSAGE_ADDED event. Now,
whenever a new message arrives at the publisher, instances of LCL_SUBSCRIBER will be
notified via the on_message() event handler method.

Event-Related Syntax

If you're looking at the event process flow in Figure 2.5, you might be wondering how
the on_message() method was fired in response to the MESSAGE_ADDED event. Unlike the
methods you've seen thus far, the on_message() method is defined as an *event handler
method*. As you might expect, event handler methods are specialized methods that register themselves as listeners for particular types of events. You can define event handler
methods within the same class that declared the event or in a completely separate class.

To declare event handler methods, you must once again enlist the aid of the METHODS
statement, as shown in Listing 2.21. Here, the FOR EVENT...OF CLASS addition links the
method with the corresponding event it's defined to handle. As you would expect, the
importing parameter list must match the exporting parameter list defined by the
event it's listening for.

```
METHODS {method_name}
  FOR EVENT {event} of CLASS {class_name}
  [IMPORTING p1 p2 ... [sender]].
```

Listing 2.21 Declaring Event Handler Methods

Once an event handler method is defined, you can register it to listen for events using
the SET HANDLER statement, whose syntax is shown in Listing 2.22. Here, the handler
tokens refer to event handler methods (without quotes) that are defined within the
class from which the SET HANDLER statement emanates. The remaining additions are
defined as follows:

- When registering event handler methods for instance events, you have a couple of
  options for identifying the scope of the event binding:
  - The optional FOR oref addition is used to bind an event handler method to a spe-
    cific object instance.
  - Alternatively, you can use the ALL INSTANCES addition to bind an event handler
    method to all object instances.
- When registering event handler methods for class events, you don't have to specify
  an object context, so neither of the FOR oref and ALL INSTANCES additions apply.
- Finally, for both instance and class event bindings, you have the option of activating
  and deactivating an event registration using the ACTIVATION addition. Here, you can
  activate an event handler method using the true ('X') value or deactivate the
  method using the false (space) value.

```
SET HANDLER handler1 handler2 ... [FOR oref|{ALL INSTANCES}]
                                  [ACTIVATION {'X'|' '}].
```

Listing 2.22 Registering Event Handler Methods

The final piece to the event syntax puzzle is the RAISE EVENT statement, whose syntax is shown in Listing 2.23. As you can see, the syntax is straightforward: You simply specify the event being raised and pass along any parameters that event handlers will use to process the event downstream.

```
RAISE EVENT evt [EXPORTING p1 = a1 p2 = a2 ...].
```

Listing 2.23 Syntax for Raising Events

Putting It All Together

To see how all this comes together in real-life ABAP code, let's look at how you might build the LCL_PUBLISHER and LCL_SUBSCRIBER classes depicted in Figure 2.5. Listing 2.24 shows the code for the LCL_PUBLISHER class. Here, you can see how we've defined the message_added event using the EVENTS keyword introduced in Section 2.1.2. This event is then triggered from within the add_message() method using the RAISE EVENT statement.

```
CLASS lcl_publisher DEFINITION.
  PUBLIC SECTION.
    METHODS:
      add_message IMPORTING iv_message TYPE string,
      confirm_receipt IMPORTING iv_subscriber TYPE string.

    EVENTS:
      message_added
        EXPORTING VALUE(ev_message) TYPE string.
ENDCLASS.

CLASS lcl_publisher IMPLEMENTATION.
  METHOD add_message.
    DATA lv_message TYPE string.
    lv_message = |Publishing message: [{ iv_message }].|.
    WRITE: / lv_message.

    RAISE EVENT message_added
      EXPORTING
        ev_message = iv_message.
  ENDMETHOD.

  METHOD confirm_receipt.
    DATA lv_message TYPE string.
    lv_message = |Message processed by { iv_subscriber }.|.
```

```
    WRITE: / lv_message.
  ENDMETHOD.
ENDCLASS.
```

Listing 2.24 Defining and Raising Events

Listing 2.25 contains the definition of the LCL_SUBSCRIBER class, which is listening for messages issued from the LCL_PUBLISHER class. We've defined an event handler method called on_message() that will be used to process publication events at runtime. The event binding takes place within the constructor() method using the SET HANDLER statement. You'll learn more about constructor methods in Chapter 4, but for now, know that this method is invoked automatically whenever an LCL_SUBSCRIBER instance is created.

```
CLASS lcl_subscriber DEFINITION.
  PUBLIC SECTION.
    METHODS:
      constructor,

      on_message FOR EVENT message_added
                  OF lcl_publisher
                IMPORTING
                  ev_message sender.

ENDCLASS.

CLASS lcl_subscriber IMPLEMENTATION.
  METHOD constructor.
    SET HANDLER on_message FOR ALL INSTANCES.
  ENDMETHOD.

  METHOD on_message.
    DATA lv_message TYPE string.
    lv_message = |Received message [{ ev_message }]|.
    WRITE: / lv_message.

    sender->confirm_receipt( 'LCL_SUBSCRIBER' ).
  ENDMETHOD.
ENDCLASS.
```

Listing 2.25 Defining and Registering an Event Handler Method

With both classes in place, you can run a test by passing a message to the add_message() method of the LCL_PUBLISHER class (see Listing 2.26). This will trigger the MESSAGE_ADDED event and allow you to see how the LCL_SUBSCRIBER class responds. Once you play around with this and learn how to interact with the event processing loop, you'll find that this feature offers many interesting possibilities.

```
DATA lo_publisher TYPE REF TO lcl_publisher.
DATA lo_subscriber TYPE REF TO lcl_subscriber.

CREATE OBJECT lo_publisher.
CREATE OBJECT lo_subscriber.

lo_publisher->add_message( 'Ping...' ).
```

Listing 2.26 Testing the Event Processing Loop

2.2.7   Working with Functional Methods

As we've stated from the outset, one of the main goals with OOP is to develop code that's intuitive and easy to read. One of the ways that object-oriented languages achieve this is by providing a syntax that resembles the sentence structure of spoken languages. For example, if you think about a method call, you have a subject (either an object or a class) and a verb (the method being called). With a little bit of creativity and proper naming, you can build statements that even people without a technical background can read and understand (at least conceptually).

To make your code flow even better, you can employ the use of *functional methods*. As the name suggests, functional methods are used to compute a single discrete value. The value in this approach is that you can plug in functional methods in the operand positions of various ABAP statements to build powerful expressions.

Listing 2.27 illustrates the basic syntax used to declare functional methods. Here, as before, you can declare IMPORTING parameters to provide inputs to the method. The lone output of the method is provided in the form of the RETURNING value parameter. As is the case with other parameter types, you're generally free to define the type of the returning parameter using the same rules that apply for EXPORTING parameters. However, type selection does play a role in determining whether a functional method can be used as an operand in selected ABAP statements.

```
METHODS func_method
   [IMPORTING parameters]
   RETURNING VALUE(rval) TYPE type
   [EXCEPTIONS...].
```

Listing 2.27 Functional Method Declaration Syntax

To demonstrate how functional methods are used in ABAP code, let's look at an example. In Listing 2.28, we've created a string tokenizer class called LCL_STRING_TOKENIZER that can be used to parse through delimited records and make it easy to access individual string tokens. This class defines two functional methods:

- The has_more_tokens() method is a Boolean method that can be used to determine whether there are more tokens in the sequence.

- The next_token() method provides a simple mechanism for accessing the next token in the sequence.

```
CLASS lcl_string_tokenizer DEFINITION.
  PUBLIC SECTION.
    METHODS:
      constructor IMPORTING iv_string TYPE csequence
                            iv_delimiter TYPE csequence,

      has_more_tokens RETURNING VALUE(rv_result) TYPE abap_bool,

      next_token RETURNING VALUE(rv_token) TYPE string.

  PRIVATE SECTION.
    DATA mt_tokens TYPE string_table.
    DATA mv_index TYPE i.
ENDCLASS.

CLASS lcl_string_tokenizer IMPLEMENTATION.
  METHOD constructor.
    SPLIT iv_string AT iv_delimiter INTO TABLE me->mt_tokens.

    IF lines( me->mt_tokens ) GT 0.
      me->mv_index = 1.
    ELSE.
      me->mv_index = 0.
    ENDIF.
  ENDMETHOD.

  METHOD has_more_tokens.
    IF me->mv_index LE lines( me->mt_tokens ).
      rv_result = abap_true.
    ELSE.
      rv_result = abap_false.
    ENDIF.
  ENDMETHOD.

  METHOD next_token.
    READ TABLE me->mt_tokens INDEX me->mv_index INTO rv_token.
    ADD 1 TO me->mv_index.
  ENDMETHOD.
ENDCLASS.
```

Listing 2.28 Working with Functional Methods (Part 1)

The code excerpt in Listing 2.29 demonstrates how you can use your string tokenizer class within regular ABAP code. Notice how we're using the has_more_rows() method as the basis of the logical expression that drives the WHILE loop that processes the string tokens. At runtime, this method will be invoked prior to the evaluation of the logical expression, and the returned value will be used to determine if the WHILE loop should continue. Not only does this save you a few lines of code, it also makes the code much more intuitive.

```
DATA lo_tokenizer TYPE REF TO lcl_string_tokenizer.
DATA lv_token TYPE string.

CREATE OBJECT lo_tokenizer
  EXPORTING
    iv_string = '09/13/2005'
    iv_delimiter = '/'.

WHILE lo_tokenizer->has_more_tokens( ) EQ abap_true.
  lv_token = lo_tokenizer->next_token( ).
  WRITE: / lv_token.
ENDWHILE.
```

Listing 2.29 Working with Functional Methods (Part 2)

Table 2.4 provides some further examples of places where you can use functional methods in common ABAP expressions.

ABAP Expression	Where It's Used
Conditional expressions (e.g., IF and WHILE statements)	As an operand in a logical expression. Example: IF oref->get_weight( ) GT 100. ... ENDIF.
CASE	As an operand in a logical expression. Example: CASE oref->get_type( ). WHEN oref->get_value1( ). ... ENDCASE.

Table 2.4 Using Functional Methods in Expressions

ABAP Expression	Where It's Used

LOOP AT, DELETE, and MODIFY	As part of the logical expression in a WHERE clause. Example:
	`LOOP AT itab`
	` WHERE field EQ oref->get_val( ).`
	` ...`
	`ENDLOOP.`

Table 2.4 Using Functional Methods in Expressions (Cont.)

Enhancements to Functional Methods

A functional method's signature supports exporting and changing parameters in addition to the singular returning parameter. Such methods can still be used inline within regular ABAP expressions; the extra exporting and changing parameters simply come along for the ride.

Predicative method calls are another useful feature. As the name suggests, these are functional method calls where the result is used as a predicate in logical expressions. To put this into perspective, consider the way that we're using predicative method calls to refactor the WHILE loop in Listing 2.29. Notice that we no longer have to compare the result of the has_more_tokens() method using a logic expression. In this context, if the returning value parameter of has_more_tokens() is initial, then the result is false; all non-initial values evaluate to true. Therefore, we can define the signature of methods using Boolean approximation types such as ABAP_BOOL or pretty much any other data type. Of course, for readability's sake, we encourage you to define your functional methods using familiar Boolean types wherever possible.

```
WHILE lo_tokenizer->has_more_tokens( ).
  ...
ENDWHILE.
```

Note that no syntactical changes are required in the implementation of methods such as has_more_tokens() to exploit this functionality. You can continue to develop functional methods as per usual, only now you can incorporate them into functional expressions in a concise and readable manner.

2.2.8   Chaining Method Calls Together

Chained method calls are a type of syntactic sugar popular among ABAP developers. They make it easy to consolidate a handful of operations into a single line of code.

To understand how chained method calls work, consider this example. In Listing 2.30, we've created a simple string utilities class called LCL_STRING. Within this class, we've defined several functional methods that perform various operations on a string value:

converting the string to upper case, trimming of leading/trailing whitespace, and replacing characters. This is all standard fare, until you get to the part where each of these methods passes back a copy of the me self-reference variable. This subtle addition to the code is what makes method chaining possible.

```abap
CLASS lcl_string DEFINITION.
  PUBLIC SECTION.
    METHODS:
      constructor IMPORTING iv_string TYPE csequence,

      trim RETURNING VALUE(ro_string) TYPE REF TO lcl_string,

      upper RETURNING VALUE(ro_string)
              TYPE REF TO lcl_string,

      replace IMPORTING iv_pattern TYPE string
                        iv_replace TYPE string
            RETURNING VALUE(ro_string)
                TYPE REF TO lcl_string,

      get_value RETURNING VALUE(rv_value) TYPE string.

  PRIVATE SECTION.
    DATA mv_string TYPE string.
ENDCLASS.

CLASS lcl_string IMPLEMENTATION.
  METHOD constructor.
    me->mv_string = iv_string.
  ENDMETHOD.

  METHOD trim.
    me->mv_string =
      condense( val = me->mv_string from = `` ).
    ro_string = me.
  ENDMETHOD.

  METHOD upper.
    me->mv_string = to_upper( val = me->mv_string ).
    ro_string = me.
  ENDMETHOD.

  METHOD replace.
    REPLACE ALL OCCURRENCES OF REGEX iv_pattern
        IN me->mv_string WITH iv_replace.
```

```
    ro_string = me.
  ENDMETHOD.

  METHOD get_value.
    rv_value = me->mv_string.
  ENDMETHOD.
ENDCLASS.
```

Listing 2.30 Working with Chained Methods (Part 1)

The code excerpt in Listing 2.31 demonstrates how to implement chained method calls from a code perspective. Here, you can see how we're taking an existing string and performing multiple operations on it in one go. This starts with the call to the trim() method. This method strips off the leading/trailing whitespace and then passes back a copy of the me self-reference. The resultant object reference is then used as the basis for the subsequent call to the upper() method, which follows the same kind of pattern. The call chain ultimately terminates with the call to get_value(), at which time we receive the formatted text "PAIGE_A_PUMPKIN".

> **Note**
> We added the line break between the calls to upper() and replace() so that the statement would fit onto a printed page in the book. Within the ABAP Editor, this sort of line break would result in a syntax error.

```
DATA lo_string TYPE REF TO lcl_string.
DATA lv_new_value TYPE string.

CREATE OBJECT lo_string
  EXPORTING
    iv_string = `  Paige A Pumpkin  `.

lv_new_value =
  lo_string->trim( )->upper( )->
    replace( iv_pattern = `\s` iv_replace = '_' )->get_value( ).

WRITE: / lv_new_value.
```

Listing 2.31 Working with Chained Methods (Part 2)

As you can see in the example, chained method calls make it easy to string together related operations in one condensed statement. For simple operations like the ones demonstrated in Listing 2.31, this makes logical sense. For more complex statements, though, chained method calls are probably a bad idea. We leave it to you as a responsible developer to know when it makes sense to sacrifice readability to save a few keystrokes.

2.3   Building Your First Object-Oriented Program

In the previous section, we looked at several examples that demonstrated how to work with objects. However, since these code excerpts were isolated, you might be wondering how all these pieces fit together in actual ABAP programs. With that in mind, this section will demonstrate the creation of a simple report program that utilizes a local class. As you'll come to find out, these concepts apply equally to the incorporation of local classes to function group definitions, module pool programs, and so on.

2.3.1   Creating the Report Program

To get things started, let's create the report program that will drive our demo. If you're new to ABAP development, this can be achieved by performing the following steps:

1. To begin, log onto the system and open the Object Navigator (Transaction SE80).

2. In the object list selection box in the **Repository Browser** on the left-hand side of the screen, choose the **Local Objects** list option (see Figure 2.6).

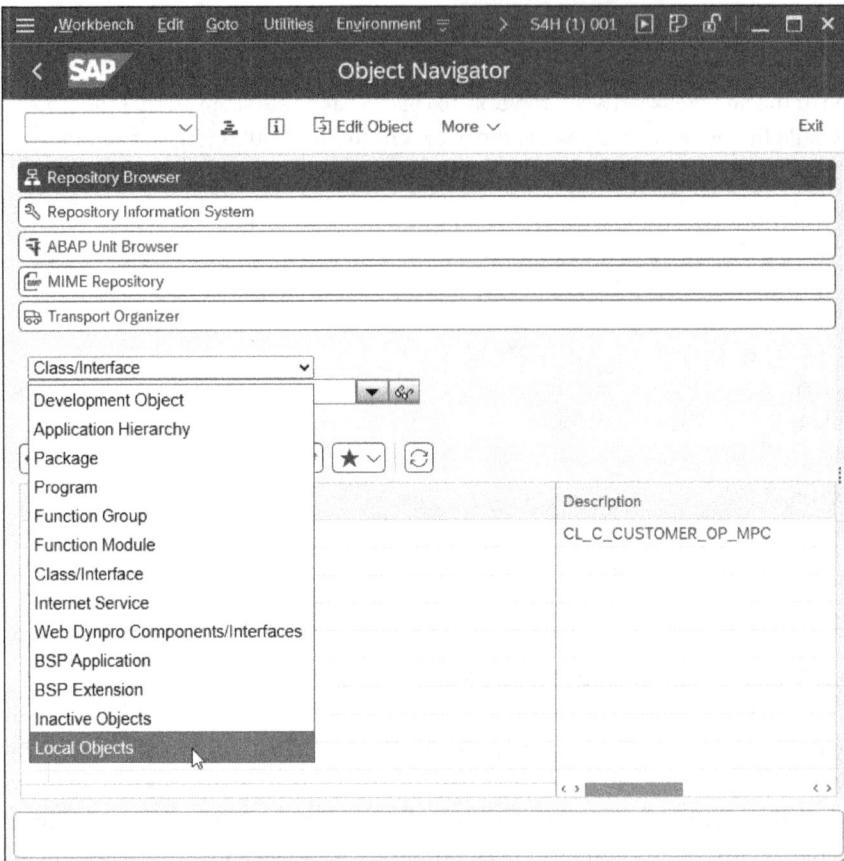

Figure 2.6 Selecting the Local Objects Repository View

3. This will pull up a tree view of locally defined development objects for your user account, as shown in Figure 2.7. To create a new report program, right-click on the top-level object node (i.e., **$TMP JWOOD** in Figure 2.7) and select **Create · Program** from the menu.

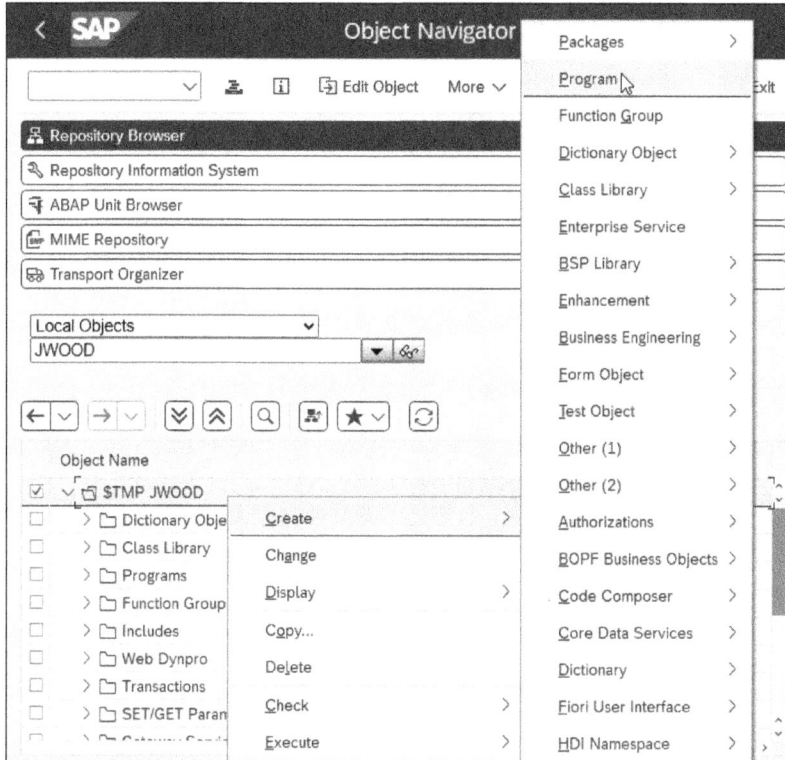

Figure 2.7 Creating a Report Program (Part 1)

4. Next, you'll be presented with the **Create Program** dialog box shown in Figure 2.8. At this step, simply specify the name of the report program (we called our report YDATE_DEMO) and press [Enter] to continue. Note that you shouldn't select the **Create with TOP Include** checkbox in this case since you're just building a simple report.

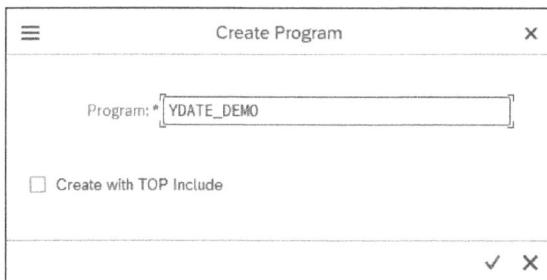

Figure 2.8 Creating a Report Program (Part 2)

5. Figure 2.9 shows the next dialog box in the creation process, where you provide a program title and additional attributes concerning the program setup. For the purpose of this simple demonstration, you can leave the default settings and click on the **Save** button to continue.

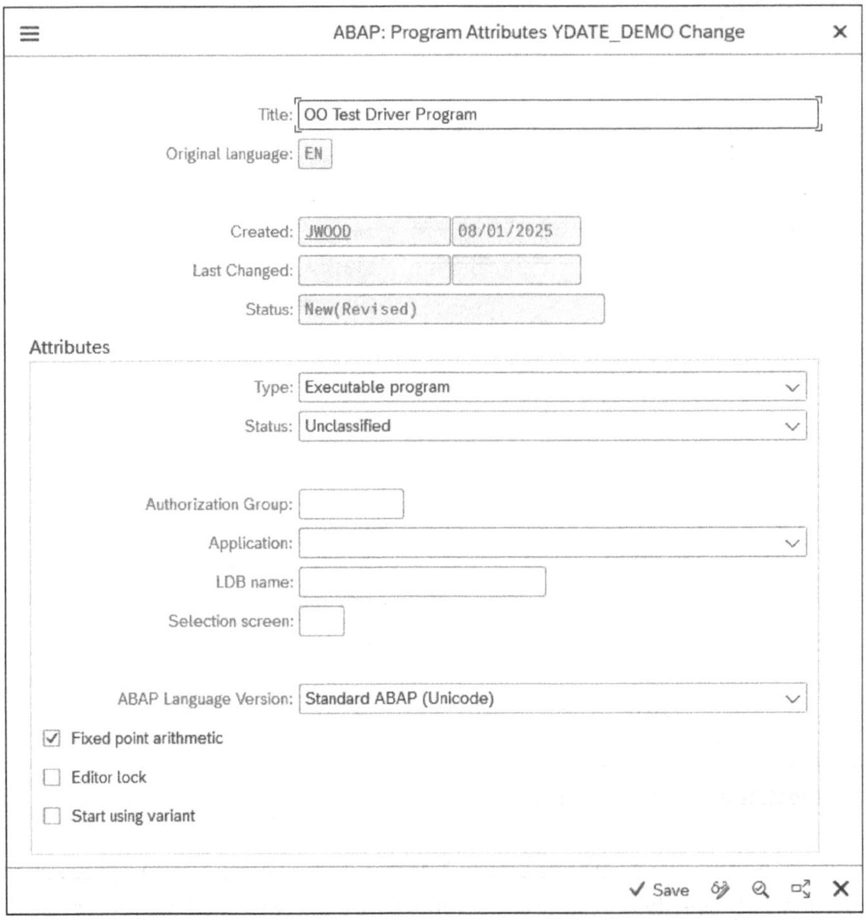

Figure 2.9 Creating a Report Program (Part 3)

6. At the next step, as shown in Figure 2.10, you'll be asked to select a package to store the object in within the ABAP Repository. Since this is a demo program, leave the default **$TMP** package selection and click the **Save** icon to continue. That way, the program will only be defined locally and can't be transported.

7. Finally, if all goes well, you should end up at an editor screen like the one shown in Figure 2.11. From there, you can get started with your coding exercise.

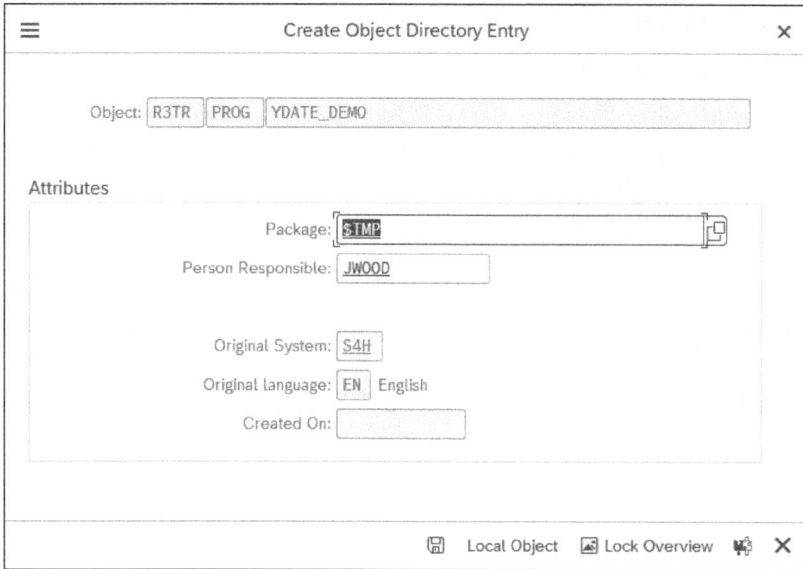

Figure 2.10 Creating a Report Program (Part 4)

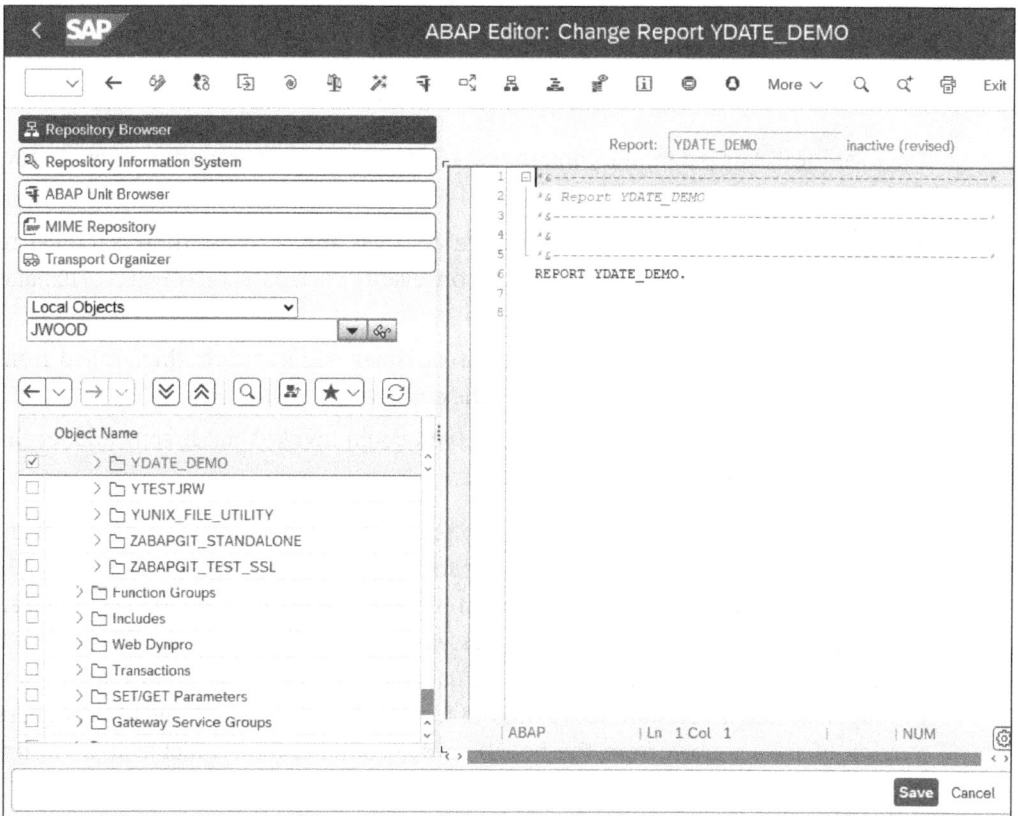

Figure 2.11 Creating a Report Program (Part 5)

2.3.2   Adding in the Local Class Definition

Once the report program is created, you can begin defining your local class in one of two ways:

- You can start keying in the class definition directly underneath the REPORT statement just like you would for other type definitions.
- Or, you can create an INCLUDE program and key in the class definition there.

The ABAP compiler doesn't care which option you choose, so it's up to you to decide how best to organize your code. For now, we'll keep things simple and define the class directly within the report program (see Listing 2.32). In Section 2.4, we'll take a closer look at some logistical implications when defining classes.

```
REPORT ydate_demo.
CLASS lcl_date DEFINITION.
  ...
ENDCLASS.

CLASS lcl_date IMPLEMENTATION.
  ...
ENDCLASS.
```

Listing 2.32 Defining Local Classes within a Report Program

Once the local class is defined, you can access it from within the report program in several different ways:

- You could define global object reference variables and then use those variables to create and use objects from within report events such as START-OF-SELECTION and END-OF-SELECTION.
- You could define local object reference variables within subroutines called from within the report program and access the objects that way.
- If the class defines a main class method, you could invoke that directly and let the class itself drive the main program logic.

Since this is a book about OOP, we tend to prefer the third option, as it frees us from having to mix-and-match programming paradigms. The code excerpt in Listing 2.33 demonstrates this approach. Here, you can see how the main program logic is driven by the main() class method, which is accessed directly within the START-OF-SELECTION event module. From here, it's OOP as per usual. Figure 2.12 shows what the program output looks like. If you want to try this out for yourself, you can download a complete version of the program from the book's source code bundle (available at *www.sap-press.com/6093*).

```
REPORT zoopbook_date_demo.
CLASS lcl_date DEFINITION.
  PUBLIC SECTION.
    CLASS-METHODS:
      main.
      ...
ENDCLASS.

CLASS lcl_date IMPLEMENTATION.
  METHOD main.
    DATA lo_birth_date TYPE REF TO lcl_date.
    DATA lv_message TYPE string.

    CREATE OBJECT lo_birth_date
      EXPORTING
        iv_date = '20030113'.

    lv_message =
      |Andersen was born on a
        { lo_birth_date->get_day_of_week( ) }.|.
    WRITE: / lv_message.

    lv_message =
      |Official birth date:
        { lo_birth_date->get_long_format( ) }.|.
    WRITE: / lv_message.
  ENDMETHOD.
ENDCLASS.

START-OF-SELECTION.
  lcl_date=>main( ).
```

Listing 2.33 Integrating Local Classes Inside Report Programs

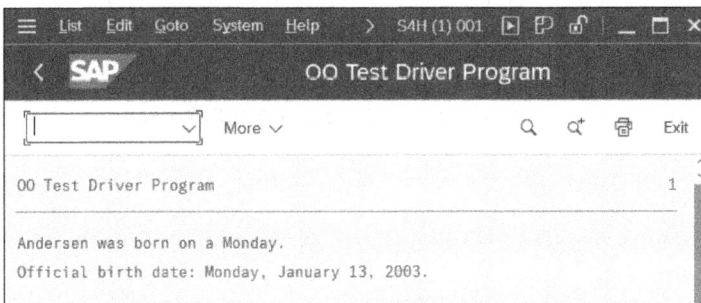

Figure 2.12 Output of the Example Program

2.4   Working with Global Classes

As we've noted at several points throughout this chapter, ABAP distinguishes between two different types of classes:

- **Local classes**
  These are the types of classes that we've looked at thus far. As you observed in Section 2.3, these classes are defined within some other ABAP module such as a report program or include program. This implies that local classes have limited visibility within the system.

- **Global classes**
  Global classes, on the other hand, are standalone ABAP Repository artifacts that are visible across the system. In this regard, you can think of global classes as being on equal footing with globally defined function modules.

From a syntax perspective, local and global classes are the same. However, as you'll learn in this section, the way that you go about developing global classes is a little bit different. With that in mind, let's jump in and take a look at global classes.

2.4.1   Understanding the Class Pool Concept

From a technical perspective, global classes are stored within the ABAP Repository inside of a special repository object called a *class pool*. Class pools are special ABAP program types that define a single global repository class as well as related local type definitions used to support the implementation of that class. Class pools are similar to function groups in the sense that they cannot be executed directly. Instead, runtime object instances are created using the CREATE OBJECT statement and then processed from there.

2.4.2   Getting Started with the Class Builder Tool

Class pools are maintained within a specialized tool in the ABAP Workbench called the *Class Builder*. To access the Class Builder from within the ABAP Workbench, choose the **Class/Interface** list option in the object list selection box of the **Repository Browser**, as shown in Figure 2.13.

Outside the ABAP Workbench, you can access the Class Builder directly using Transaction SE24 or via the menu path shown in Figure 2.14: **Tools · ABAP Workbench · Development · SE24 – Class Builder**.

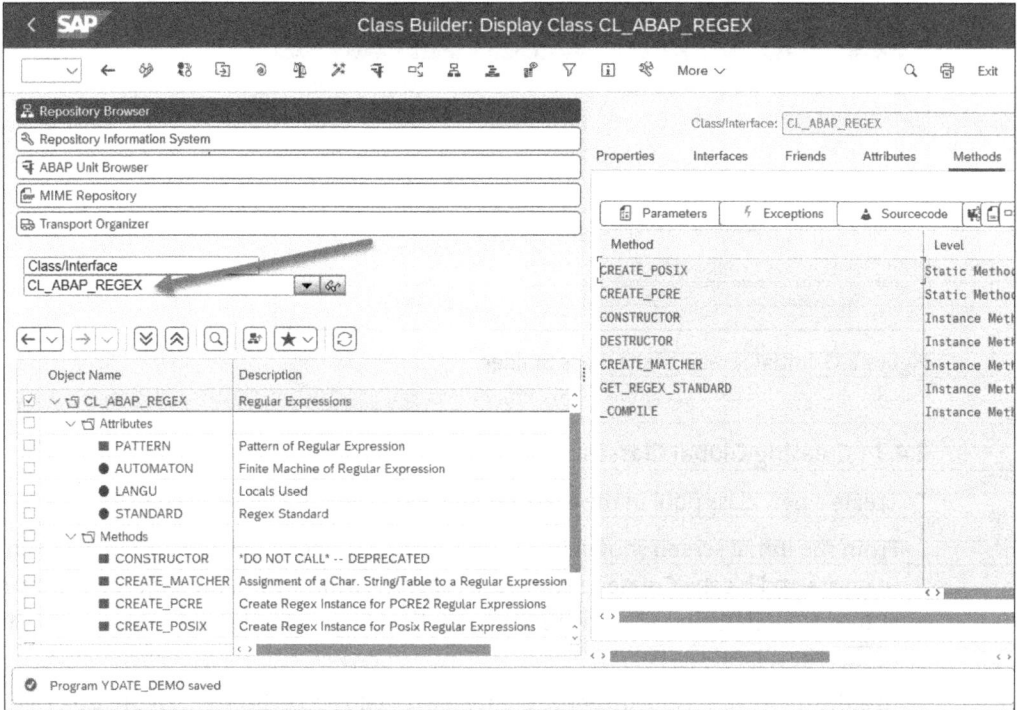

Figure 2.13 Accessing the Class Builder from the ABAP Workbench

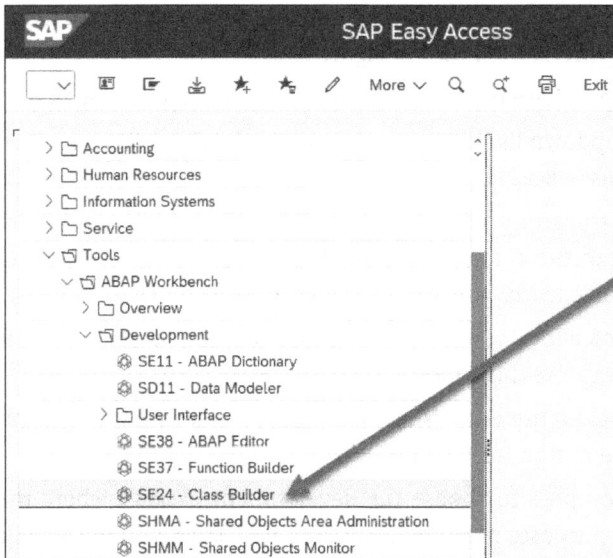

Figure 2.14 Accessing the Class Builder from the SAP Easy Access Menu

Figure 2.15 shows what the initial screen of the Class Builder looks like when accessed in standalone mode.

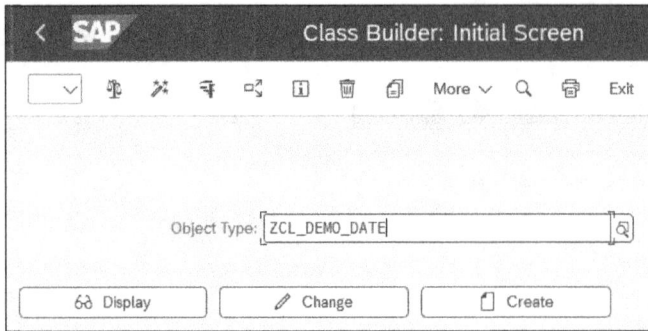

Figure 2.15 Initial Screen of the Class Builder

2.4.3   Creating Global Classes

To create a new class pool in the Class Builder, perform the following steps:

1. From the initial screen shown in Figure 2.15, type in the name of the class you want to create and hit the **Create** button. Alternatively, from within the Object Navigator, you can key in the class name in the input field shown in Figure 2.13 and press the ⌈Enter⌋ key.

2. In either case, you'll end up at the **Create Class** dialog box shown in Figure 2.16. Here, you must fill in the following attributes:

   – In the **Class** field, you confirm the name of the class that you're creating. Since this class is part of the overall ABAP Repository, you must give it a unique name using the {Namespace}CL_ prefix, as shown in Figure 2.16.

   – In the **Description** field, provide a brief description of what the class is used for.

   – The **Inst. Generation** dropdown list is used to determine the class's *instantiation context*. We'll explore this concept in Chapter 4.

   – The **Class Type** radio button group allows you to specify what type of class you're building. For now, stick to the default **Usual ABAP Class** option. The remaining class types will be described in Chapter 8 and Chapter 9, respectively.

   – Finally, the **Final** checkbox allows you to determine whether the class is closed off from inheritance. We'll explore this concept in Chapter 5.

3. Once the class definition details are established, you can proceed with the creation process by clicking the **Save** button (see Figure 2.16).

4. At this point, you'll be prompted to choose the development package where the class will be stored. For the purposes of our demo classes, we'll stick with the default $TMP local package.

5. If all goes well, you should end up at the form editor screen shown in Figure 2.17. From here, you can begin rounding out the class definition by defining attributes, methods, and so on.

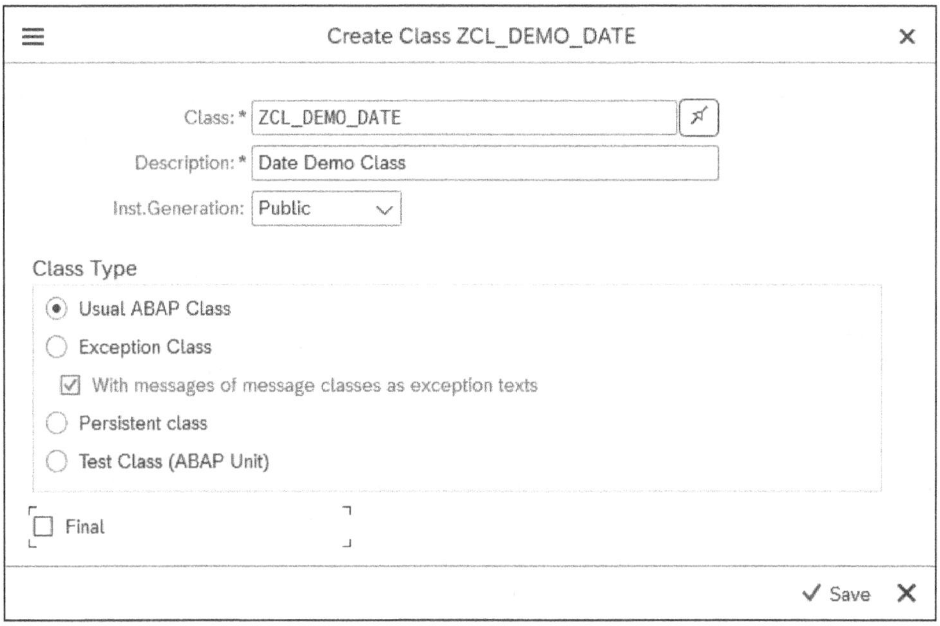

Figure 2.16 Creating a Class Pool

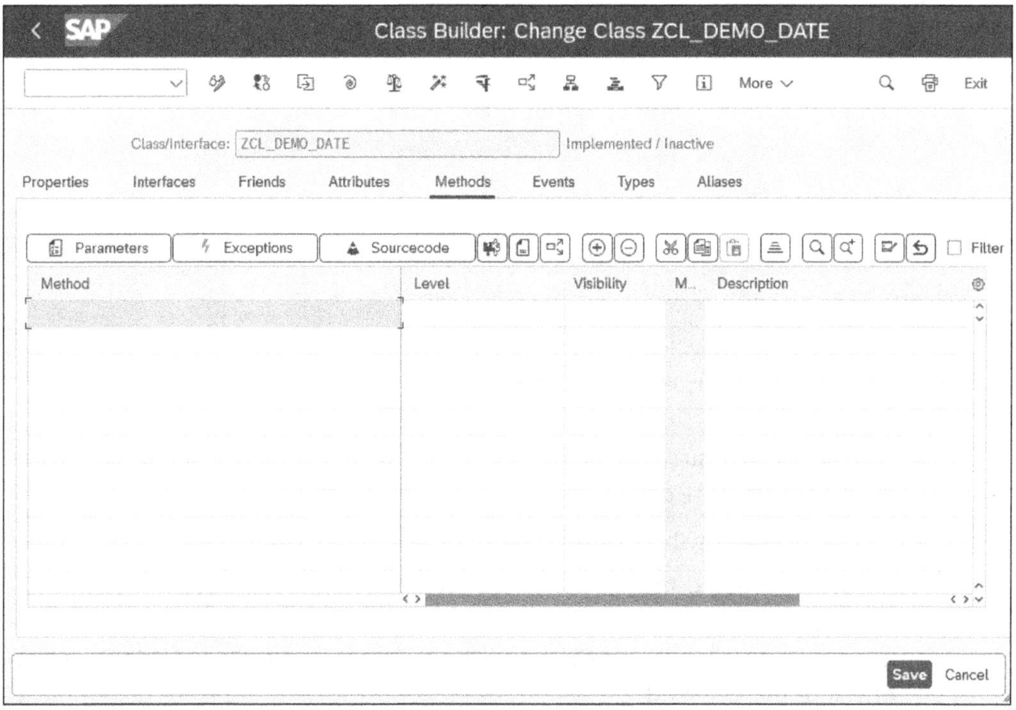

Figure 2.17 Main Editor Screen of the Class Builder

2.4.4   Using the Form-Based Editor

As shown in Figure 2.17, the default view of the Class Editor tool provides you with a form-based view for defining various types of components. For the most part, these forms simply provide an input mask for entering the component declaration details described in Section 2.1.2. In the following sections, we'll take a look at these forms and show how they're used to declare class components.

Defining Attributes

Attributes are defined on the **Attributes** tab of the Class Editor. This tab of the Class Editor provides an entry table in which you can specify all the various attributes for a class (see Figure 2.18).

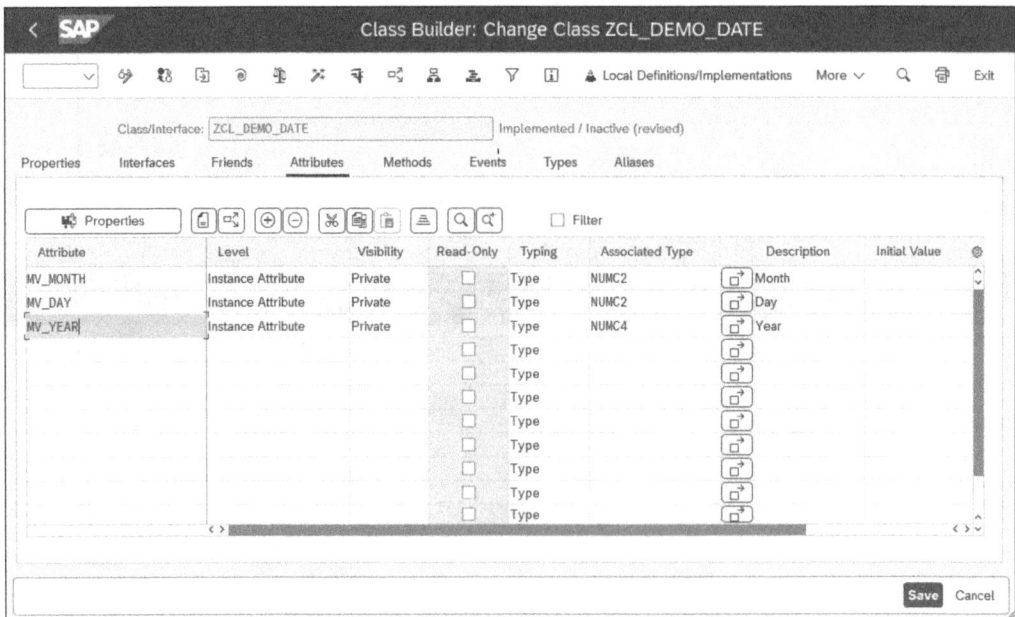

Figure 2.18 Defining Attributes in the Class Editor

As you can see in Figure 2.18, the attributes table defines several columns that you can use to specify attribute definitions:

- The name of an attribute is specified in the **Attribute** column.
- The **Level** column is used to define the declaration type of the attribute. Figure 2.19 contains a list of the possible declaration types available in the input value help for this column. Global class attributes can be declared as an **Instance Attribute**, a **Static Attribute**, or a **Constant**.

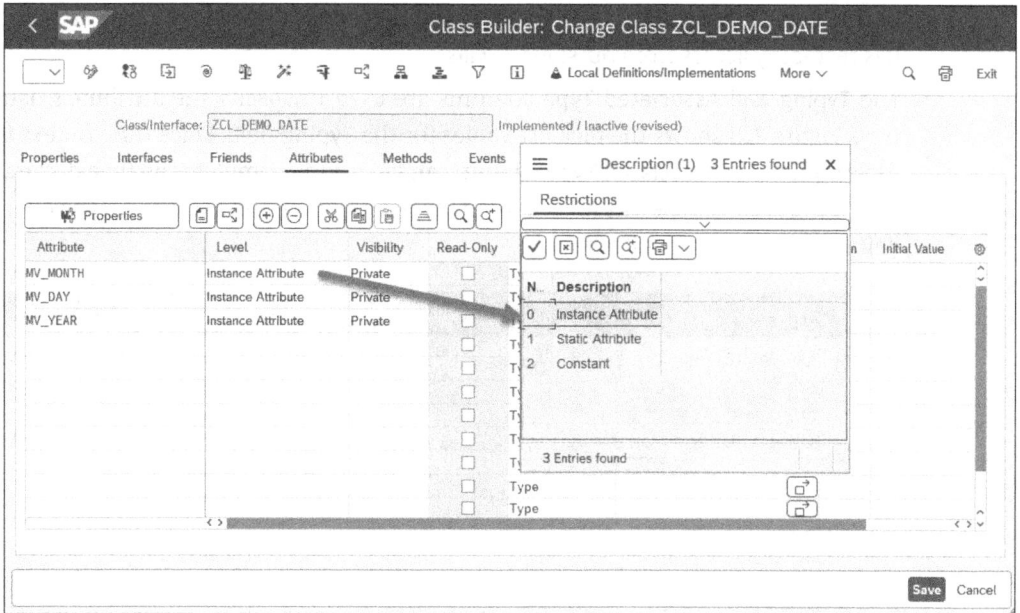

Figure 2.19 Setting the Declaration Type for an Attribute

- The **Visibility** column is used to assign the attribute to a specific visibility section within the class. Figure 2.20 shows the possible values: **Public**, **Protected**, and **Private**.

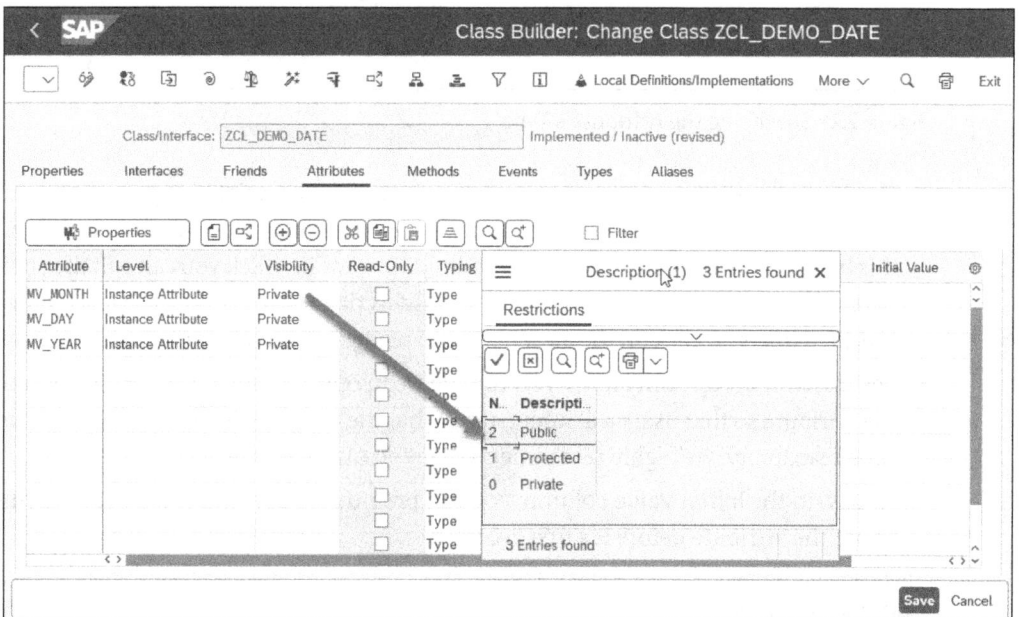

Figure 2.20 Assigning the Visibility Section for an Attribute

- In the **Read-Only** column, you can select the checkbox to specify that the attribute has read-only access from outside the class.

- The **Typing** and **Associated Type** columns are used to specify the attribute's data type. Figure 2.21 shows the pick list values for the **Typing** field. Once this context is determined, the **Associated Type** column can be used to complete the type declaration using built-in types, ABAP Dictionary types, or even custom types or type pools defined within the class itself.

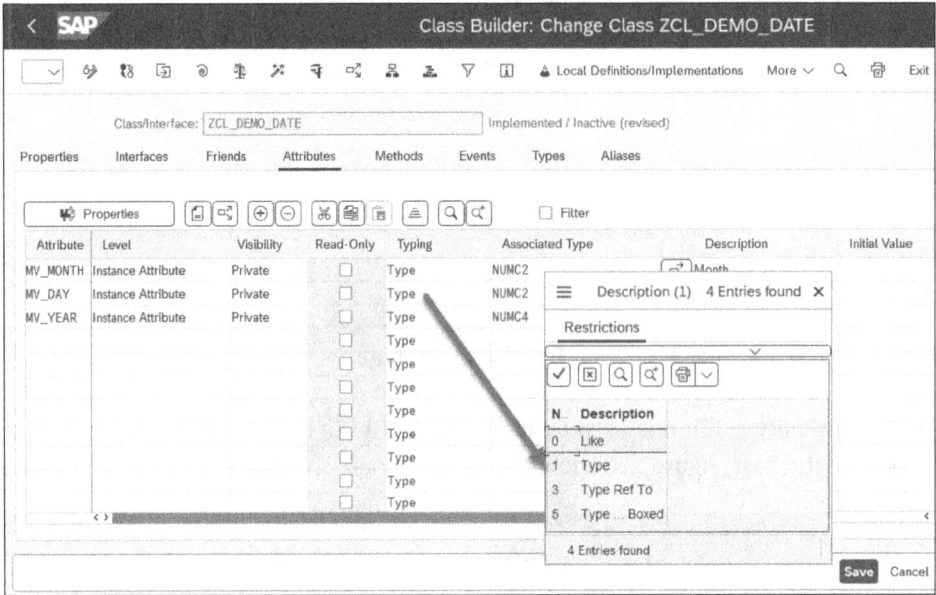

Figure 2.21 Specifying the Attribute's Type

- The **Direct Type Entry** button directly to the right of the **Associated Type** column provides you with another option for specifying an attribute's type. When you click on this button, you'll be taken to a source code editor view in which you can edit the attribute definition directly using ABAP code, as described in Section 2.1.2. Once you've specified the type, click on the **Back** button to navigate back to the **Attributes** tab.

- The **Description** column allows you to type in an optional short text description of the attribute so that users will understand what the attribute is used for. To improve code readability, we highly recommend that you utilize this feature.

- Finally, in the **Initial Value** column, you can provide a value that will be used to initialize the attribute before it's first accessed.

Defining Methods

Methods are defined on the **Methods** tab in the Class Editor. Here, much like you saw on the **Attributes** tab, the Class Editor provides you with an input table for defining the methods of a class. As you can see in Figure 2.22, most of the attributes here are self-explanatory. The lone exception to this is the **Level** field. This field is used to determine if the method is defined at the instance level or the class level (see Figure 2.23).

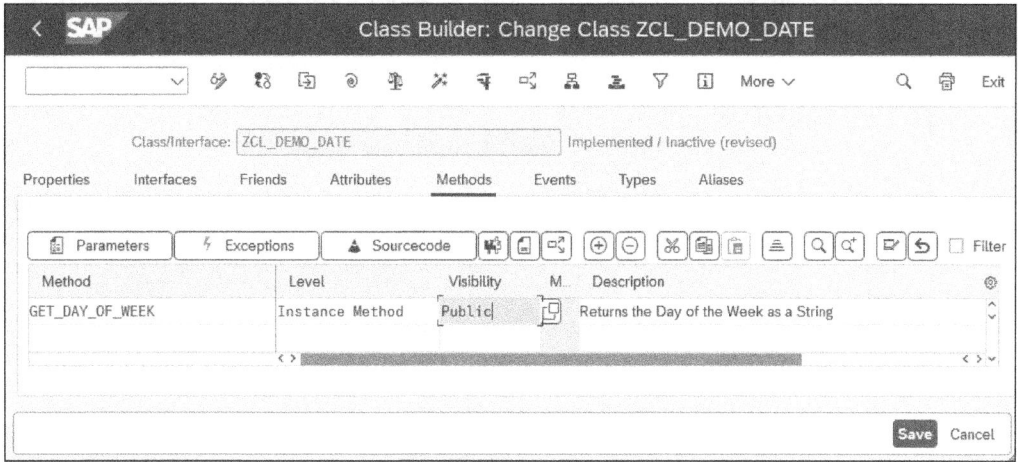

Figure 2.22 Defining Methods in the Class Editor

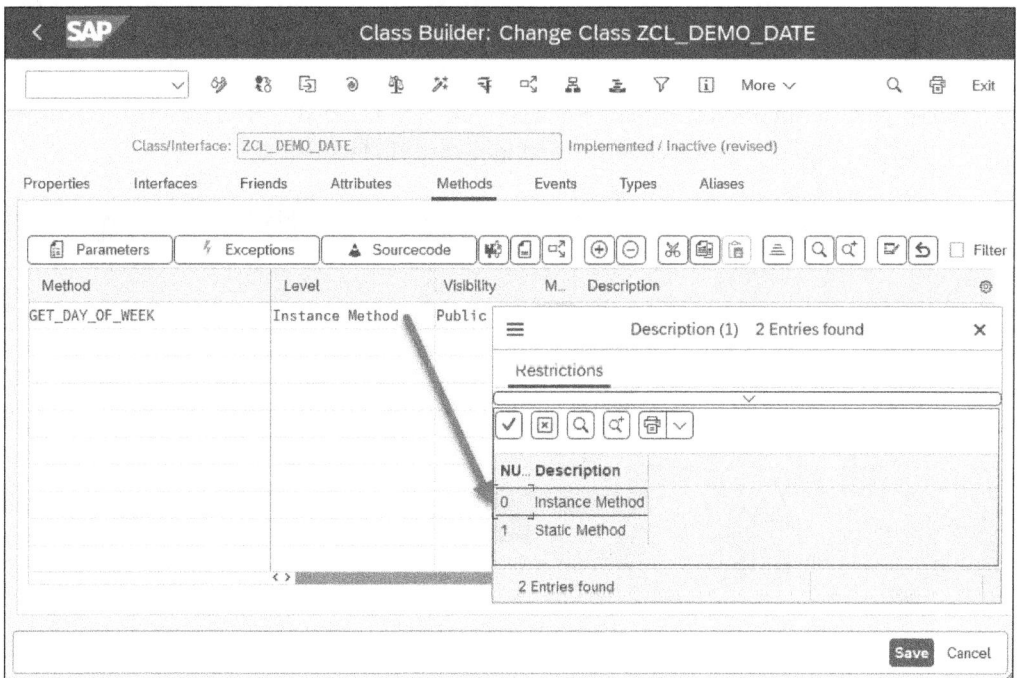

Figure 2.23 Setting the Declaration Type for a Method

To define the parameters for a method, place your cursor on the name of the method you wish to edit in the **Method** column and click on the **Parameters** button in the toolbar directly above the method input table. This will open the **Method Parameters** input screen shown in Figure 2.24. From there, you can fill in parameter details, similar to how you defined attributes on the **Attributes** tab.

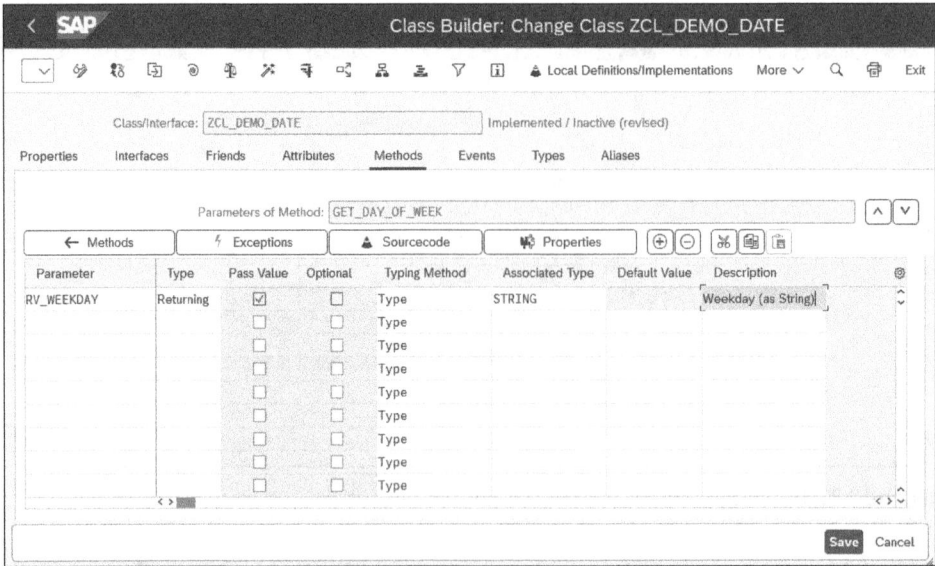

Figure 2.24 Defining Method Parameters

To edit the method's implementation, you can either double-click on the method name or click the **Sourcecode** button highlighted in Figure 2.25.

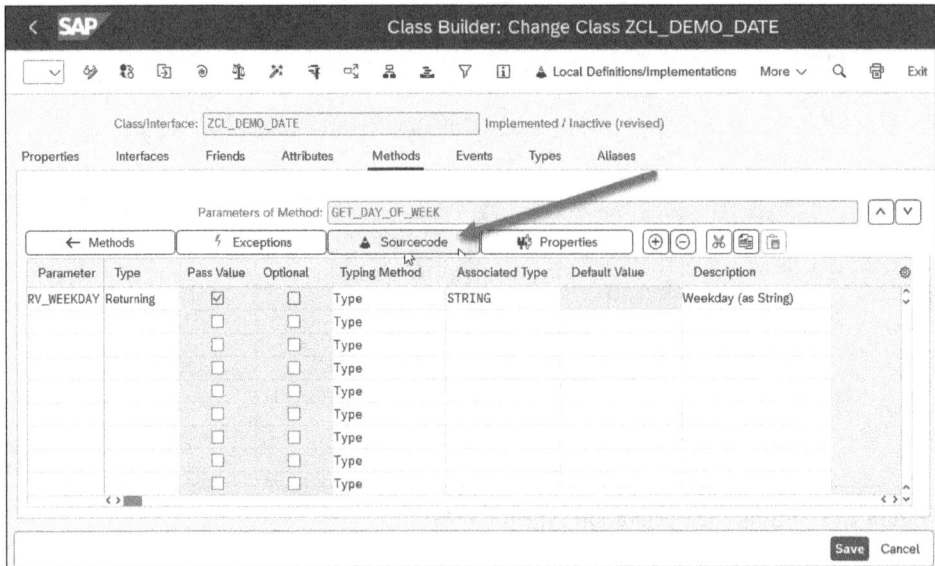

Figure 2.25 Navigating to the Method Source Code Editor

In the source code editor, you can implement the method inside of a METHOD...END-METHOD processing block within the ABAP Editor as per usual (see Figure 2.26).

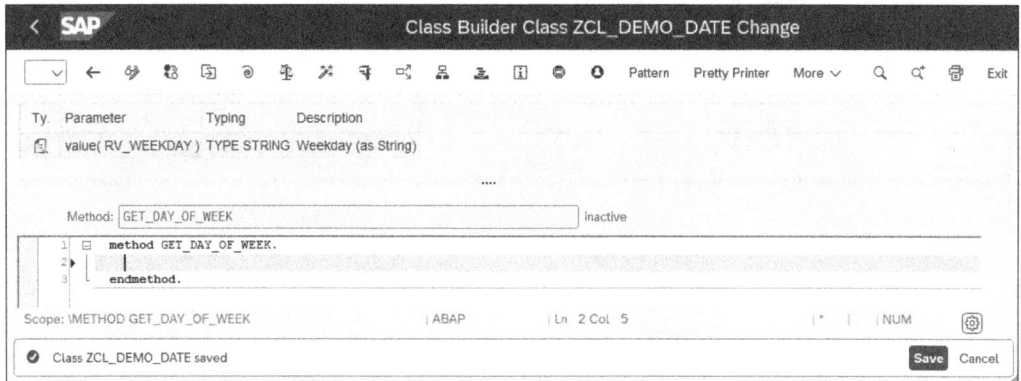

Figure 2.26 Implementing a Method in the ABAP Editor

Defining Events

You define events on the **Events** tab of the Class Editor. Here, you are provided with an input table similar to the ones used to specify components on the **Attributes** and **Methods** tabs (see Figure 2.27).

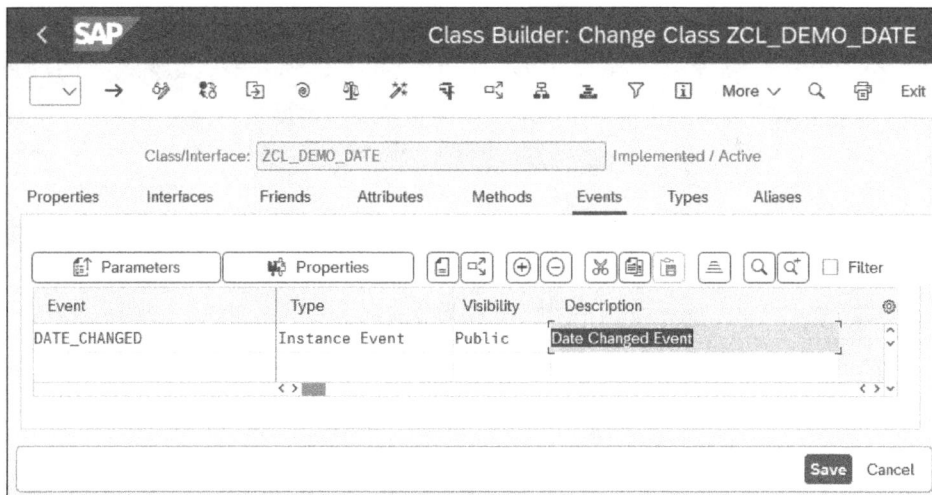

Figure 2.27 Defining Events in the Class Editor

To declare parameters for an event, place your cursor on the name of the target event in the **Event** column and click the **Parameters** button in the toolbar above the input table. This will open the **Event Parameters** input screen shown in Figure 2.28. Here, you can specify the exporting parameters of the event, similar to how you defined method parameters in the previous section.

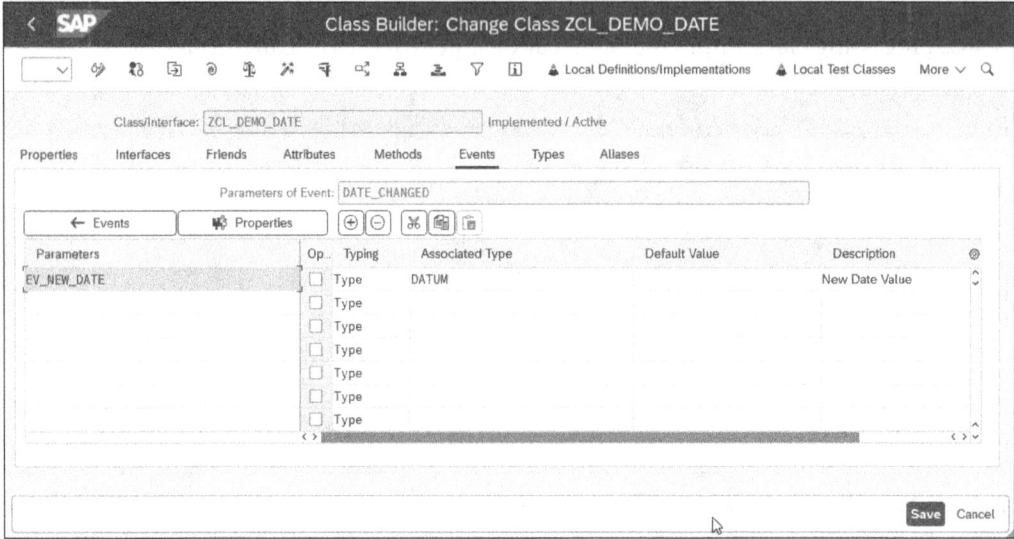

Figure 2.28 Defining Event Parameters

Defining Custom Data Types

You define custom data types on the **Types** tab of the Class Editor. As you can see in Figure 2.29, the **Types** tab has a similar look and feel to the **Attributes** tab shown earlier in Figure 2.18.

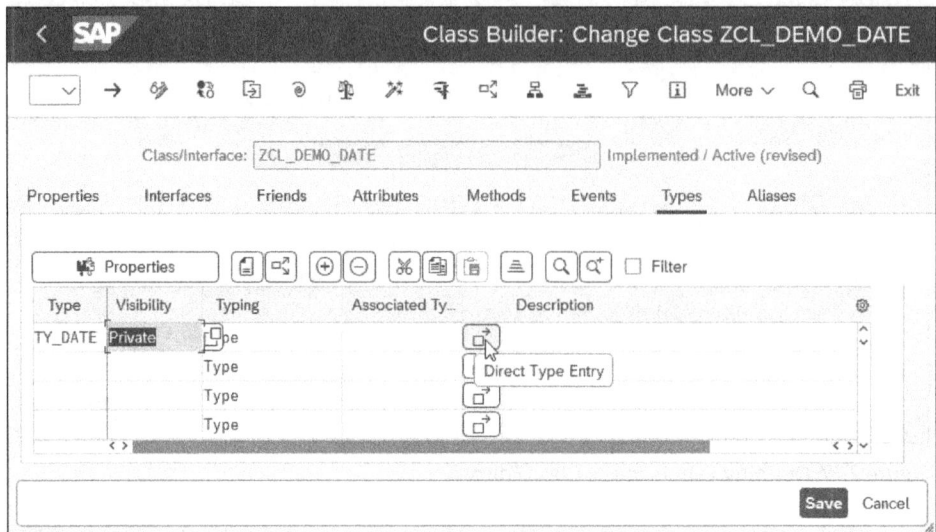

Figure 2.29 Defining Custom Types in the Class Editor

Once again, you can use the **Direct Type Entry** icon (shown in Figure 2.29) to jump into the ABAP Editor and refine the type definition using the TYPES statement as necessary (see Figure 2.30).

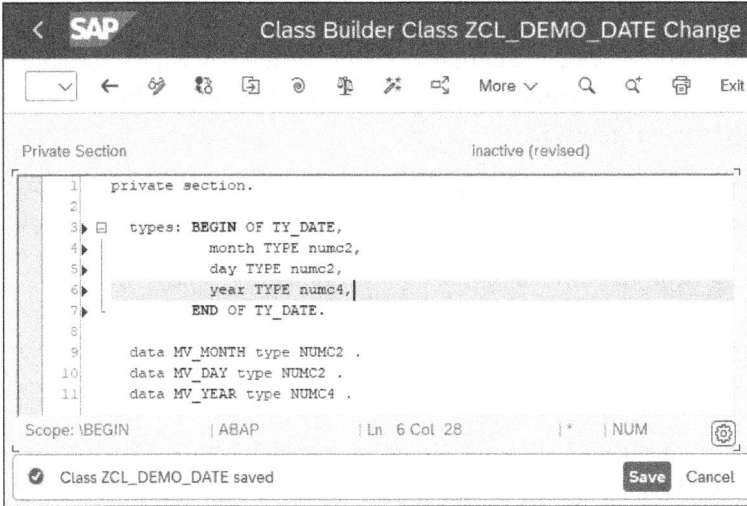

Figure 2.30 Defining Custom Types Using Direct Type Entry

Local Definitions/Implementations

In addition to the components considered thus far, class pools can also be defined with local types used to aid in global class implementation. These local types can be used to define local variables in components such as methods. To access this section of the code, click on the **Local Definitions/Implementations** button shown in Figure 2.31. From there, you'll be taken to a code-based editor in which you can define local data types and local class types as desired. In the latter case, you define the local classes using the same syntax demonstrated in Section 2.1.

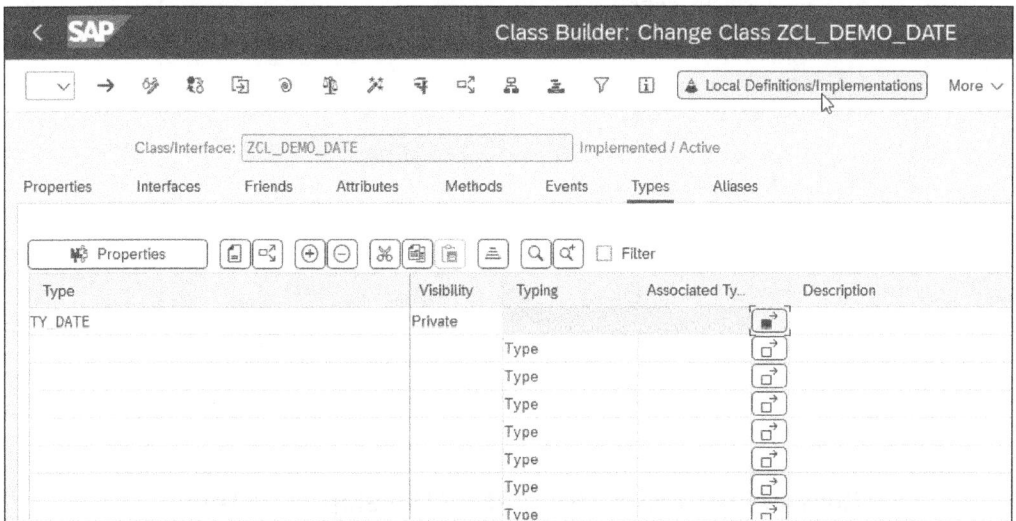

Figure 2.31 Accessing the Local Definitions Include of a Class Pool

2.4.5 Using the Source Code Editor

The Class Builder also comes with an alternative to the form-based editor: the source code-based editor. To access this mode, click on the **Source Code-Based** button in the main toolbar, as shown in Figure 2.32.

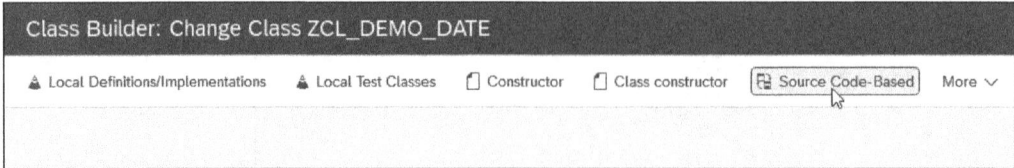

Figure 2.32 Accessing the Source Code-Based Editor (Part 1)

This will bring you to a normal ABAP editor screen, as shown in Figure 2.33. From here, you can work with the basic ABAP Objects syntax covered over the course of this chapter. You can stick with this mode to write your code or toggle back to the form-based view by clicking the **Form-Based** button in the editor toolbar.

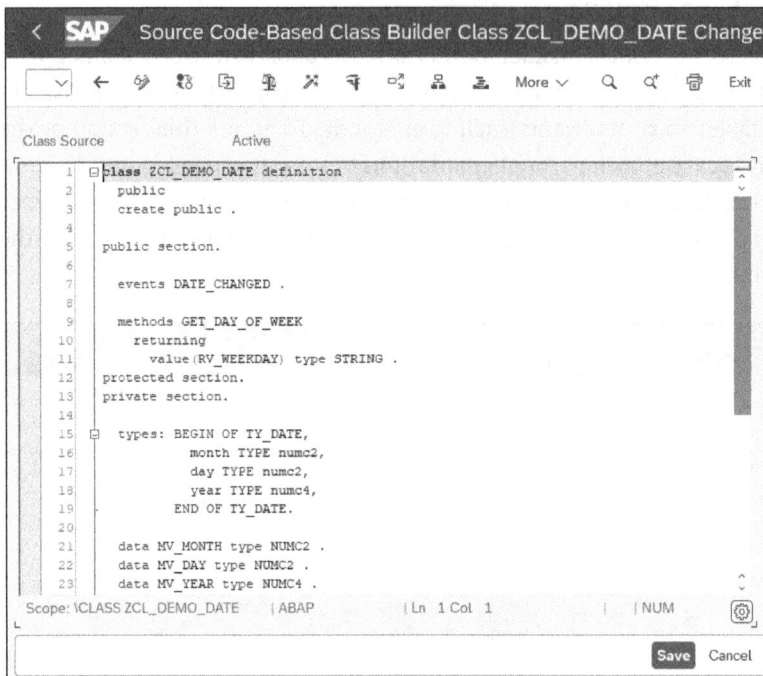

Figure 2.33 Accessing the Source Code-Based Editor (Part 2)

2.5 Developing Classes Using ABAP Development Tools

In the early days of ABAP development, the only way to develop classes in ABAP Objects was to use the ABAP Workbench tools. With the SAP NetWeaver 7.03 release, you now have a new and exciting toolset to work with: *ABAP development tools* (formerly called ABAP development tools in Eclipse). In this section, we'll take an introductory look at some of the basic features of ABAP development tools.

2.5.1   What Is Eclipse?

Before we dive into ABAP development tools, we'll give you a brief introduction to Eclipse. Eclipse is a flexible, Java-based integrated development environment (IDE) initially developed by IBM and then subsequently donated to the open-source community in 2001. From there, the project picked up a lot of steam as other major vendors (or "stewards") jumped on board and added their contributions to the project. These efforts culminated in the launch of the Eclipse Foundation in 2004.

Eclipse is like any other IDE: It provides functionality to edit source code, organize code into projects, and so forth. What sets Eclipse apart is its flexible plug-in system. Software vendors and individuals alike can create plug-ins to extend the core functionality of the IDE in all kinds of interesting ways. If you look online, you can find plug-ins that allow you to develop in just about any programming language imaginable using Eclipse. With the advent of ABAP development tools, that list of programming languages now includes ABAP.

From a strategy perspective, moving toward an Eclipse-based IDE makes a ton of sense for SAP. Technology-wise, it's fair to say that SAP has reached the limits of what can be achieved using classic Web Dynpro technology in the ABAP Workbench. Indeed, you might be stunned by some of the Eclipse-based features demonstrated in Section 2.5.3. These capabilities simply weren't possible using the old technology stack! Plus, as new developers come onto the scene, it's in SAP's best interest to offer a familiar development platform that's on par with competing IDEs, such as Microsoft Visual Studio Code.

Note

Even though we think most users will never want to go back to the ABAP Workbench once they get their hands on ABAP development tools, we should point out that the ABAP Workbench is not going away anytime soon. The great news for developers is that you now have more choices to work with. You can stick with the ABAP Workbench, jump over to ABAP development tools, or toggle back and forth between the two. Since the two IDEs can be used simultaneously, developers can dip their toes in both ponds and decide which IDE works best for them over time.

2.5.2   Setting Up the ABAP Development Tools Environment

Since the ABAP development tools are bundled as a series of plug-ins, they can generally be installed on top of most any recent Eclipse installation. This could include Eclipse installations used to develop software outside of the SAP landscape, or other

SAP-based Eclipse solutions such as SAP HANA tools. If you're an experienced Eclipse developer, you can find instructions for setting up the ABAP development tools plug-ins at *https://tools.hana.ondemand.com/#abap*. Otherwise, check out for step-by-step instructions that walk you through the installation process.

> **Release Compatibility**
>
> At the time of writing, ABAP development tools requires that the application server (AS) ABAP backend must be on SAP kernel 7.20 or higher and Basis version 7.31 with support pack stack 04. Though this may change over time, you're probably not going to be able to use ABAP development tools with an old AS ABAP system running on SAP NetWeaver 7.0.

Once you have your Eclipse installation in place, you can access ABAP development tools by launching the IDE and performing the following steps:

1. From the top-level **File** menu, choose the **New • ABAP Project** menu option as shown in Figure 2.34.

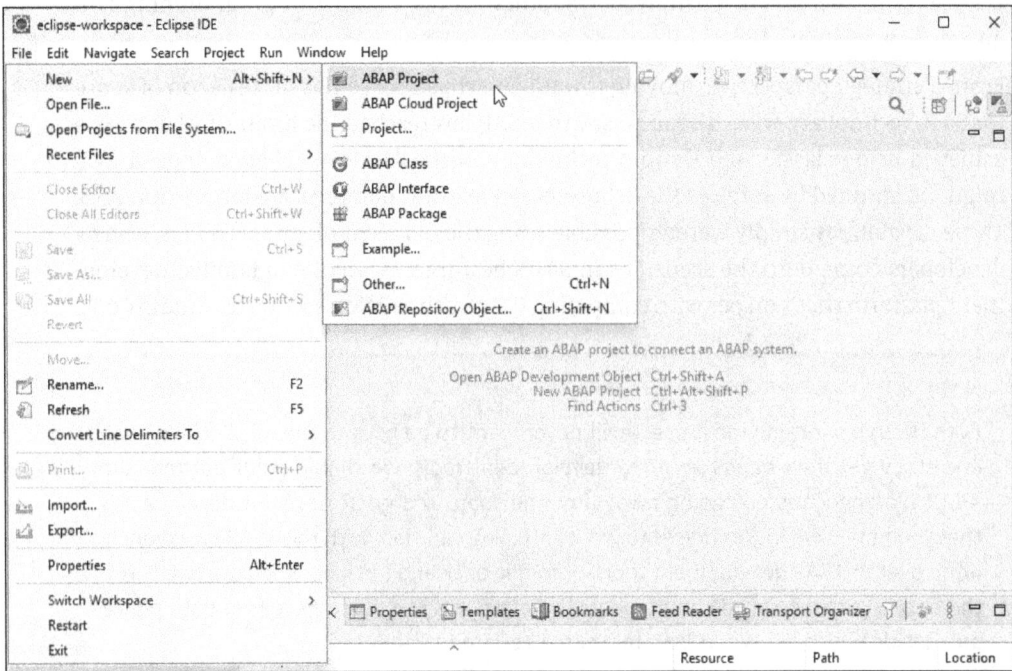

Figure 2.34 Creating an ABAP Development Tools Project (Part 1)

2. This will launch the **New ABAP Project** wizard. As you can see in Figure 2.35, the list of available system connections is loaded from your SAP logon pad. Choose the target system connection and click the **Next** button to continue.

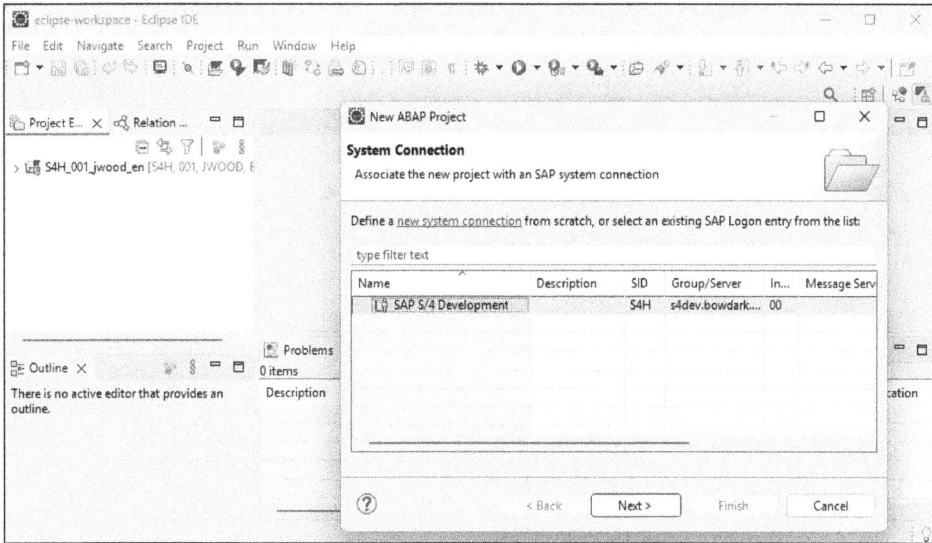

Figure 2.35 Creating an ABAP Development Tools Project (Part 2)

3. At the **Connection Settings** step shown in Figure 2.36, you can confirm and adjust the system connection settings, as well as turn on options like single sign-on with secure network communication (SNC). Click the **Next** button to continue.

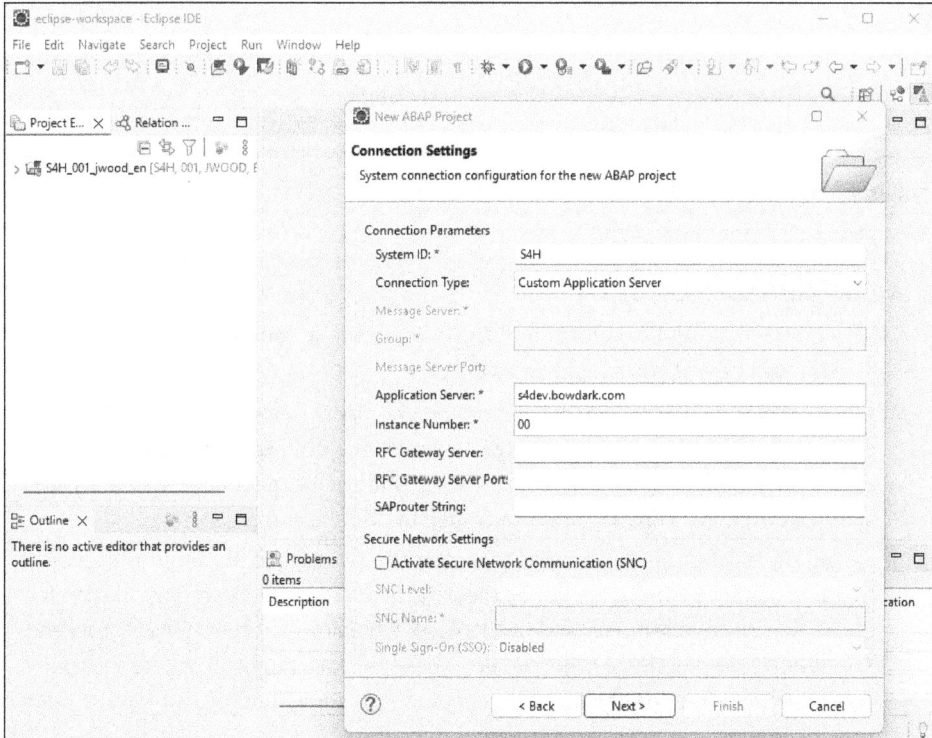

Figure 2.36 Selecting the Target AS ABAP System to Connect To

4. At the next step, you'll be prompted to provide login credentials so the ABAP development tools plug-ins can log into the AS ABAP backend (see Figure 2.37). Click the **Finish** button to create the project.

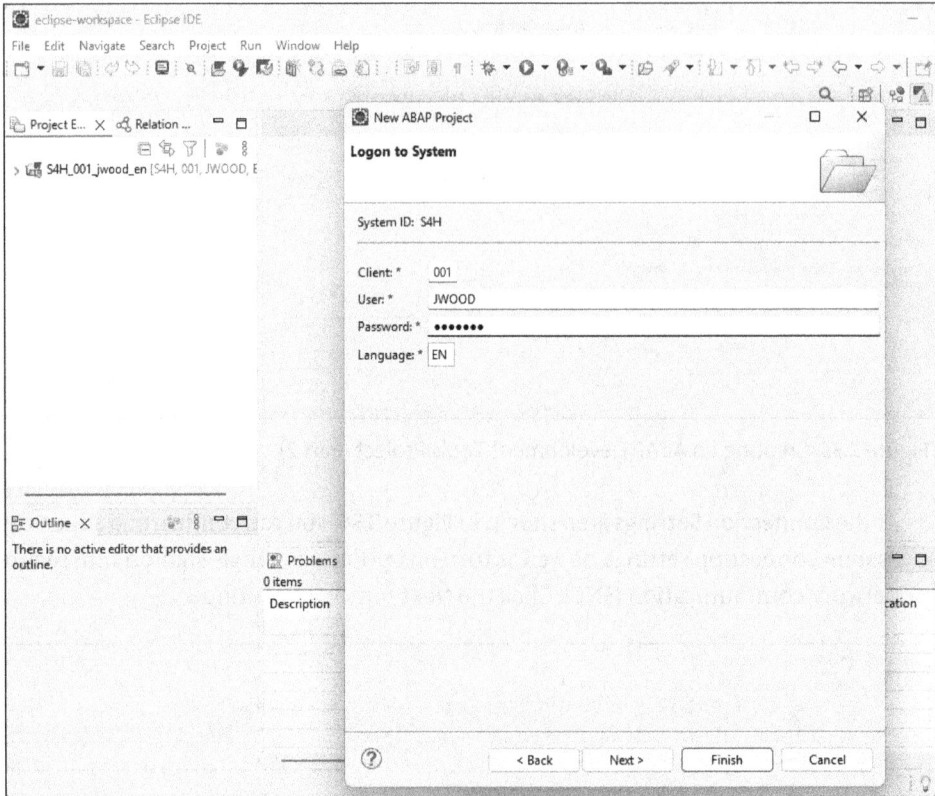

Figure 2.37 Authenticating with the AS ABAP Backend

A Word on Workspaces and Projects

As you create new ABAP projects in Eclipse, there are a couple of things going on behind the scenes that you should be aware of:

- Eclipse stores project files in a directory on your local machine, referred to as the Eclipse *workspace*. You can find this directory from within Eclipse by choosing **Window · Preferences** in the top-level menu bar and then expanding **General · Workspace**. From there, a link should be provided to find the current settings.

- For ABAP-based projects, the files stored in the workspace are limited to project files and metafiles. These are used to keep track of your project settings and preferences. Source code remains within the ABAP Repository on the AS ABAP backend. This is different than most other project types (e.g., Java projects), where the source code is maintained locally and then checked into some kind of software revision control system such as Git or Perforce.

2.5.3   Working with the ABAP Development Tools Class Editor

Once an ABAP project is created and a connection is established within the AS ABAP backend, you can begin creating and editing ABAP-related artifacts, as shown in Figure 2.38. The editor screen within the ABAP perspective is split into four editor panes:

- On the top left-hand pane, you have the **Project Explorer** window, which resembles the tree-based repository browser that allows you to search for development objects from within the ABAP Workbench.

- The top right-hand pane consists of a tabbed pane containing one or more editor views. For example, you can open a class, a report, and a function module on different tabs and toggle back and forth between them. Eclipse also supports split-screen views so that you can perform side-by-side comparisons.

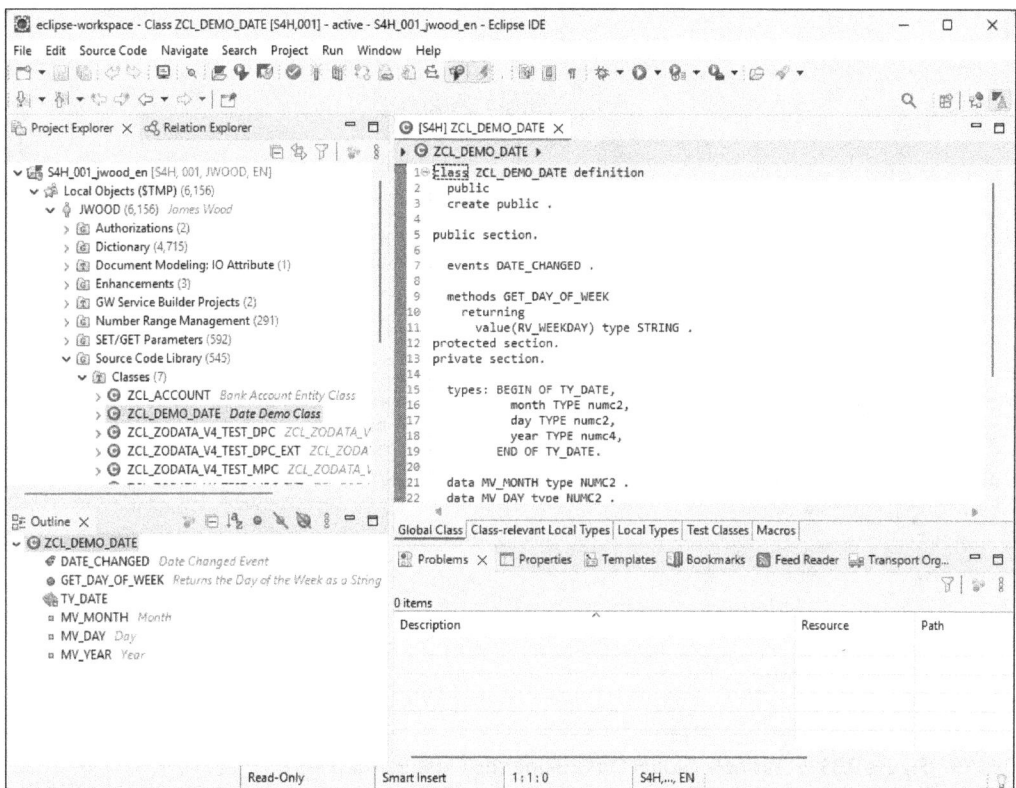

Figure 2.38 Working with the ABAP Development Tools Class Editor

- In the lower left-hand pane, you have the **Outline** view, which makes it easy to navigate directly to specific modules within the development object you're editing.

- Finally, the lower right-hand pane is generally reserved for secondary views, such as properties views.

Since each of these editor panes are contained within a splitter pane, you can shrink or expand particular panes by dragging the separator bars with your mouse. You can also expand or hide panes by clicking the corresponding icons in the top right-hand corner of the pane.

Once you become acquainted with the ABAP perspective, you can experiment with the various Eclipse-based editors by opening an existing object (by double-clicking on it in the project explorer) or creating a new one from the provided context menus. Since we're most interested in editing ABAP classes, the steps required to create a new class are as follows:

1. Within the desired ABAP development package, right-click on the **Source Code Library** folder and choose **New • ABAP Class** from the menu (see Figure 2.39).

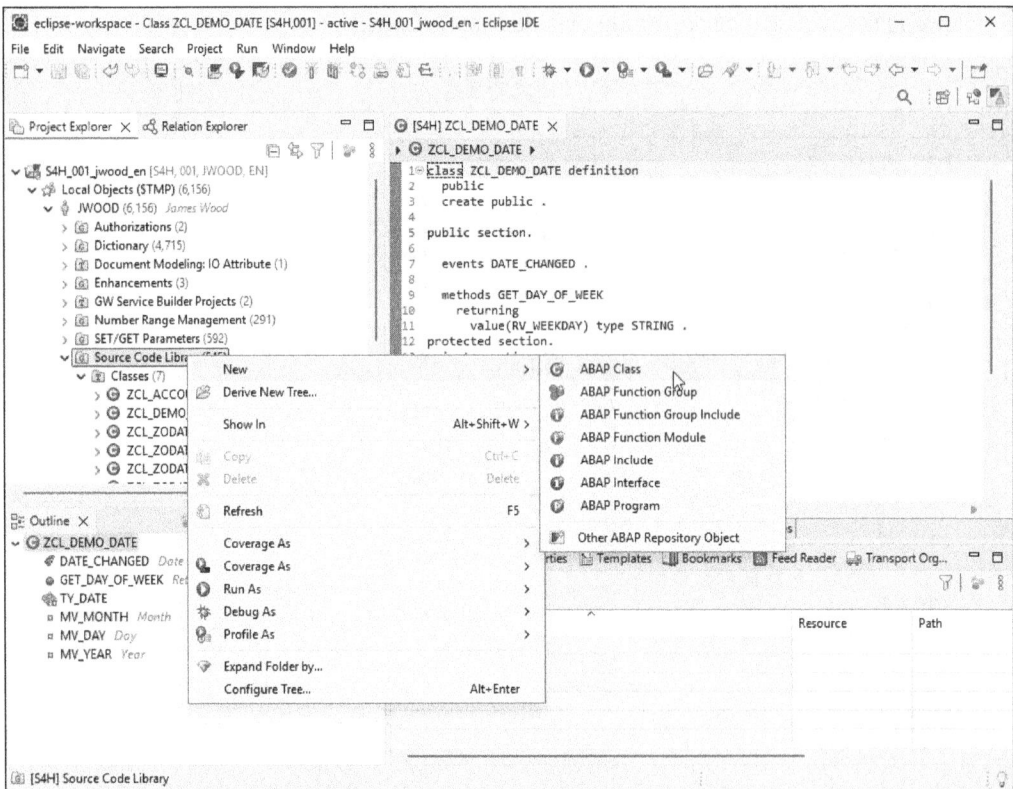

Figure 2.39 Creating a Global Class Using the Class Editor (Part 1)

2. This will open the **New ABAP Class** dialog box shown in Figure 2.40. Specify the class's **Name** and a brief **Description**. From there, you can follow the wizard through to completion, accepting all the default settings to create the new class.

Once you open a class, you can edit it using normal ABAP Objects syntax. To maximize the editor area, double-click on the top-level tab for the class you're editing. Along the top of the editor screen, you can access all the familiar toolbar functions available in the normal Class Builder tool: syntax check, save, activate, and so on.

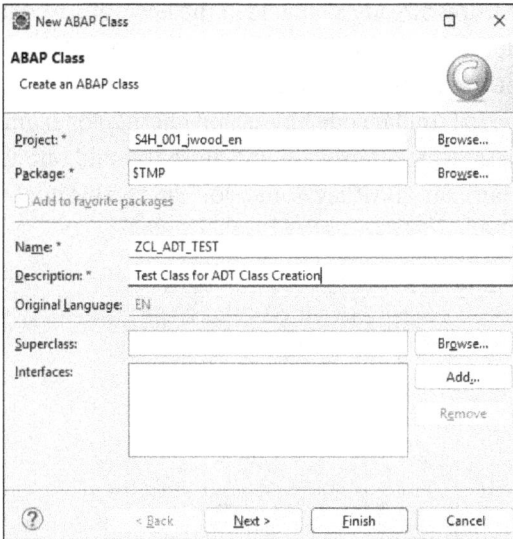

Figure 2.40 Creating a Global Class Using the Class Editor (Part 2)

There are quite a few useful features within the text editor itself. For example, in Figure 2.41, you can see how the editor is equipped with intelligent code completion features. These features can be accessed on demand as you're coding by pressing [Ctrl]+[Space].

Figure 2.41 Working with Code Assistant Features in ABAP Development Tools

Besides the normal code-completion features already available in the new source code editor of the ABAP Workbench, ABAP development tools also taps into a powerful feature of the Eclipse IDE: *templates*. As the name suggests, templates refer to predefined code templates that make it easy to drop in familiar code blocks and idioms. For example, in Figure 2.42, you can see how we're using a template in Eclipse to build out an ABAP CASE statement. As you're typing familiar ABAP keywords, you can try this out by once again pressing [Ctrl]+[Space].

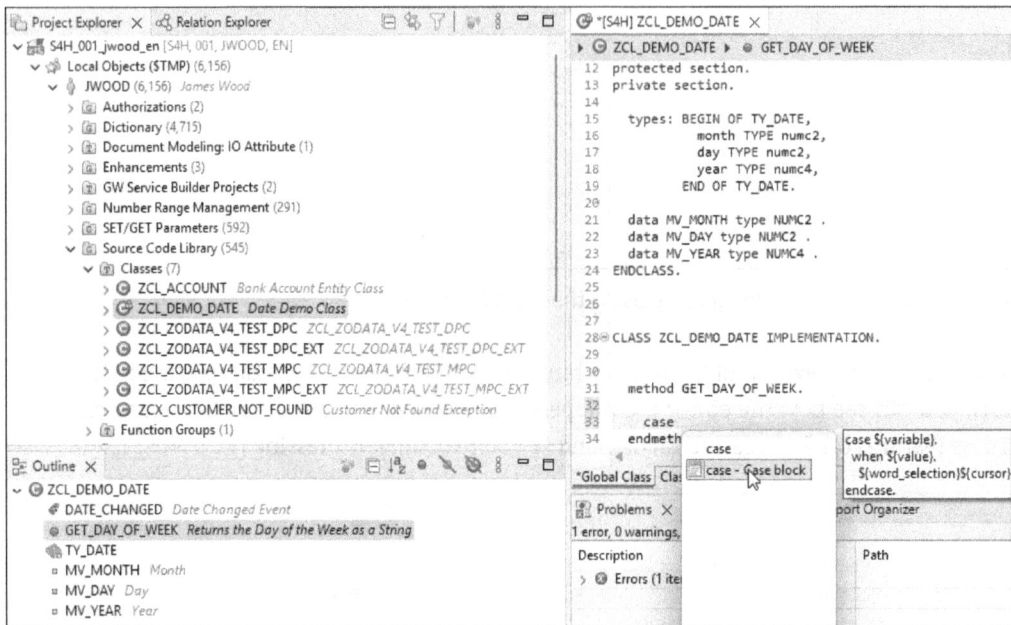

Figure 2.42 Working with ABAP Templates (Part 1)

If you look closely at the template being inserted in Figure 2.43, you can see how the template code contains several variables contained within the ${} blocks (e.g., ${variable}). As you insert a template, these variables can help further define a code block and give you more than just a simple piece of boilerplate code.

To see which templates are available, select the **Window** • **Preferences** menu option as shown in Figure 2.44. From there, select **ABAP Development** • **Editors** • **Source Code Editors** • **Templates** to access the template editor table shown in Figure 2.45. Here, you can view or edit existing templates or create new ones as desired.

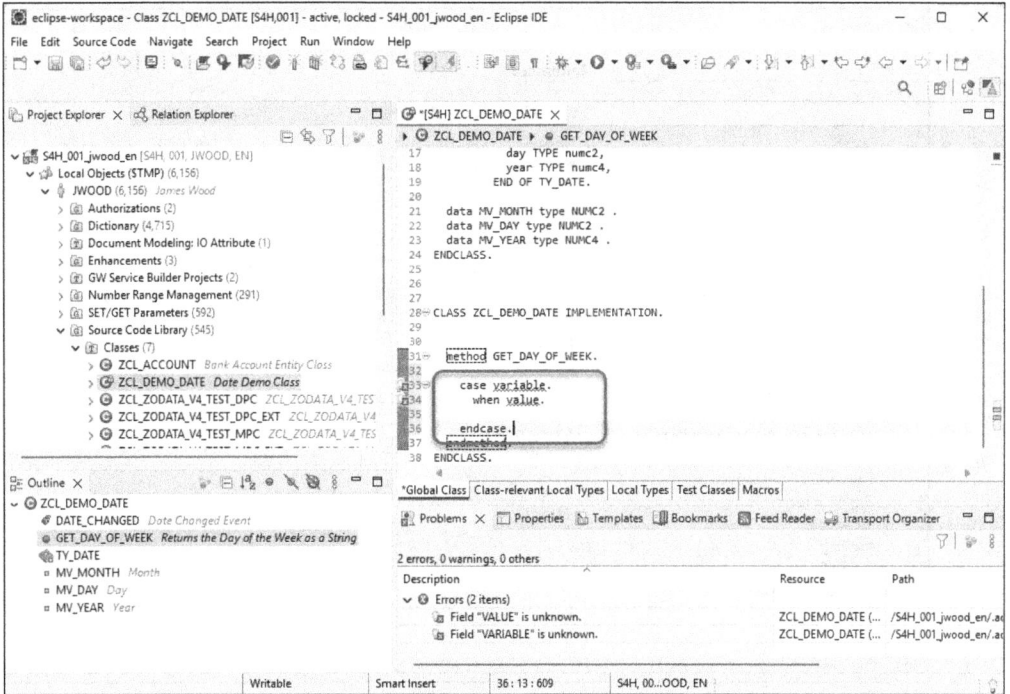

Figure 2.43 Working with ABAP Templates (Part 2)

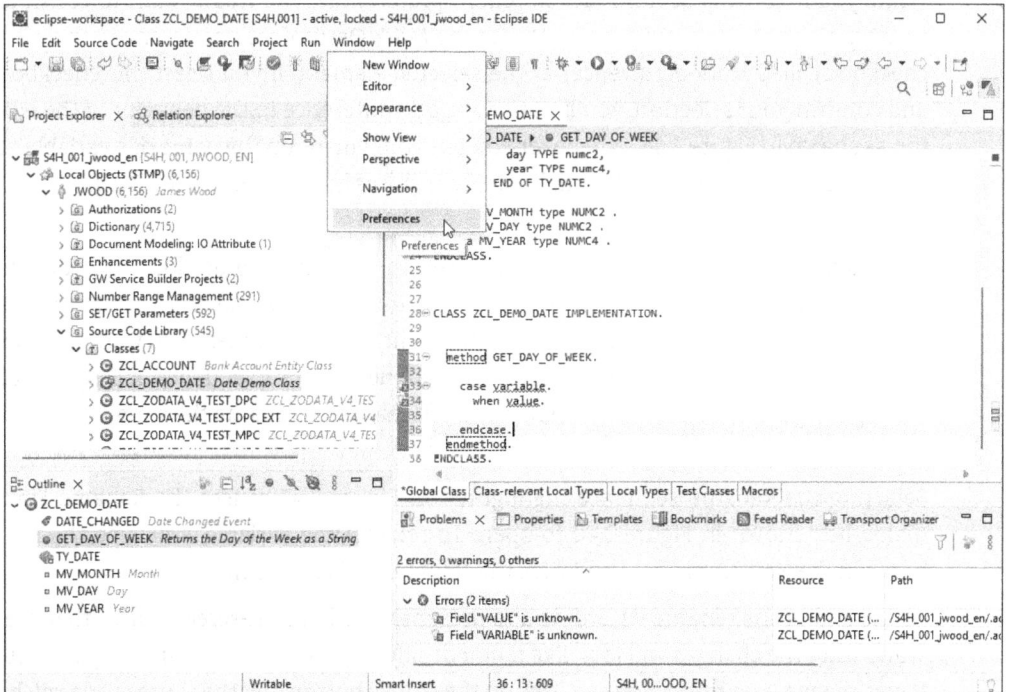

Figure 2.44 Customizing ABAP Templates (Part 1)

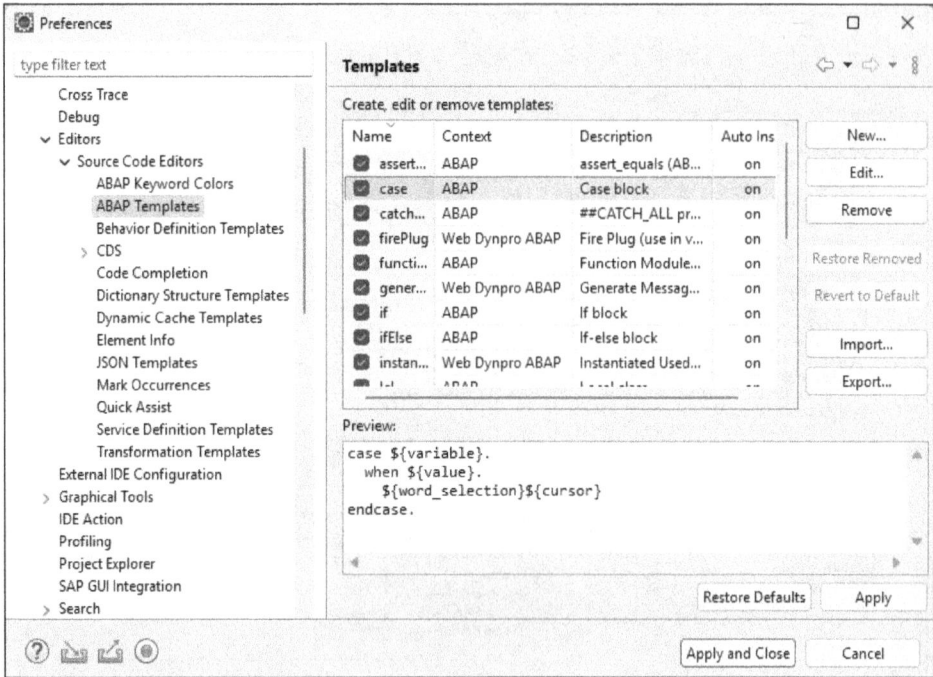

Figure 2.45 Customizing ABAP Templates (Part 2)

While you're looking at ABAP development preferences, you might want to consider turning on another useful feature. Under the **Mark Occurrences** preferences, there's a checkbox called **Mark occurrences of the selected element**. If you select this checkbox and confirm your selection, you'll notice a subtle difference in the behavior of the editor, as shown in Figure 2.46. Whenever you put your cursor on the name of a variable or code module, the editor will highlight all occurrences of that element with a gray background.

Besides making it easy to visualize where certain elements are used in the code, the mark occurrences feature also allows you to tap into the powerful refactoring features built into editor. To access these features, highlight the code element you want to refactor and press Ctrl+1. (This sequence is called the Quick Fix sequence in Eclipse.) For example, Figure 2.47 shows how we're using this feature to rename the variable lt_day_names.

Selecting the **Rename** sequence highlighted in Figure 2.47 causes the **Rename Field** dialog box shown in Figure 2.48 to open. To rename the variable, you type the new name in the **New Name** field and click the **Next** button to continue.

At the end of the rename wizard, you're presented with a split-screen editor that previews what the changes will look like if you decide to proceed, as shown in Figure 2.49. If you change your mind, you can click on the **Cancel** button and the source code will be left untouched. Otherwise, confirm the selection by clicking on the **Finish** button.

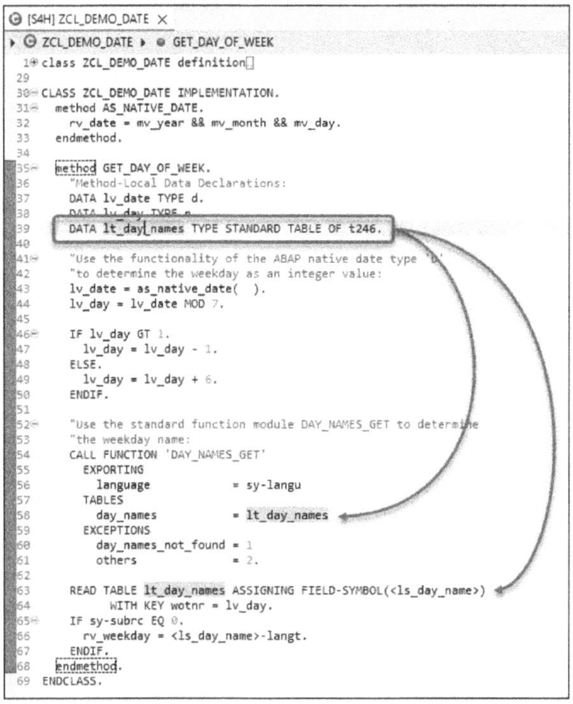

Figure 2.46 Working with the Mark Occurrences Feature

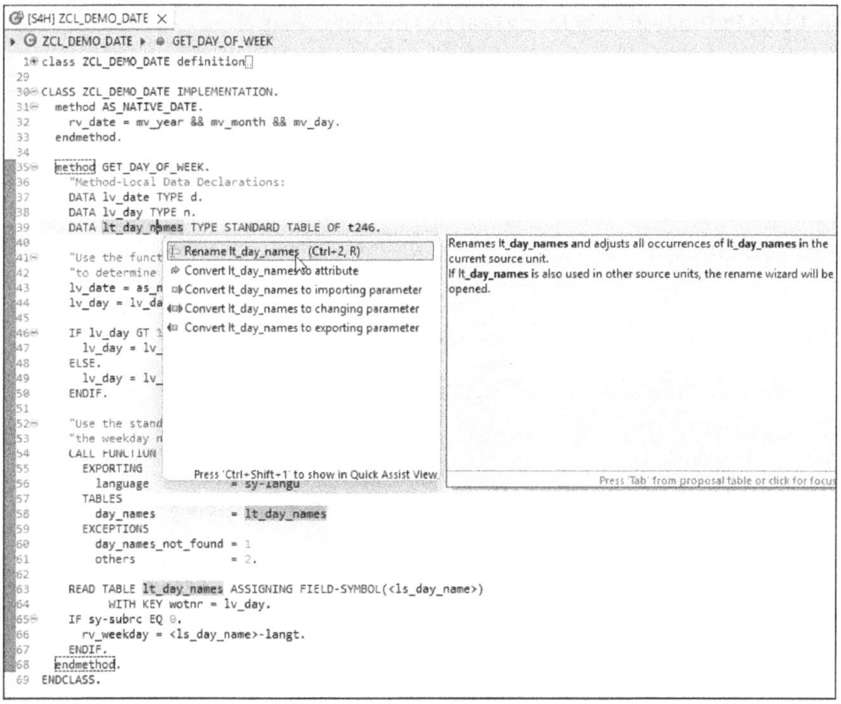

Figure 2.47 Using the Built-In Refactoring Features in Eclipse (Part 1)

```
[S4H] ZCL_DEMO_DATE  ✕
▶ ⊙ ZCL_DEMO_DATE  ▶  ⊙ GET_DAY_OF_WEEK
 1⊖ class ZCL_DEMO_DATE definition□
29
30⊖ CLASS ZCL_DEMO_DATE IMPLEMENTATION.
31⊖   method AS_NATIVE_DATE.
32      rv_date = mv_year && mv_month && mv_day.
33    endmethod.
34
35⊖   method GET_DAY_OF_WEEK.
36      "Method-Local Data Declarations:
37      DATA lv_date TYPE d.
38      DATA lv_day TYPE n.
39      DATA lt_day_names TYPE STANDARD TABLE OF t246.
40
41⊖      "Use the functionality of the ABAP native date type 'D'
42       "to determine the weekday as an integer value:
43       lv_date = as_native_date( ).
44       lv_day = lv_date MOD 7.
45
46⊖      IF lv_day GT 1.
47         lv_day = lv_day - 1.
48       ELSE.
49         lv_day = lv_day + 6.
50       ENDIF.
51
52⊖      "Use the standard function module DAY_NAMES_GET to dete
53       "the weekday name:
54       CALL FUNCTION 'DAY_NAMES_GET'
55         EXPORTING
56           language          = sy-langu
57         TABLES
58           day_names         = lt_day_names
59         EXCEPTIONS
60           day_names_not_found = 1
61           others            = 2.
62
63       READ TABLE lt_day_names ASSIGNING FIELD-SYMBOL(<ls_day_
64               WITH KEY wotnr = lv_day.
65⊖      IF sy-subrc EQ 0.
66         rv_weekday = <ls_day_name>-langt.
67       ENDIF.
68     endmethod.
69  ENDCLASS.
```

Rename Field

Choose Name

New Name: * [lt_day_names]   18 remaining

(?)   < Back   Next >   Finish   Cancel

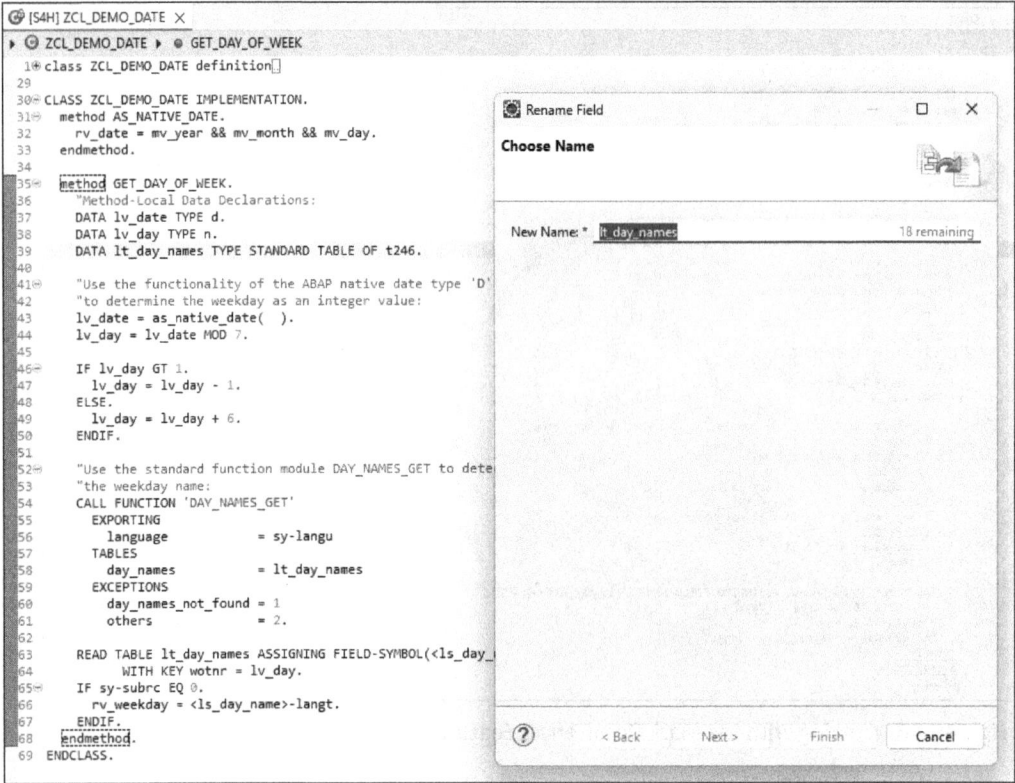

Figure 2.48 Using the Built-In Refactoring Features in Eclipse (Part 2)

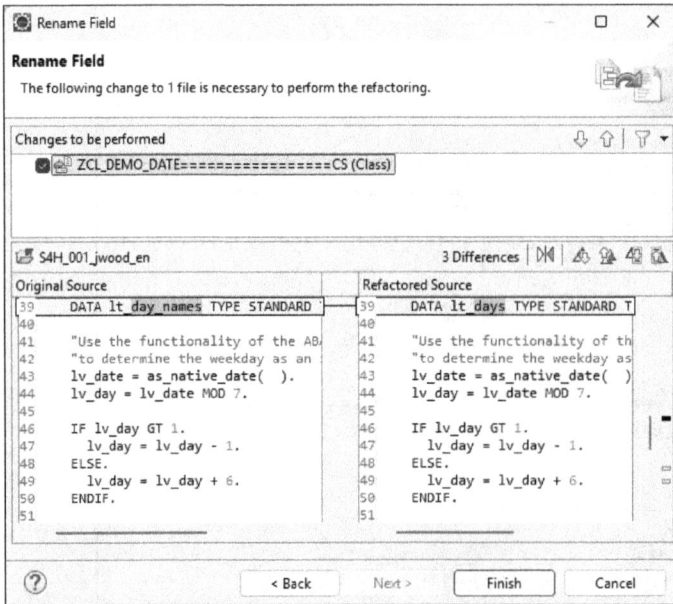

Rename Field

Rename Field
The following change to 1 file is necessary to perform the refactoring.

Changes to be performed   ⬇ ⬆ | 🔽 ▾
   ☑ ZCL_DEMO_DATE================CS (Class)

S4H_001_jwood_en   3 Differences

Original Source	Refactored Source
39 DATA lt_day_names TYPE STANDARD	39 DATA lt_days TYPE STANDARD T
40	40
41 "Use the functionality of the AB	41 "Use the functionality of th
42 "to determine the weekday as an	42 "to determine the weekday as
43 lv_date = as_native_date( ).	43 lv_date = as_native_date( )
44 lv_day = lv_date MOD 7.	44 lv_day = lv_date MOD 7.
45	45
46 IF lv_day GT 1.	46 IF lv_day GT 1.
47 lv_day = lv_day - 1.	47 lv_day = lv_day - 1.
48 ELSE.	48 ELSE.
49 lv_day = lv_day + 6.	49 lv_day = lv_day + 6.
50 ENDIF.	50 ENDIF.
51	51

(?)   < Back   Next >   Finish   Cancel

Figure 2.49 Using the Built-In Refactoring Features in Eclipse (Part 3)

You can find menu-based access to these features and more in the **Source Code** menu shown in Figure 2.50. For example, you have the option to clean up unused variables, format portions of the code according to preferences or project standards, and much, much more.

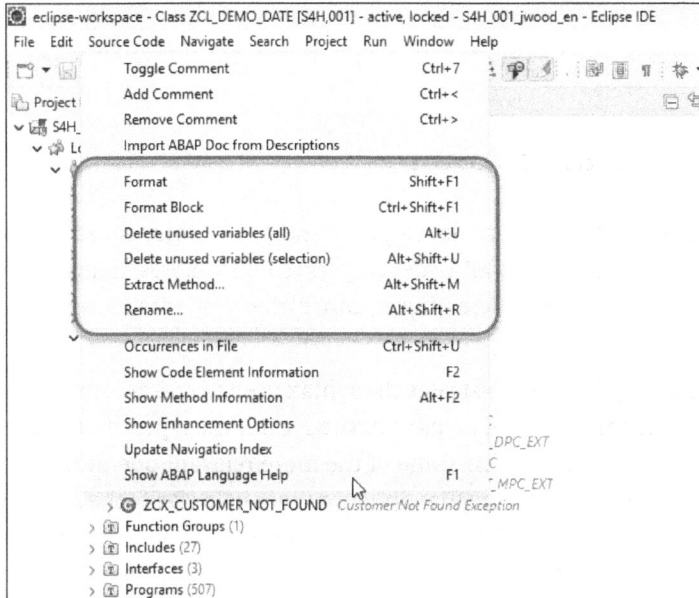

Figure 2.50 Accessing Other Refactoring Features in Eclipse

As you're editing development objects with these features, keep in mind that most of the changes are made in-memory. This means that the changes are not synchronized with the ABAP Repository until you save and/or activate the code. At that point, the changes are committed and available for subsequent editing in Eclipse or the backend ABAP Workbench.

2.6 Working with Constructor Expressions

In the early days of ABAP, one of the primary complaints of developers coming to ABAP Objects from other contemporary object-oriented languages was that ABAP Objects syntax was simply too verbose. For example, consider the code excerpts in Listing 2.34 and Listing 2.35. Both of these code excerpts carry out the task of instantiating a "point" object. In Listing 2.34, you can see how the Java programming language allows you to accomplish this in one line of code using the Java new operator. When compared with the ABAP-based equivalent in Listing 2.35, it's clear that several more keystrokes are involved. First, you have to declare an object reference variable, and then you have to instantiate the object using the familiar CREATE OBJECT statement. Both approaches get us to the same place, but the ABAP way leads to early onset carpal tunnel syndrome.

```
Point p = new Point(2, 3);
```

Listing 2.34 Instantiating an Object in Java

```
DATA lo_point TYPE REF TO lcl_point.
CREATE OBJECT lo_point
  EXPORTING
    x = 2
    y = 3.
```

Listing 2.35 Instantiating an Object in ABAP Objects

In an effort to make ABAP Objects more developer-friendly, SAP introduced a new operator type in release 7.40 that allows ABAP developers to build expressions similar to those they might be accustomed to creating in other object-oriented languages: *constructor expressions*.

The syntax diagram in Listing 2.36 demonstrates the syntax used to create constructor expressions. To understand this syntax, it helps to have a feel for the types of constructor operators available. Table 2.5 highlights some of the more relevant operator types offered with release 7.40. Here, you can see that operator types help clarify the context of a particular constructor expression. For example, the presence of the NEW operator type tells you that the constructor expression is used to instantiate a data object. The rest of the expression syntax is then used to specify the type of the object being created and any parameters needed to carry out the instantiation process.

```
... operator [type|#]( ... ) ...
```

Listing 2.36 Syntax Diagram for Constructor Expressions

Operator	Description
Instance operator (NEW)	This operator type is used to create or instantiate objects. Here, the term "objects" refers to object instances in an object-oriented context as well as anonymous data objects.
Value operator (VALUE)	This operator is used to construct complex values that are used in initialization expressions, as method arguments, and so forth.
Reference operator (REF)	This operator is used to create data references on the fly. For example, if a method expects a data reference, you can use the REF operator to create that reference without having to use the GET_REFERENCE OF dobj INTO... statement to copy the data reference into a helper variable first.

Table 2.5 Selected Constructor Operator Types Introduced with Release 7.40

Operator	Description
Conversion/casting operators (CONV and CAST)	These operators are used to perform type conversions on the fly to reduce the number of helper variables needed to work with constructs like application programming interface (API) signatures.
Conditional operators (COND and SWITCH)	These operators are used to streamline conditional initialization statements into a single expression.

Table 2.5 Selected Constructor Operator Types Introduced with Release 7.40 (Cont.)

So, what do real world constructor expressions look like? Well, consider the code excerpt in Listing 2.37. We've refactored the ABAP code excerpt from Listing 2.35 using the NEW operator as well as inline declaration (i.e., the DATA(variable) part). From a compilation and runtime perspective, both code excerpts work exactly the same. However, it goes without saying that the newer approach is much more concise.

```
DATA(lo_point) = NEW lcl_point( x = 2, y = 3 ).
```

Listing 2.37 Working with the NEW Operator

From a pure OOP perspective, the NEW operator is perhaps the most compelling option. However, while the other operator types were not designed exclusively for OOP, they certainly make for more streamlined coding. For instance, consider the code excerpt in Listing 2.38. Here, we're using syntax features to reduce the number of lines it takes to build a SQL query using the object-oriented *ABAP Database Connectivity* (ADBC) library. Without these features, the number of lines of code required to implement this example would nearly double to accommodate all the helper variable declarations.

```
DATA(lv_program_id) = 'R3TR'.
DATA(lv_object_type) = 'CLAS'.
DATA(lv_pattern) = 'CL_ABAP%'.

DATA(lo_stmt) = NEW cl_sql_statement( ).
lo_stmt->set_param( REF #( lv_program_id ) ).
lo_stmt->set_param( REF #( lv_object_type ) ).
lo_stmt->set_param( REF #( lv_pattern ) ).

DATA(lo_rs) =
  lo_stmt->execute_query(
    `SELECT obj_name, author ` &&
      `FROM tadir ` &&
     `WHERE mandt = ? AND pgmid = ? AND object = ? ` &&
       `AND obj_name LIKE ?` ).
```

```
DATA(lt_rsmd) = lo_rs->get_metadata( ).
DATA(lr_line_type) = lo_rs->get_struct_ref( md_tab = lt_rsmd ).
DATA(lo_line_descr) = CAST cl_abap_structdescr(
  cl_abap_structdescr=>describe_by_data_ref( lr_line_type ) ).
DATA(lo_table_descr) = CAST cl_abap_tabledescr(
  cl_abap_tabledescr=>create( p_line_type = lo_line_descr ) ).

DATA lr_results_tab TYPE REF TO data.
CREATE DATA lr_results_tab TYPE HANDLE lo_table_descr.
lo_rs->set_param_table( lr_results_tab ).
lo_rs->next_package( ).
lo_rs->close( ).

FIELD-SYMBOLS <lt_results> TYPE INDEX TABLE.
FIELD-SYMBOLS <ls_result> TYPE any.

ASSIGN lr_results_tab->* TO <lt_results>.
LOOP AT <lt_results> ASSIGNING <ls_result>.
  WRITE: / <ls_result>.
ENDLOOP.
```

Listing 2.38 Using Constructor Expressions to Simplify an ADBC Query

2.7 UML Tutorial: Object Diagrams

In Chapter 1, Section 1.6, you learned how to use Unified Modeling Language (UML) class diagrams to specify the static architecture of an object-oriented system. Most of the time, these diagrams are straightforward and easy to interpret. However, the relationship between certain classes is not always so intuitive. In these cases, *object diagrams* can be used to depict a simulation of the actual objects created in reference to these classes at runtime. Oftentimes, just seeing an example of how the actual objects are configured at runtime can shed some light on the nature of complex class relationships.

Figure 2.51 illustrates a portion of a class diagram that shows the recursive aggregation relationship between a bill of materials (BOM) document and its components/items. The diamond on the MaterialBOM side of the association is used to indicate that the BOM is an *aggregate*, containing zero or more items.

The static class diagram in Figure 2.51 is useful for identifying the basic relationship between a BOM and its items, but it doesn't really help you visualize just how complex the relationships might be between BOMs and items at runtime. For example, in complex engineering scenarios, it's not uncommon for a BOM to contain items that are also complex assemblies. Depending on the nature of what you're building, this hierarchy can be nested arbitrarily deep.

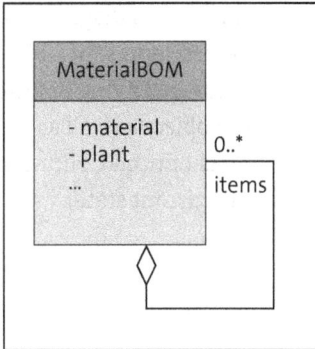

Figure 2.51 UML Class Diagram Showing Material BOM Aggregation

To visualize what's going on at runtime, what you really need is a diagram that shows the relationship between the actual object instances that will be created. In UML, this task is performed by the *object diagram*.

As you can see in Figure 2.52, object diagrams depict a snapshot of object instances and their interrelationships. Here, you can see how the BOM object for a laptop computer expands into a multi-level hierarchy. If you're developing algorithms to process this hierarchy, it can be useful to have this view of the data to visualize how you might go about traversing nodes.

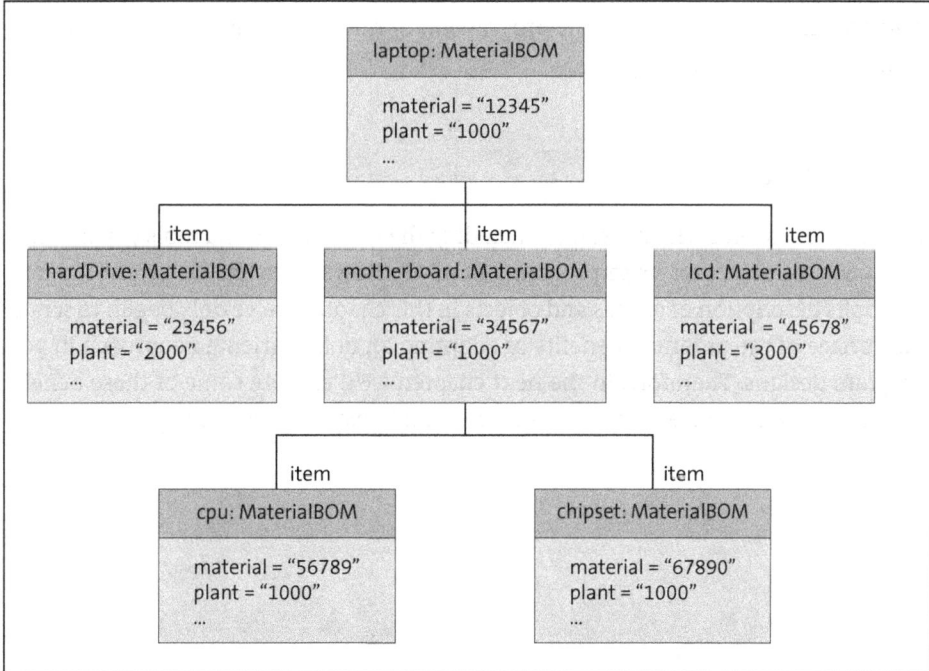

Figure 2.52 Object Diagram Showing the Expanded BOM for a Laptop Computer

As you can see from Figure 2.52, object diagrams are very similar to class diagrams in many respects. However, in object diagrams, the rectangular boxes represent object instances instead of classes. Figure 2.53 shows the basic notation for specifying objects in an object diagram. In the top box, you provide the name of the object as well as the type of class the object is created in reference to. The lower box is optional, allowing you to provide additional runtime details about the object (i.e., its current state).

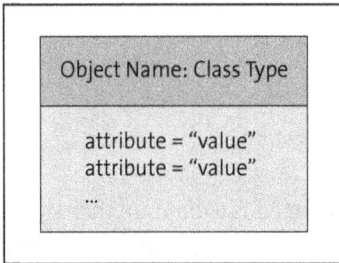

Figure 2.53 Object Instance Notation in Object Diagrams

Object diagrams can display as many objects as needed to illustrate the class/object relationships. The diagram functions as a viewport into the system at a particular point in time. Objects are created and destroyed in programs all the time, so it's important not to get hung up on trying to illustrate every possible object that will be created in the system at runtime. If one diagram cannot fully describe the relationship, additional diagrams can be used to show the progression of the object configuration as the program continues to run.

2.8   Summary

This chapter covered a lot of ground, including the basic syntax and tools you'll need to begin writing object-oriented programs in ABAP. However, while we've given a nuts-and-bolts description of classes and objects in this chapter, we've only begun to scratch the surface of the potential benefits of adopting an object-oriented approach in your program designs. Therefore, in the next chapter, we'll explore some of these benefits by examining the concepts of *encapsulation* and *implementation hiding*.

Chapter 3
Encapsulation and Implementation Hiding

*Classes are abstractions used to extend the functionality of a program-
ming language by introducing user-defined types. Encapsulation and
implementation hiding techniques simplify the way that developers
interface with these user-defined types, making object-oriented program designs easier to understand, maintain, and enhance.*

One of the most obvious ways to speed up the software development process is to
leverage preexisting code. However, while most projects strive to create reusable
source code artifacts, few succeed in delivering modules that can be classified as *reusable*. In most cases, this lack of reusability can be traced back to the fact that the mod-
ule(s) become too tightly coupled with their surrounding environment. With so many
"wires" getting in the way, it's hard to pick up these modules and drop them somewhere else. Therefore, to improve reusability, you should cut the cords and learn to
build autonomous objects (or components) that can *think* and *act* on their own.

In this chapter, you'll learn how to breathe life into your objects by exploring the benefits of combining data and behavior together under one roof. Along the way, we'll
explore the use of access control mechanisms and show how they can be used to define
and shape the *interfaces* of classes, making them easier to modify and reuse in other
contexts.

3.1 Lessons Learned from Procedural Programming

Contrary to popular belief, many core object-oriented programming (OOP) concepts
are based on similar principles rooted in the procedural programming paradigm. In
both paradigms, the basic goal is to provide developers with the tools they need to
translate requirements from the physical world into software-based solutions. However, while both programming models share in this goal, they go about achieving it in
very different ways. In this section, we'll take a closer look at the procedural approach
and consider some of the limitations that ultimately caused language designers to
move in the direction of an object-oriented approach.

3.1.1 Decomposing the Functional Decomposition Process

Typically, procedural developers formulate their program designs using a process called *functional decomposition*. This term is taken from the world of mathematics, where mathematical functions are broken down into a series of smaller discrete functions that are easier to understand on their own. From a development perspective, functional decomposition refers to the process of *decomposing* a complex program into a series of smaller modules (e.g., functions, procedures, or subroutines).

One common approach for discovering these procedures is to scan through the functional requirements and highlight all the verbs used to describe the steps a program must take to meet its objectives. After all the steps have been identified, they are then *composed* into a main program that's responsible for making sure that the procedures are executed in the right sequence. This process of organizing and refining the main program is sometimes called *stepwise refinement*.

For small to medium-sized programs, this strategy works well. However, as programs start to branch out and grow in complexity, the design tends to become unwieldy and the main program is saddled with too many responsibilities. In addition to keeping track of all the different procedures and making sure that they're processed in the right order, the main program is also normally responsible for managing all the data used by the various procedures. For this reason, such programs are often referred to as *god programs*. These types of issues become even more pronounced when we look at modern program designs like web apps and web services.

> **Note**
>
> In their book *Design Patterns Explained: A New Perspective on Object-Oriented Design, Second Edition* (Addison-Wesley, 2004), Alan Shalloway and James Trott suggest that the term "god program" stems from the fact that only God can understand these programs.

As noted earlier, with functional decomposition, the level of abstraction is the *subroutine*. Within a given subroutine definition, you can implement logic to perform a particular task using data provided from one of two places:

- Parameters that are passed into the subroutine from the calling program
- Global variables that are visible from within the subroutine

Regardless of the approach you use to supply subroutines with data, the reality is that there's no clean way of doing this without introducing some undesirable dependencies. For example, if you make liberal use of global variables, you open yourself up to the possibility of data corruption errors. Imagine the impact of switching out the call sequence of a pair of subroutines that make changes to the same global variables. If subroutine b depends on subroutine a to initialize the data and the call sequence gets flipped due to functional requirement changes, it's very likely that you'll see strange data-related errors in the processing (see Figure 3.1).

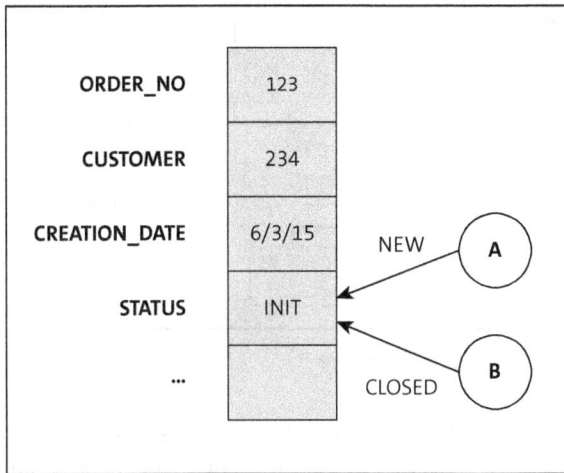

Figure 3.1 Data Collision Errors Between Subroutines and Global Variables

Conversely, replacing global variables by passing around parameters places an additional burden on the main program to keep track of those parameters. You end up cluttering the subroutine's parameter interface, which in turn leads to the tight coupling problem described earlier.

Ideally, your modules should assume more responsibilities internally so that they are less reliant on controlling programs and modules when carrying out their tasks. Think of it this way: If you were to compare the organization of a software program with organizational (org) structures in an enterprise, which of the two org structures depicted in Figure 3.2 and Figure 3.3 would you want your programs to look like? In the case of the flat org structure depicted in Figure 3.2, you have one centralized module that's responsible for micromanaging lots of submodules. On the other hand, the verticalized org structure shown in Figure 3.3 is much more balanced, with higher level modules delegating responsibilities down to specialized submodules.

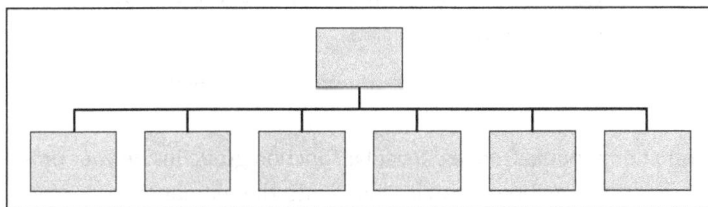

Figure 3.2 Example of a Flat Organizational Structure

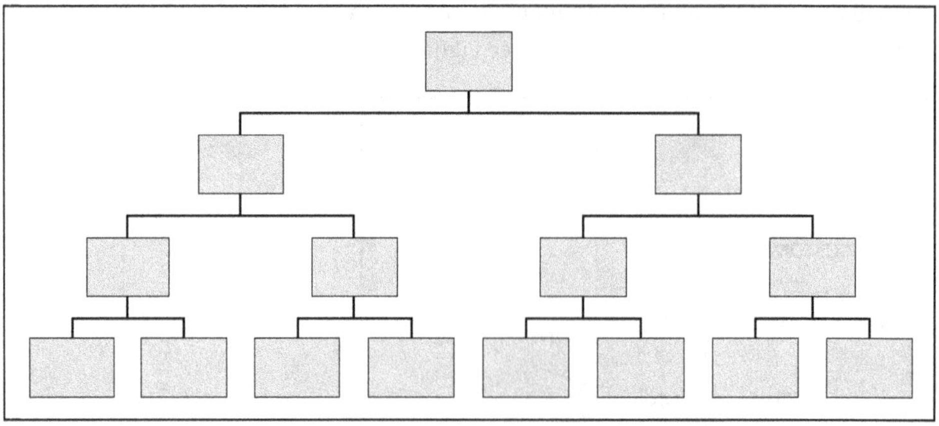

Figure 3.3 Example of an Expanded Organizational Structure

In programming, just like in business, it's important to delegate responsibilities so that your programs remain flexible. For that to happen, submodules need to be smart enough to figure certain things out on their own—and that requires data. In the sections to come, you'll find that combining data and behavior together within a class helps us develop modules that can attain the kind of autonomy we're looking for.

3.1.2   Case Study: A Procedural Code Library in ABAP

To better illustrate some of the procedural programming challenges noted in Section 3.1.1, let's consider an example. In this section, we'll sketch out the development of a date utility library using ABAP function modules.

If you've worked with function modules before, then you know that they're defined within the context of a *function group*. In some respects, function groups bear similarities to classes in that you can use them to define data and behaviors together within a self-contained unit (called a *function pool*). However, this analogy breaks down when you consider the fact that you cannot load multiple instances of a function group inside your program. This limitation makes it difficult for developers to work with the global data inside of a function group since additional logic is required to partition the data into separate work areas (or instances).

> **Note**
> Whenever you call a function module from a particular function group inside your program, the global data from the function group is loaded into the memory of the internal session of your program. Any subsequent calls to function modules within that function group will share the same global data allocated whenever the first function module was called.

Because of this shortcoming, most function module developers tend to design their functions as *stateless* modules that operate on data that's maintained elsewhere. In this context, the term "stateless" implies that the function modules have no recollection of prior invocations and don't maintain any sort of internal state. As a result, function module developers need only worry about implementing the procedural logic—keeping track of the data and sessions is someone else's problem.

Blame It on the BAPIs

The stateless approach to function module development increased in popularity quite a bit in the late 1990s and early 2000s when SAP introduced *Business Application Programming Interfaces (BAPIs)*. At that time, SAP rolled out loads of function modules that promoted a stateless architecture. To call these BAPIs, you would generally have to define a slew of global variables that would be used to process BAPI calls. This is illustrated with BAPI_USER_GET_DETAIL, which is commonly used to read user details. In the function signature shown in Figure 3.4, you can see that there's quite a bit of data about a user that must be maintained outside of the function module. It's also interesting to note that the same variables would be needed to perform other operations on users such as create, change, and so forth.

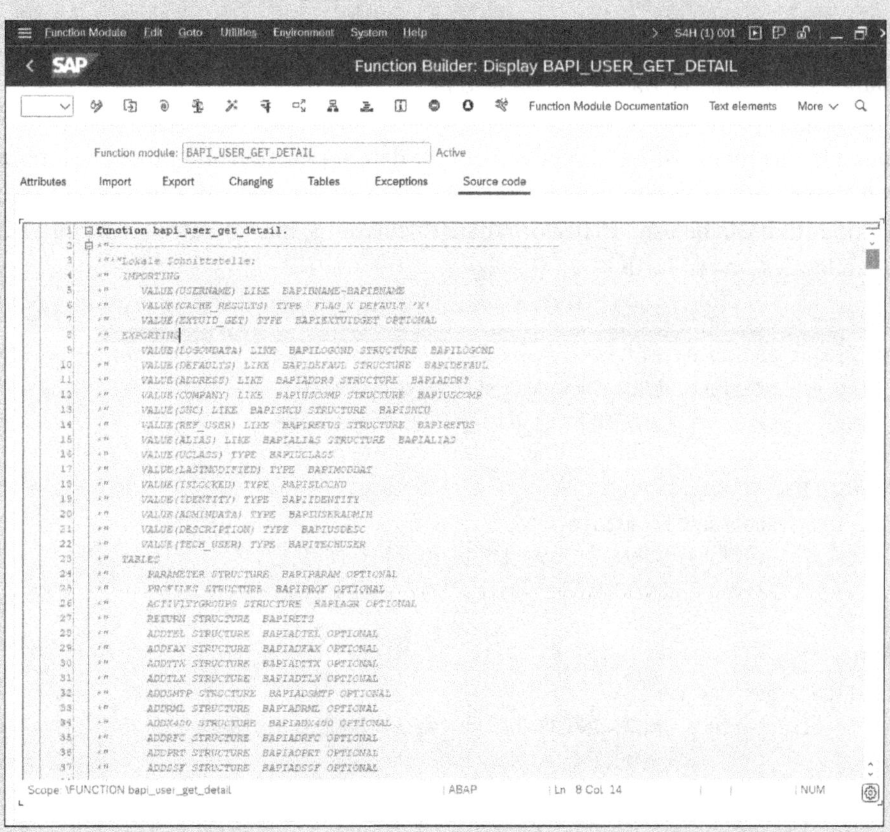

Figure 3.4 An Example of a Stateless BAPI Function

For the purposes of our date library example, we'll build our utility functions as stateless function modules. Within these functions, we'll operate on a date value represented by the SCALS_DATE structure shown in Figure 3.5. Though we could've just as easily used the internal ABAP date (D) type, we elected to use a structure type so that we could clearly address the individual components of a date (e.g., month, day, or year) without using offset semantics.

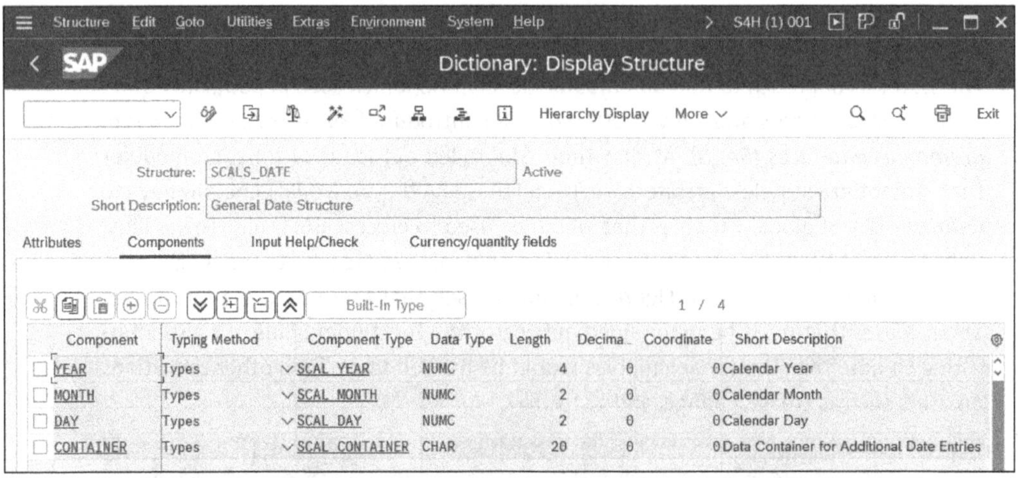

Figure 3.5 Modeling the Data Used for the Date Library

The code excerpt in Listing 3.1 sketches out the date application programming interface (API) in a function group called ZDATE_API. Here, we've defined a handful of utility methods that can be used to perform date calculations, format dates according to different locales, and so forth.

```
FUNCTION-pool zdate_api.
FUNCTION z_add_to_date.
  * Local Interface IMPORTING VALUE (iv_days) TYPE i
  *                 CHANGING (cs_date) TYPE scals_date
  ...
ENDFUNCTION.
FUNCTION z_subtract_from_date.
  * Local Interface IMPORTING VALUE (iv_days) TYPE i
  *                 CHANGING (cs_date) TYPE scals_date
  ...
ENDFUNCTION.
FUNCTION z_get_day_name.
  * Local Interface IMPORTING VALUE (is_date) TYPE scals_date
  *                 EXPORTING ev_day TYPE string
  ...
ENDFUNCTION.
FUNCTION z_get_week_of_year.
```

```
* Local Interface IMPORTING VALUE (is_date) TYPE scals_date
*                  EXPORTING ev_week TYPE i
  ...
ENDFUNCTION.
FUNCTION z_format_date.
* Local Interface IMPORTING VALUE (is_date) TYPE scals_date
*                          VALUE (iv_format) TYPE csequence
*                  EXPORTING ev_formatted TYPE string
  ...
ENDFUNCTION.
```

Listing 3.1 Building a Date Utility Library Using Function Modules

Within an ABAP program, you might use functions in the ZDATE_API function group to operate on date values being evaluated as part of a data processing routine like the contrived reporting example in Listing 3.2. With this kind of scenario in mind, in the upcoming sections, we'll consider how our date API might stand up to the maintenance requests that pop up over time. This analysis will set the stage for Section 3.1.3, when we shift our focus to objects.

```
REPORT zsome_report.
START-OF-SELECTION.
  PERFORM get_data.

FORM get_data.
  DATA ls_date TYPE scals_date.
  DATA lt_itab TYPE STANDARD TABLE OF ...
  FIELD-SYMBOLS <ls_wa> LIKE LINE OF lt_itab.

  SELECT *
    INTO TABLE lt_itab ...

  LOOP AT lt_itab ASSIGNING <ls_wa>.
    ls_date = ...

    CALL FUNCTION 'Z_ADD_TO_DATE'
      EXPORTING
        iv_days = <ls_wa>-work_days
      CHANGING
        cs_date = ls_date.
    ...
    CALL FUNCTION 'Z_SUBTRACT_FROM_DATE'
      EXPORTING
        iv_days = <ls_wa>-offset
      CHANGING
        cs_date = ls_date.
```

```
  ...
  CALL FUNCTION 'Z_FORMAT_DATE'
    EXPORTING
      is_date = ls_date
      iv_format = `MM/DD/YYYY`
    IMPORTING
      ev_formatted = lv_formatted.
  ...
  ENDLOOP.
ENDFORM.
```

Listing 3.2 Incorporating the Date API Into an ABAP Report Program

Expanding the Scope of the Date API

For the first scenario, imagine that you discover a need to expand the date API to also keep track of time. While this seems easy enough in principle, it could become challenging, since the structure used to model the date value doesn't contain components to capture a time stamp.

If you look at the SCALS_DATE structure in the ABAP Dictionary (in Figure 3.6), you'll see that this structure cannot be enhanced or appended to. You might get away with using the unused CONTAINER field, but this wouldn't be obvious to developers who aren't intimately familiar with the internal workings of the date API.

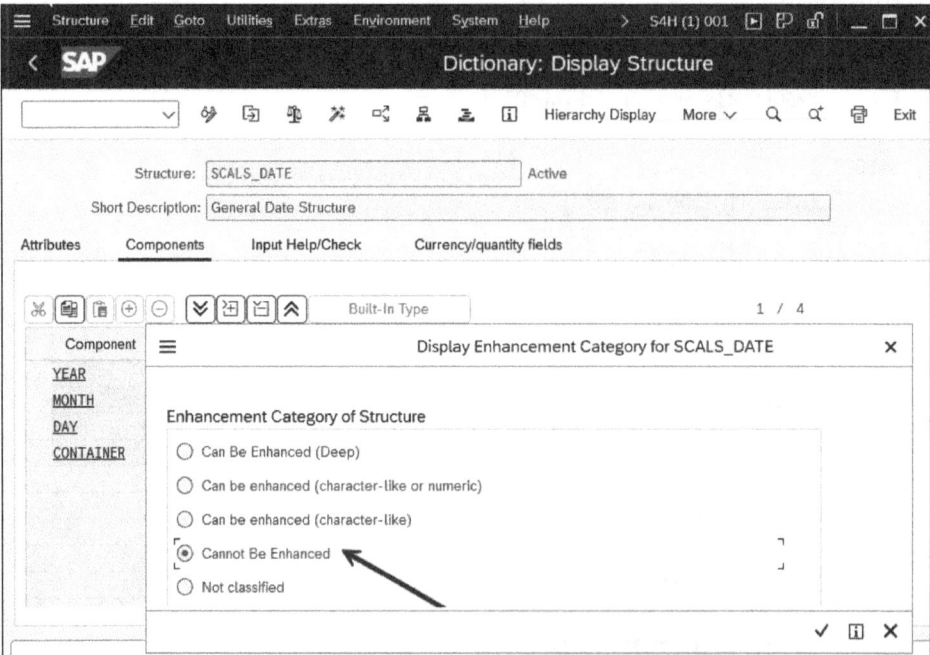

Figure 3.6 Looking at the Enhancement Category of the SCALS_DATE Structure

To implement this change correctly, you'd probably have to change the signature of your function modules to utilize a new structure. Besides requiring a fair amount of rework within the functions themselves, this also requires that you make wholesale changes to the programs that call them.

Though you might be saying to yourself that the choice of the SCALS_DATE structure for the date API's data model was a poor one (and you'd be right to say so), that's not the issue here. The point of this demonstration is to illustrate the fact that the date API exposes way too much information about its internal representation. Consumers of the date API shouldn't know (or care) whether you use the native ABAP date type (D), a structure, or something else entirely.

By exposing this kind of information in the function signatures, you've effectively locked yourself into a brittle design. For better or worse, you must stick with the design choices you've made and try your best to enhance around them. With stateless modules, this is about the best you can hope for.

Dealing with External Data Corruption

For the next scenario, imagine that you receive a defect report that indicates the Z_FORMAT_DATE function is producing invalid output. After much investigation, you determine that the invalid output isn't caused by the logic in Z_FORMAT_DATE but rather by an invalid day value specified in the SCALS_DATE structure's DAY field. You discover that the invalid value is set within the calling report program, which is accessing the SCALS_DATE structure outside of the ZDATE_API function group.

Though such errors might be easy to fix once you identify them, they can be difficult to find. Since the ZDATE_API function group doesn't technically own the data, there's nothing stopping other modules from overwriting and/or corrupting the API's data model. In a perfect world, *all* updates to the date API's data model would go through functions in the ZDATE_API function group so that you can isolate them and enforce the necessary validation rules (e.g., you can't have a date value of 20260231). However, this is something the procedural model simply can't guarantee. To enforce these rules and control access, you need support from the underlying language implementation.

3.1.3   Moving Toward Objects

The ZDATE_API function group introduced in Section 3.1.2 is an example of an *abstract data type* (ADT). As the name suggests, ADTs are data types that provide an abstraction around some entity or concept (e.g., a date). Included in this abstraction is the data itself, as well as a set of operations that can be performed on that data.

For ADTs to be effective, you must keep the data and operations as close to one another as possible. As you observed in Section 3.1.2, such cohabitation is virtually impossible to achieve with procedural programming techniques. Because of this divide, the date API (though admittedly contrived) was awkward to use and quite error prone. These problems become even more pronounced as the size and complexity of such code libraries expand.

In many ways, all the problems we'll consider in this section can be traced back to one central theme: poor support for data. While it would seem obvious that data is the foundation upon which any successful computer program runs, the stark reality is that data takes a back seat to actions in the procedural programming paradigm. As a result, procedural programs tend to decay at a much faster pace than programs built using models that place a greater emphasis on the data.

3.2   Data Abstraction with Classes

Recognizing many of the limitations outlined in Section 3.1, software researchers developed the OOP paradigm from the ground up with a strong emphasis on data *and* behavior. As you've already learned, classes are the vehicle that drives this equilibrium, encapsulating data (attributes) and behavior (methods) together inside a self-contained unit.

Encapsulation improves the organization of the code, making object-oriented class libraries much easier to understand and use than their procedural counterparts. To put this into perspective, let's reflect on the clumsiness of the function module-based date library we created in Section 3.1.2. Each time you accessed one of the API functions, you had to pass in an externally managed structure that contained all of the date information needed to handle the request. Plus, if you wanted to work with multiple dates, then you had to define multiple variables and track those variables manually outside of the function group.

Let's compare that experience with a reimagined date API built using an ABAP Objects class. In Listing 3.3, we've created a class called LCL_DATE that provides the same functionality of the ZDATE_API function group. As you look over the class definition, notice the simplification in the signature of the API methods. Instead of passing around an SCALS_DATE structure, the date information is stored internally in an instance attribute called MS_DATE_INFO. Besides simplifying the interface, this design change also allows us to get out of the business of tracking date information externally. Now, the date API is truly an ADT that provides a *complete* abstraction around a date value, as opposed to a loosely associated set of stateless function modules.

```
CLASS lcl_date DEFINITION.
  PUBLIC SECTION.
    DATA ms_date_info TYPE scals_date.
    METHODS:
      add IMPORTING iv_days TYPE i
          RETURNING VALUE(ro_date) TYPE REF TO lcl_date,
      subtract IMPORTING iv_days TYPE i
               RETURNING VALUE(ro_date) TYPE REF TO lcl_date,
```

```
    get_day_name RETURNING VALUE(rv_day) TYPE string,
    get_week_of_year RETURNING VALUE(rv_week) TYPE i,
    format IMPORTING iv_pattern TYPE csequence
          RETURNING VALUE(rv_date) TYPE string.
    ...
ENDCLASS.
```

Listing 3.3 Reimagining the Date Utilities API as an ABAP Objects Class

The code excerpt in Listing 3.4 demonstrates how you can work with the refactored date library. Once an `LCL_DATE` instance is created, you no longer have to worry about handling the date value. Instead, you can use methods like `add()` and `subtract()` to apply the changes in-place. From a code readability standpoint, this is much easier to follow, because the context of an operation like `add()` is clearly the object referenced by `lo_date`.

```
DATA lo_date TYPE REF TO lcl_date.
DATA lv_message TYPE string.

CREATE OBJECT lo_date
  EXPORTING
    iv_date = '20261207'
lo_date->add( 30 ).
lv_message = |{ lo_date->subtract( 15 )->format( 'YYYYMMDD' ) }|.
```

Listing 3.4 Working with an Object-Oriented API

Ultimately, objects created in reference to encapsulated classes take on their own *identity*, allowing developers to start thinking about their designs in more conceptual terms (e.g., a date). Consumers of these classes don't have to worry about low-level implementation details; to the end user, the `LCL_DATE` class is like a black box that performs various date manipulations. You don't have to supply the `LCL_DATE` class with data, context, or instructions; it intrinsically *knows* how to do its job.

In the next section, you'll learn how to round out ADTs like the `LCL_DATE` class by closing off access to internal components such as the `MS_DATE_INFO` attribute. This safeguard ensures that *all* operations on date values are mediated through API methods that rigorously validate incoming requests to ensure that the integrity of date values is maintained. As you'll see, this approach offers several important benefits.

3.3   Defining Component Visibilities

The term *encapsulation* refers to the idea of enclosing something inside of a *capsule*. The imagery associated with a word like capsule implies that you're setting some kind of boundary between the internal components of a class and the outside world. The purpose of this boundary is to protect (or hide) the inner mechanisms of the object that are sensitive to change. Most of the time, the most vulnerable parts of an object are its attributes, since these define the object's state. However, in this book, we'll look at ways to hide *any* design decisions that are subject to change.

In this section, we'll describe the ABAP Objects language constructs that you can use to establish boundaries within your classes. Then, in the section that follows, you'll learn how to use these boundaries to build robust classes that can easily be adapted to ever-changing functional requirements.

3.3.1   Working with Visibility Sections

ABAP Objects provides three visibility sections for controlling access to the components defined within a class: the PUBLIC SECTION, the PROTECTED SECTION, and the PRIVATE SECTION. Within a CLASS DEFINITION statement, all component declarations must be defined within one of these three visibility sections. The code excerpt in Listing 3.5 demonstrates the syntax used to define components within these sections.

```
CLASS lcl_visibility DEFINITION.
  PUBLIC SECTION.
    DATA x TYPE i.
  PROTECTED SECTION.
    DATA y TYPE i.
  PRIVATE SECTION.
    DATA z TYPE i.
ENDCLASS.
```

Listing 3.5 Working with Visibility Sections

As you might expect, components defined within the PUBLIC SECTION of a class are publicly accessible from any context in which the class itself is visible (i.e., anywhere you can use the class type to declare an object reference variable). These components make up the *public interface* of the class.

Components defined within the PRIVATE SECTION of a class are only accessible from within the class itself. Note that this is more than just a mere suggestion; this is something that's strictly enforced by the ABAP compiler/runtime environment. For example, the code excerpt in Listing 3.6 would produce a compilation error because the z attribute of the LCL_VISIBILITY class is defined as a private attribute. The *only* way to update the value in z would be to go through a method defined in the LCL_VISIBILITY class.

```
DATA lo_visible TYPE REF TO lcl_visibility.
CREATE OBJECT lo_visible.
IF lo_visible->z GT 0.
  ...
ENDIF.
```

Listing 3.6 Attempting to Access Private Components of a Class

We'll defer discussion of the PROTECTED SECTION until after we cover inheritance in Chapter 5. For now, simply note that components defined in the PROTECTED SECTION are only accessible within a class and its subclasses.

When working in the form-based view of the Class Builder, you can assign components of global classes to visibility sections using the **Visibility** column highlighted in Figure 3.7.

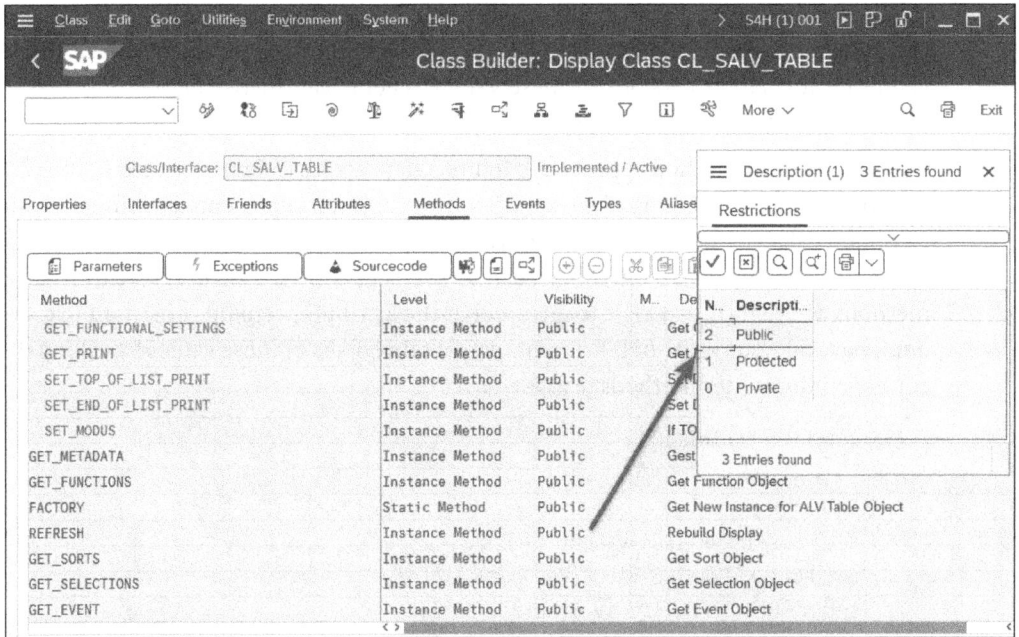

Figure 3.7 Setting the Visibility of Components Using the Form-Based View of the Class Builder

Designing Across Multiple Dimensions

Choosing the right visibility section for a given component can be tricky, and it requires a fair amount of thought. Instead of thinking about the individual components, you must think in terms of the class's overall interface. If you want to make your class simple and easy to use, then you'll need to strip down the public interface to just the essentials. This makes the interface less busy and therefore easier to understand and consume.

In general, clients of a class should be on a need-to-know basis. In other words, if a client doesn't require direct access to a component, then there's no need for them to even be aware of its existence. Declaring such components within the PRIVATE SECTION of a class makes life easier for everyone: Clients get to work with a simplified interface and the owners of the class have the freedom to change or improve the internal implementation of a class without the fear of breaking existing client code.

With this concept in mind, we recommend that you define most attributes within the PRIVATE SECTION of a class. The primary reason for hiding attributes is to ensure that the state of the object cannot be tampered with haphazardly. If a client needs to update the state of an object, they can do so through a method defined in the PUBLIC SECTION. The advantage of this kind of indirection is that you can control the assignment of the attribute using business rules that are defined inside the method. This eliminates a lot of the guesswork in troubleshooting data-related errors, since you know that all changes to an attribute are routed through a single method. Methods that update the value of private attributes are sometimes called *setter* (or *mutator*) methods. To access these values (or formatted versions of these values), clients can invoke *getter* (or *accessor*) methods that broker access in the other direction.

This getter/setter method approach to indirect data access is demonstrated in the LCL_TIME class in Listing 3.7. Here, the state of the time object is represented by three private attributes called mv_hour, mv_minute, and mv_second. Any updates to these attributes are controlled through setter methods such as set_hour() or set_minute(). Within these methods, we've included logic to ensure that the attributes remain consistent (e.g., we don't have an hour value of 113). Clients can obtain copies of these values by calling the corresponding getter methods (e.g. get_hour()).

```
CLASS lcl_time DEFINITION.
  PUBLIC SECTION.
    METHODS:
      set_hour IMPORTING iv_hour TYPE i,
      get_hour RETURNING VALUE(rv_hour) TYPE i,
      set_minute IMPORTING iv_minute TYPE i,
      get_minute RETURNING VALUE(rv_minute) TYPE i,
      set_second IMPORTING iv_second TYPE i,
      get_second RETURNING VALUE(rv_second) TYPE i.

  PRIVATE SECTION.
    DATA: mv_hour TYPE i,
          mv_minute TYPE i,
          mv_second TYPE i.
ENDCLASS.

CLASS lcl_time IMPLEMENTATION.
  METHOD set_hour.
    IF iv_hour BETWEEN 0 AND 23.
      me->mv_hour = iv_hour.
    ELSE.
      "TODO: Error handling...
    ENDIF.
  ENDMETHOD.
```

```
  METHOD get_hour.
    rv_hour = me->mv_hour.
  ENDMETHOD.
  ...
ENDCLASS.
```

Listing 3.7 Working with Getter and Setter Methods

As an alternative to the getter method approach, ABAP also allows us to define read-only attributes within a class definition. This is achieved using the READ-ONLY addition to the DATA keyword. The code excerpt in Listing 3.8 demonstrates how you might refactor the LCL_TIME class from Listing 3.7 to use this feature.

```
CLASS lcl_time DEFINITION.
  PUBLIC SECTION.
    DATA: mv_hour TYPE i READ-ONLY,
          mv_minute TYPE i READ-ONLY,
          mv_second TYPE i READ-ONLY.
  ...
ENDCLASS.
```

Listing 3.8 Defining Read-Only Attributes in a Class

While this feature can come in handy for simple classes that are primarily used for transferring data, we encourage you to use this option sparingly, since it exposes the internal implementation details of your class.

3.3.2   Understanding the Friend Concept

In the previous section, you learned that components defined within the private and protected sections of a class are not visible outside of that class (or subclasses, in the case of protected components). However, in certain cases, it's advantageous to grant special access to classes of your choosing. Such classes are called *friends* of the class that grants them access.

Listing 3.9 illustrates the syntax used to create friend relationships between a defining class CL_SOME_CLASS and its friends: C1, C2, and so on. Here, the FRIENDS addition is added to a CLASS DEFINITION statement to declare this relationship up front to the ABAP compiler. You can specify multiple friend classes after the FRIENDS addition (not to mention interfaces, which are covered in Chapter 6).

```
CLASS cl_some_class DEFINITION FRIENDS c1 c2 i3 i4.
  ...
ENDCLASS.
```

Listing 3.9 Defining Friend Relationships in Classes

To demonstrate how friend relationships work between classes, consider the example code in Listing 3.10. Here, we have a pair of classes called LCL_PARENT and LCL_CHILD that have a friend relationship. The LCL_CHILD class takes advantage of this relationship by accessing the LCL_PARENT class's mv_credit_card_no attribute in a method called buy_toys(). Since mv_credit_card_no is defined as a private attribute, the only way for LCL_CHILD to access this value is through the friend relationship. Without this addition, the code in Listing 3.10 would produce a syntax error.

```
CLASS lcl_child DEFINITION DEFERRED.
CLASS lcl_parent DEFINITION FRIENDS lcl_child.
  PRIVATE SECTION.
    DATA mv_credit_card_no TYPE string.
ENDCLASS.

CLASS lcl_child DEFINITION.
  PUBLIC SECTION.
    METHODS buy_toys.
ENDCLASS.

CLASS lcl_child IMPLEMENTATION.
  METHOD buy_toys.
    DATA: lo_parent TYPE REF TO lcl_parent,
          lo_store TYPE REF TO lcl_toy_store.
    lo_parent = ...
    lo_store = ...

    lo_store->checkout( lo_parent->mv_credit_card_no ).
  ENDMETHOD.
ENDCLASS.
```

Listing 3.10 Bypassing Access Control Using Friends

You can achieve the same effect for global classes maintained in the form-based view of the Class Builder by plugging in the target friend classes on the **Friends** tab, as shown in Figure 3.8.

As you begin working with friend relationships, there are a couple of important things to consider. First, it's important to note the direction and nature of the friend relationship. In Listing 3.10, class LCL_PARENT explicitly granted friend access to class LCL_CHILD. This relationship definition is not reflexive. For example, it would not be possible for class LCL_PARENT to access the private components of class LCL_CHILD without the LCL_CHILD class granting friend access to LCL_PARENT first. Second, notice that classes cannot arbitrarily declare themselves friends of another class. For instance, it would not be possible for class LCL_CHILD to surreptitiously declare itself a friend of class LCL_PARENT. If this were the case, access control would be a waste of time, since any class could bypass this restriction by simply declaring themselves a friend of whatever class they were trying to access.

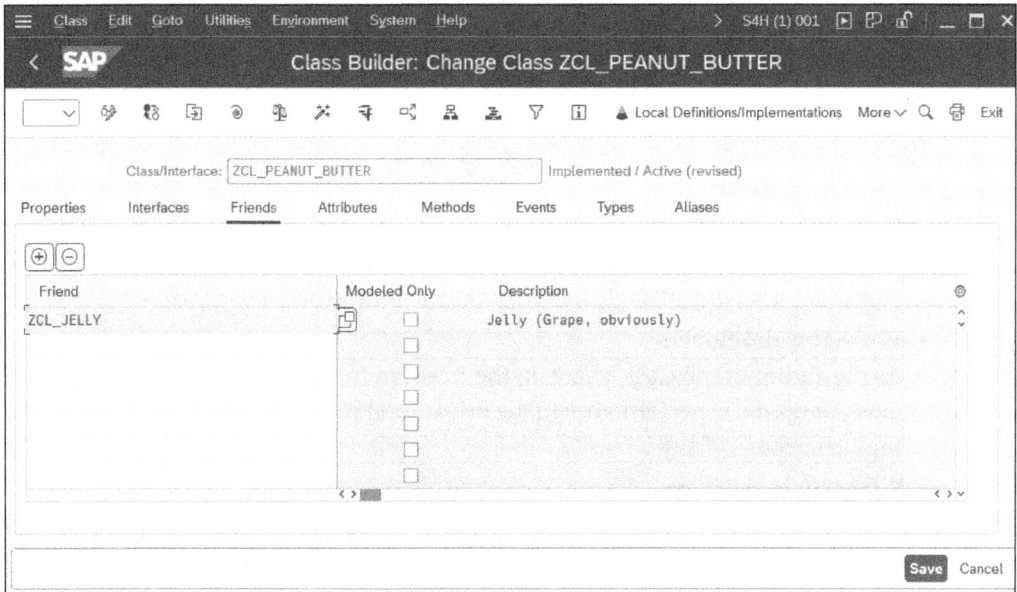

Figure 3.8 Defining Friendship Relationships Between Global Classes

The example code in Listing 3.10 also introduced a new addition to the CLASS DEFINITION statement that we have not covered before: the DEFERRED addition. In a scenario like this, the DEFERRED addition used in the first CLASS DEFINITION statement for LCL_CHILD is needed to instruct the compiler of the existence of the LCL_CHILD class in the CLASS DEFINITION statement for the LCL_PARENT class. Without this clause, the compiler would say that class LCL_CHILD is unknown whenever you try to establish the friend relationship in the definition of class LCL_PARENT.

To Friend or Not to Friend

Many purists argue that the use of friends should not be allowed in object-oriented languages since they bypass traditional access control mechanisms. Whether or not you agree with this sentiment, we recommend that you use friend relationships sparingly in your designs, because it truly is rare that you would need to open up access like this.

3.4   Hiding the Implementation

As you've seen, class visibility sections determine which parts of the class are accessible to other code. While it may be tempting to declare everything public for ease of access, that quickly leads to fragile systems. The real power of these visibility sections lies in their ability to hide implementation details.

Private attributes and methods, defined in the PRIVATE SECTION, act like the inner workings of a machine. These details are vital for making the machine run but not something an end user should be tampering with. By shielding these details, you create a clean and dependable contract. Other developers (and maybe even your future self) can rely on the class's public interface without needing to know or worry about the moving parts inside.

This separation has several practical benefits:

- **Protecting clients**
  Consumers of your class interact only with stable, documented methods and attributes in the PUBLIC SECTION. They don't risk introducing bugs by reaching into sensitive areas or relying on logic that may change tomorrow.

- **Protecting developers**
  As the author of the class, you gain the freedom to refine algorithms, rename variables, or optimize performance in the private and protected sections without breaking client code.

- **Supporting evolution**
  ABAP apps often live for years, sometimes decades. Hiding the implementation means your classes can evolve safely as requirements and technologies change.

At the end of the day, defining private attributes and methods isn't about being secretive—it's about setting boundaries that make your classes safer to use, easier to maintain, and more adaptable in the long run.

3.5   Designing by Contract

As you've learned, encapsulation and implementation hiding techniques can be used to define very precise public interfaces for a class. These interfaces form a *contract* between the developer of a class and users of that class. The contract metaphor is taken from the business world, where customers enter into contractual agreements with suppliers providing goods or services. In his book, *Object-Oriented Software Construction* (Prentice Hall, 1997), Bertrand Meyer describes how this concept could be adapted into object-oriented software designs in order to improve the reliability of software components that are "...implementations meant to satisfy well-understood specifications."

In this context, objects are subject to a series of *invariants* (or constraints) that specify the valid states for the object. To maintain these invariants, methods are defined using *preconditions* (what must be true before the method is executed) and *postconditions* (what must be true after the method is executed). In Chapter 8, we'll look at ways to deal with exceptions to these rules.

The primary goal when applying the *design by contract* approach in your software designs is to produce components that deliver *predictable* results. The boundaries set by the visibility sections ensure that *loopholes* are not introduced into the contract. For instance, the procedural-based date library that we first introduced in Section 3.1.2 had many loopholes that made it possible to bypass the business rules implemented inside the function modules. The encapsulation techniques we applied in the class-based reimplementation of this library closed these loopholes by encapsulating the date data as a private attribute that's cut off from external tampering.

Client programmers using classes based on these principles know what to expect from the class based on the provided public interface. Similarly, class developers are free to change the underlying implementation so long as they continue to honor the contract outlined in the public interface. Over time, the dual nature of this relationship helps to increase trust as you accumulate reusable modules that clients know will work.

3.6   UML Tutorial: Sequence Diagrams

So far, our study of the Unified Modeling Language (UML) has been focused on diagrams that are used to describe the static architecture of an object-oriented system. In this section, we'll introduce the first of several *behavioral diagrams* that are used to illustrate the behavior of objects at runtime. The *sequence diagram* depicts a message sequence chart between objects that are interacting inside a software system.

Figure 3.9 shows a simple sequence diagram that illustrates a cash withdrawal transaction in an ATM. A sequence diagram is essentially a graph in two dimensions. The objects involved in the interaction are aligned along the horizontal axis. The vertical axis represents time. Sequence diagrams are initiated by a request message from some kind of external source. In the example in Figure 3.9, the external source is a user interfacing with the ATM. This initial message is called a *found message*. In object-oriented terms, a message is analogous to a method call. Messages are sent to objects (depicted as object boxes on the object diagrams we described in Chapter 2). The dashed line protruding from underneath the object box represents the object's *lifeline*.

The intersection of a message and an object's lifeline is depicted with a thin rectangular box called an *activation bar*. The activation bar shows when an object is active during the interaction. Objects are activated via messages (i.e., method calls). Messages can include parameters that help clarify the operation to be performed by the object. However, it's not a good idea to try to fully specify the method interface in a sequence diagram—that's what a class diagram is for. You should only use parameters for emphasis or clarity. Synchronous method calls can have a *return* message that can also have optional parameters.

In some cases, a method might need to call other local helper methods to complete its task. In this case, you can illustrate a *self call* by drawing a circuitous arrow to another activation bar that is stacked on top of the current activation bar. For example, in Figure 3.9, messages `dispenseCash` and `printReceipt` are both represented as self calls on the `atm` object inside method `withdraw`.

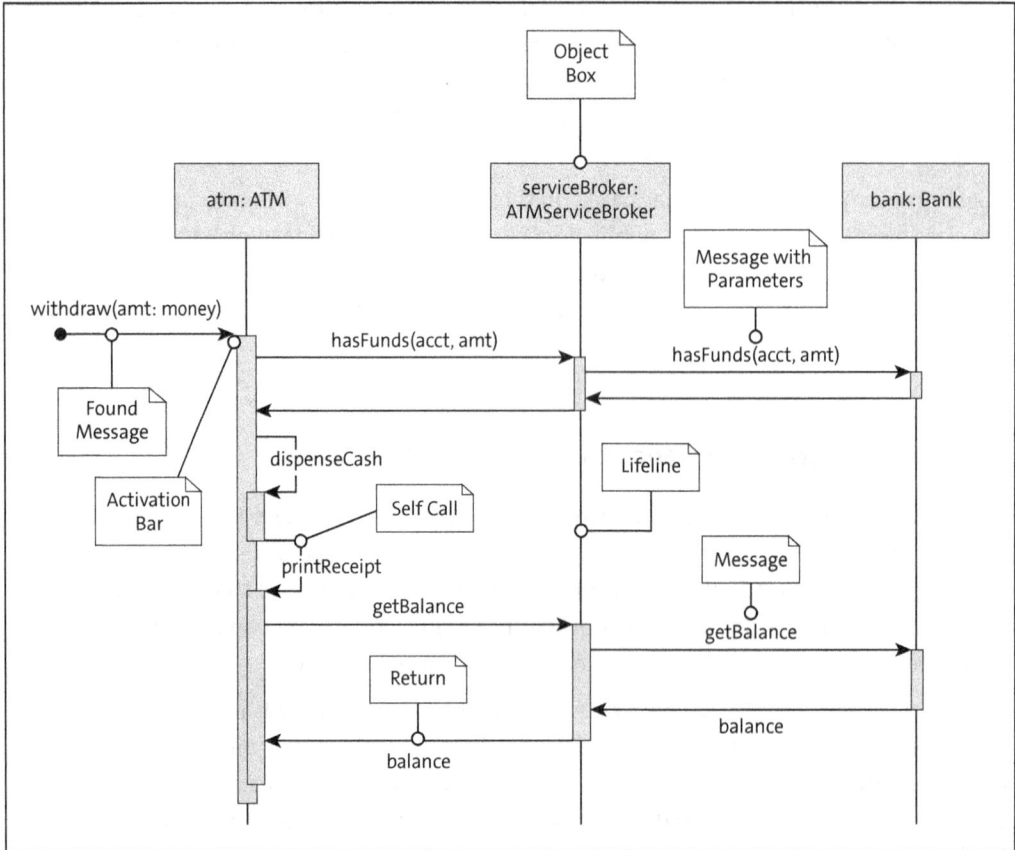

Figure 3.9 Sequence Diagram for Withdrawing Cash from an ATM

Sequence diagrams are very useful for explaining complex interactions where the order of operations is difficult to follow. One of the reasons that sequence diagrams are so popular is that the notation is very intuitive and easy to read. To maintain this readability, it's important to avoid cluttering a sequence diagram with too many interactions. In the coming chapters, we'll look at other types of interaction diagrams that illustrate fine-grained behavior within an object or more involved interactions that span multiple use cases.

3.7 Summary

In this chapter, you learned about the many advantages of applying encapsulation and implementation hiding techniques to your class designs. Encapsulating data and behavior in classes simplifies the way that users and clients work with classes. Hiding the implementation details of these classes strengthens the design even further, making classes much more resistant to change and/or data corruption. The combination of these two design techniques helps you design intelligent classes that are highly self-sufficient. Such classes are easy to reuse in other contexts since they are loosely coupled to the outside world.

In the next chapter, we'll examine the basic lifecycle of an object. We'll also learn about special methods called *constructors* that can be used to ensure that object instances are always created in a valid state.

3

Chapter 4
Object Initialization and Cleanup

Some of the most elusive bugs to trap are the ones that can be traced back to missing or invalid variable initializations. With that in mind, this chapter takes a deep dive into the lifecycle of objects and explores techniques for ensuring that objects are properly initialized before consumers ever get their hands on them.

In the previous chapter, you learned how to use encapsulation and implementation hiding techniques to protect the integrity of an object. Such objects produce consistent and reliable results, freeing developers from constantly worrying about data correctness issues in their programs. However, all these measures are wasted if you fail to properly initialize the object in the first place.

In this chapter, we'll consider some techniques for ensuring that objects are always created in a valid state. We'll also examine the overall object lifecycle, paying particular attention to how object resources are managed by the automatic memory management functionality of the ABAP runtime environment.

4.1 Understanding the Object Creation Process

One of the primary goals of the object-oriented design process is to identify ways to delegate responsibilities to objects. This approach transfers complexity from the main program into objects that are intelligent enough to handle the tasks they're assigned.

To coordinate these efforts, the main program needs to be able to create and destroy objects on demand. While the ABAP runtime environment takes care of most of the low-level technical details related to object allocation, there are some performance costs associated with creating objects dynamically. To recognize how these costs can affect the performance of your programs, it's important to understand what's going on behind the scenes whenever you request the creation of an object using the CREATE OBJECT statement.

To put all this into perspective, let's consider an example of a simple ABAP report program that needs to create objects at runtime. For the purposes of our discussion, we'll assume that this report is running in the foreground in an SAP GUI session. However, the basic principles remain the same for background processes.

As you can see in Figure 4.1, the report program is running inside of a logical memory frame called an *internal session*. Conceptually, you can think of internal sessions as part of a call stack that gets created as ABAP programs call other programs (e.g., using the CALL TRANSACTION or SUBMIT statements). Each internal session manages the data objects of the program that's running, as well as the data objects of other programs (e.g., function pools and class pools) that the program is using. The call stack itself is contained within a *main session* that gets allocated whenever you open SAP GUI or create a new session.

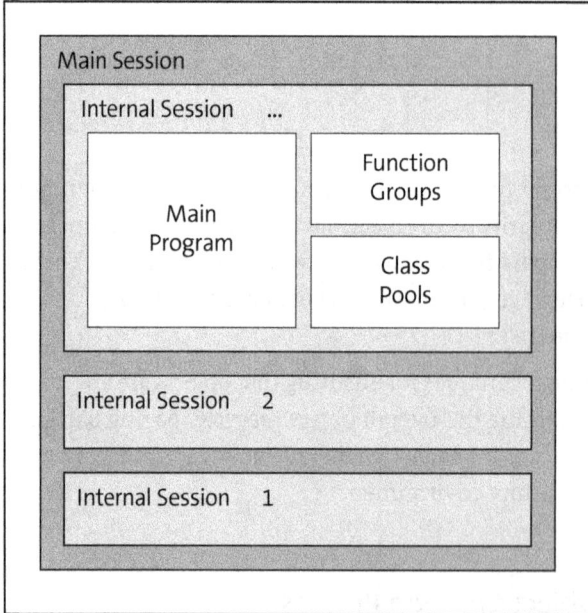

Figure 4.1 Logical Memory Areas of a User Session

If we zoom out a bit more, you can see that main sessions are stored within another logical memory area called a *user session* (also sometimes referred to as a *context*). As the name suggests, user sessions contain state information about users that are logged on to the system: basic user attributes, assigned authorizations, etc. They also keep track of the programs a user is running along with the data objects used by those programs.

Internally, user sessions are stored with a special section of shared memory on the application server (AS) ABAP host called the *roll buffer*. As you can see in Figure 4.2, the roll buffer is separated from the local memory area allocated to service work processes, such as the one that's processing our report program. Why implement such separation? Well, since the AS ABAP is a time-sharing system designed to support multiple users, there must be a way to swap users in and out between tasks to maximize work process utilization. In SAP terms, this swap process is referred to as a *rollout* (when a user is evicted from a work process) or a *roll-in* (when a user request is assigned to a work process).

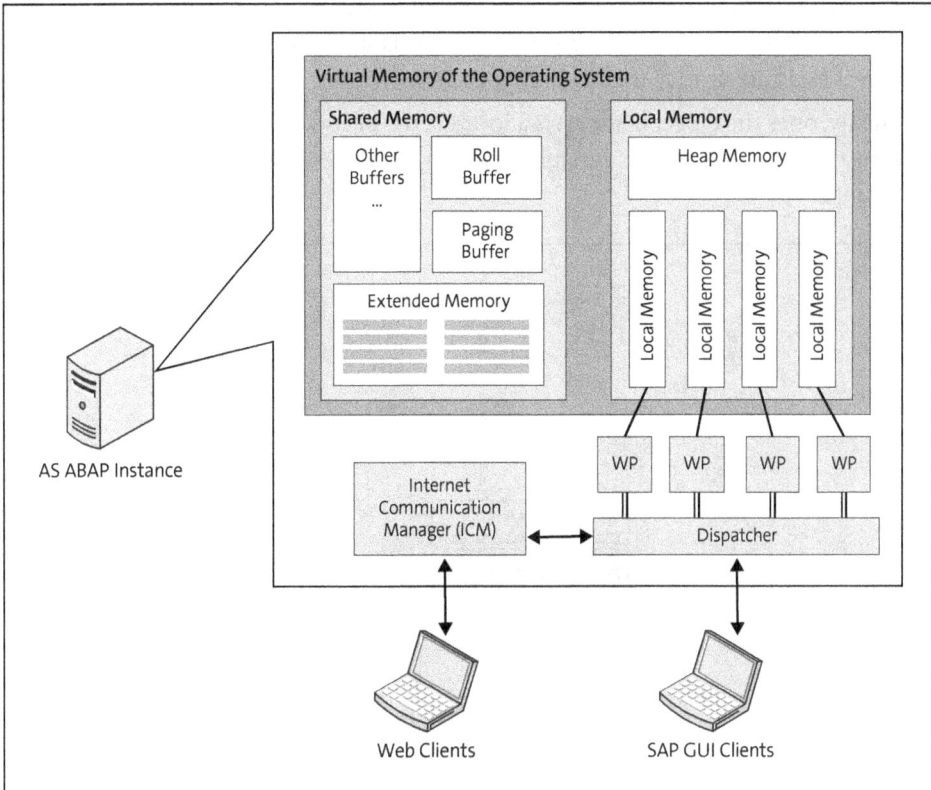

Figure 4.2 Basic Memory Structure of an AS ABAP Instance

Since rollouts and roll-ins happen all the time during normal operation, they must execute swiftly to avoid system latency. This means that SAP must keep the user session itself small so that it can be copied quickly and easily. The upshot of all this from an object management perspective is that SAP can't afford to carry around lots of object instances within a user session. To get around this limitation, SAP created a layer of separation between object instances and the programs that allocate them. This is where the object reference variables we introduced in Chapter 2 come into play.

Figure 4.3 illustrates how this separation is achieved from a memory allocation perspective. Here, our ABAP report is operating within a work process and our user session has been copied (rolled-in) to the roll area of the work process.

Whenever you request the creation of an object at runtime (using the CREATE OBJECT statement), the ABAP runtime environment will carry out the allocation request as follows:

1. First, it looks at the class definition referenced by your object reference variable and determines how much memory it needs to allocate an object instance of this type.

2. Then, once it knows how much memory it needs, it will scan through the extended memory area to find a chunk of memory large enough to store the object and its data.

3. During this memory allocation process, the runtime environment will also set aside some additional memory to create a *header* data structure that keeps track of the object's administrative details (see Figure 4.3).

4. Finally, once the object instance is allocated, the header structure's address is copied into the object reference variable that the CREATE OBJECT statement is performed against.

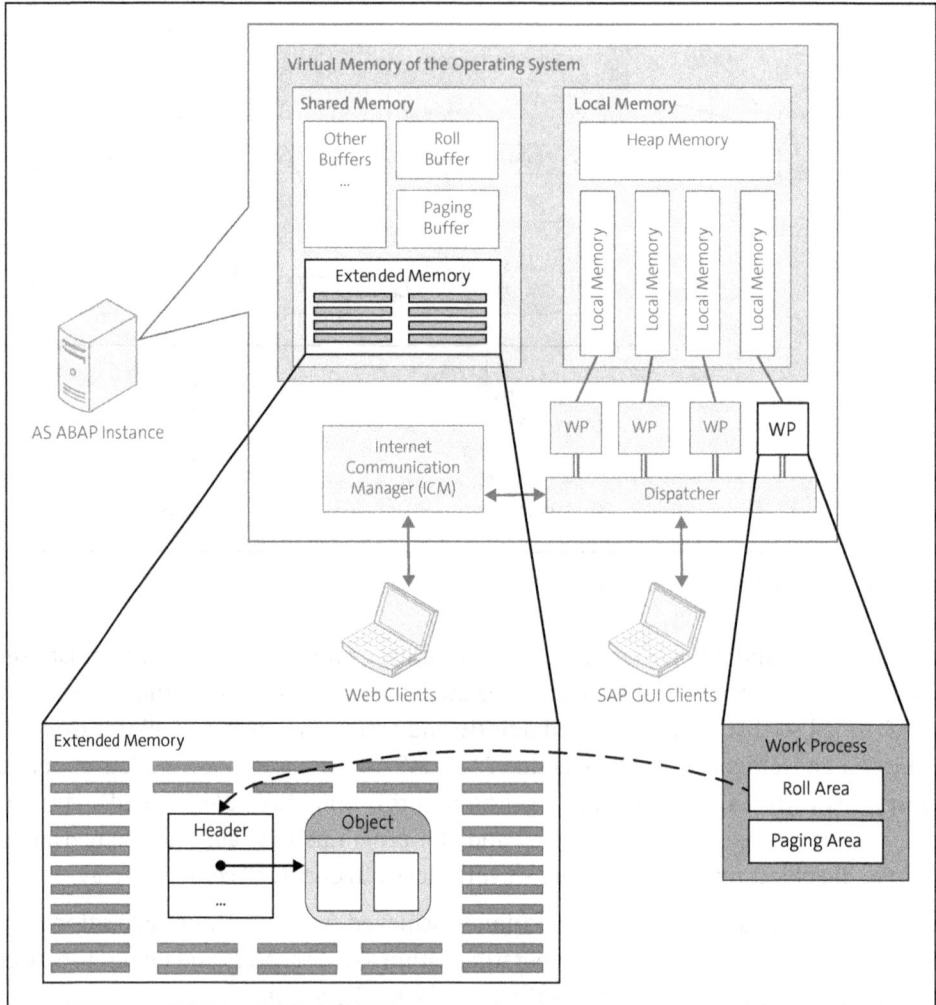

Figure 4.3 Memory Allocation of Objects

The primary consequence of this approach for dynamically generating objects is the additional time required to allocate the proper amount of memory for an object. As multiple programs create and destroy objects, the extended memory area can become fragmented, making it difficult to locate a contiguous chunk of memory large enough to hold an object. Skeptics sometimes point to these performance costs as a reason for not using objects in their programs, claiming that they can't afford the additional over-head at runtime.

However, if you look carefully at your existing ABAP programs, you'll likely find that you're already using many types of dynamic data objects. For example, internal tables are dynamic data objects that require additional memory to be dynamically allocated as additional rows are appended to them. Most design decisions involve some kind of trade-off, and in the case of objects, you might have to sacrifice some performance to realize the many benefits of object-oriented programming. Fortunately, SAP has opti-mized the performance of the ABAP runtime environment such that these perfor-mance issues are rarely a concern. Still, we'll investigate some basic guidelines for tun-ing performance in Section 4.5.

> **Note**
>
> For an excellent description of dynamic data objects, check out Horst Keller's blog post, "ABAP Geek 12 - The Deep," at *http://s-prs.co/v609300*.

4.2   Working with Constructors

Encapsulated objects rely on data stored in private or hidden attributes to keep track of their internal state so that they can respond to method requests intelligently. There-fore, it's crucial that you ensure an object's attributes are properly supplied with the data it needs to perform its duties when called upon. Otherwise, you end up back in procedural hell where methods are stateless, and you have to go to great lengths to fetch data before you can do any useful work.

In order to avoid these kinds of situations, you need a way to *guarantee* that the attri-butes of an object are properly initialized before *any* calls are made to methods that depend on these attributes. Of course, you could try to be disciplined in your approach and make sure that you call all the appropriate setter methods before you use the object, but then you must remember to do it every time you instantiate an object. In the best-case scenario, you've introduced a lot of redundant code. In the worst-case sce-nario, you forget to call a method here and there and create an even bigger problem for yourself. Clearly, you'll need a better method for initializing objects. To take the guess-work out of this process, object-oriented languages such as ABAP Objects allow you to define special initializer methods within a class definition called *constructors*. These specialized callback methods are invoked automatically by the ABAP runtime environ-ment *after* an object instance is allocated but *before* control is handed back to a con-sumer. Therefore, they're an ideal place for injecting the relevant logic needed to ini-tialize an object.

4.2.1   Defining Constructors

Constructors are defined using essentially the same syntax you use to define regular instance methods (see Listing 4.1). The notable difference is that you can only define importing parameters in the method signature. If you think about it, this makes sense: Since the constructor is called by the ABAP runtime environment in response to a CREATE OBJECT statement, as opposed to the normal CALL METHOD statement, there's really nothing to pass back at this stage. To make this distinction clear, callers are not permitted to invoke constructors indirectly using the CALL METHOD statement.

```
METHODS constructor
          IMPORTING [VALUE(]i1  i2 ...[)]
               TYPE type [OPTIONAL]...
          EXCEPTIONS ex1 ex2.
```

Listing 4.1 Syntax for Defining an Instance Constructor

You can create constructors in global classes by clicking the **Constructor** button in the Class Editor as shown in Figure 4.4.

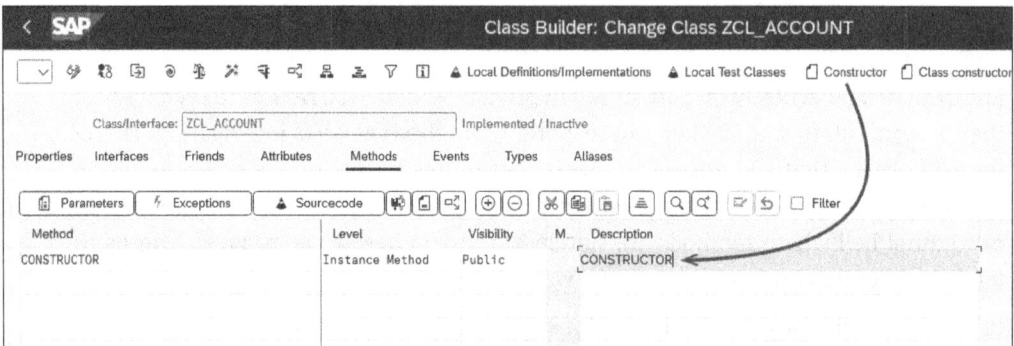

Figure 4.4 Creating Constructors for Global Classes

4.2.2   Understanding How Constructors Work

To demonstrate how constructors work, let's consider an example based on an account class used to model a bank account. In Listing 4.2, we've created a local class called LCL_ACCOUNT that allows us to view the current balance of the account and withdraw or deposit funds. The account details are stored in private attributes that do not have associated setter methods. Instead, we'll assume that these attributes are initialized via a database lookup inside the constructor() method. This implies that consumers must supply a valid account number whenever they create an account object. In Chapter 8, you'll see how class-based exceptions can be used to enforce this rule; for now, we'll simply assume that this is the case.

```
CLASS lcl_account DEFINITION.
  TYPE-POOLS: abap.
  PUBLIC SECTION.
    METHODS:
      constructor IMPORTING iv_account_no
                      TYPE string,
      get_balance RETURNING VALUE(rv_balance)
                      TYPE bapicurr_d,
      deposit     IMPORTING iv_amount
                      TYPE bapicurr_d,
      withdrawal  IMPORTING iv_amount
                      TYPE bapicurr_d
                  RETURNING VALUE(rv_result)
                      TYPE abap_bool.
  PRIVATE SECTION.
    DATA: mv_account_no TYPE string,
          mv_balance    TYPE bapicurr_d.
ENDCLASS.

CLASS lcl_account IMPLEMENTATION.
  METHOD constructor.
*   Query database tables to retrieve account details:
*   SELECT FROM ...
*     WHERE account_no = iv_account_no.
  ENDMETHOD.

  METHOD get_balance.
    rv_balance = me->mv_balance.
  ENDMETHOD.

  METHOD deposit.
    me->mv_balance = me->mv_balance + iv_amount.
  ENDMETHOD.

  METHOD withdrawal.
    IF iv_amount LE me->mv_balance.
      me->mv_balance = me->mv_balance - iv_amount.
      rv_result = abap_true.
    ELSE.
      rv_result = abap_false.
    ENDIF.
  ENDMETHOD.
ENDCLASS.
```

Listing 4.2 Working with Constructors (Part 1)

The code excerpt in Listing 4.3 shows how we access our new constructor using the CREATE OBJECT statement. Notice how the addition of the constructor() method has effectively put a lock on the front door of our LCL_ACCOUNT class. Since the mv_balance instance attribute is initialized before the object is ever handed over to consumers, you no longer need to worry about making sure that the account balance is up to date: It gets initialized internally, and updates can only occur through the deposit() and withdraw() methods that have built-in logic to guard against overdraws. You also don't need to worry about passing around account numbers since this detail is established up front during the object allocation process.

```
DATA lo_checking TYPE REF TO lcl_account.
CREATE OBJECT lo_checking
  EXPORTING
    iv_account_no = '1234567890'.

"Assume the opening balance is zero:
lo_checking->deposit( '100.00' ).
IF lo_checking->withdraw( '200.00' ) NE abap_true.
  WRITE: / 'Insufficient funds.'.
ENDIF.
lo_checking->deposit( '300.00' ).
```

Listing 4.3 Working with Constructors (Part 2)

4.2.3   Class Constructors

Most of the time, when we talk about constructors, we're talking about *instance constructors* that are used to initialize an instance of an object that is being created. However, it's also possible to define a *class constructor* for a class. Class constructors provide a mechanism for initializing class/static attributes and are implicitly called by the system before any accesses are made to the class inside a program.

> **Note**
>
> In this sense, class constructors are functionally equivalent to static initializer blocks from Java or C#.

Class constructors are defined using the syntax shown in Listing 4.4. As you can see, you're not allowed to define parameters for class constructors since they're implicitly called by the system.

```
CLASS-METHODS class_constructor.
```

Listing 4.4 Syntax for Defining a Class Constructor

You can create class constructors for global classes by clicking the **Class constructor** button on the Class Editor screen (see Figure 4.5).

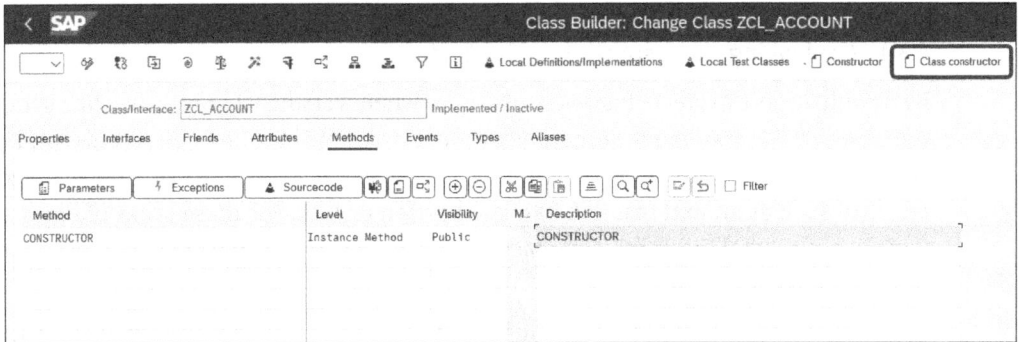

Figure 4.5 Creating Class Constructors for Global Classes

From an implementation perspective, class constructors are defined just like any other class method. You can preallocate and initialize shared resources to eliminate overhead and speed up the object instantiation process. You'll see some practical benefits of these capabilities in Section 4.3.

> **Note**
>
> Since class constructors are defined at the class level, you cannot access instance components within a class constructor since no instances of the class exist in that context. Of course, it *is* possible to instantiate an object of the class within the class constructor and then use that object reference to access instance components. You'll see examples of this technique in Section 4.3.

4.3   Taking Control of the Instantiation Process

Most of the classes that we've looked at up to this point are trivial and easy to instantiate. However, as you get into more complex designs, you may run into situations where object instantiation becomes a bit trickier. For example, you may find that you want to constrain the number of object instances users can create, or you may run into situations where there might be multiple ways to construct an object.

In this section, we'll look at a few common object-creational patterns that you can use to deal with complex object creation requirements. These patterns were originally documented in the classic software engineering text *Design Patterns: Elements of Reusable Object-Oriented Software* (Addison-Wesley, 1995). For an excellent ABAP-based treatment on these concepts, we highly recommend that you pick up a copy of *Design Patterns in ABAP Objects* (SAP PRESS, 2017).

4.3.1   Controlling the Instantiation Context

By default, users can create instances of classes whenever and wherever they like using the CREATE OBJECT statement. Sometimes, though, this kind of wide-open access is undesirable. For example, imagine that you have an object that needs to acquire a shared resource such as an external database connection or lock. For these types of objects, you need to control the number of object instances that are created to prevent resource contention issues.

To assert this kind of control, you must modify the instantiation context at the class definition level using the CREATE addition highlighted in Listing 4.5. Table 4.1 describes each of the possible instantiation contexts in detail.

```
CLASS lcl_db_connection DEFINITION
  CREATE {PUBLIC | PROTECTED | PRIVATE}.
  ...
ENDCLASS.
```

Listing 4.5 Specifying the Instantiation Context of an ABAP Objects Class

Instantiation Context	Visibility
PUBLIC	Classes with this default instantiation context can be instantiated anywhere the class itself is visible without restrictions. This means that instances of the class can be created internally via instance/class methods or externally from any normal ABAP programming context: subroutines, functions, report programs, and so forth.
PROTECTED	Instances of these classes can only be created inside methods of the class itself and its subclasses.
PRIVATE	Instances of these classes can only be created inside methods of the class itself.

Table 4.1 Instantiation Contexts for Classes

You can set the instantiation context for global classes by opening the Class Builder tool and navigating to the **Properties** tab. Here, you can set the instantiation context using the **Instance Generation** dropdown list shown in Figure 4.6.

In the upcoming sections, we'll look at a couple of widely used creational patterns that utilize this feature to control the way that object instances are created.

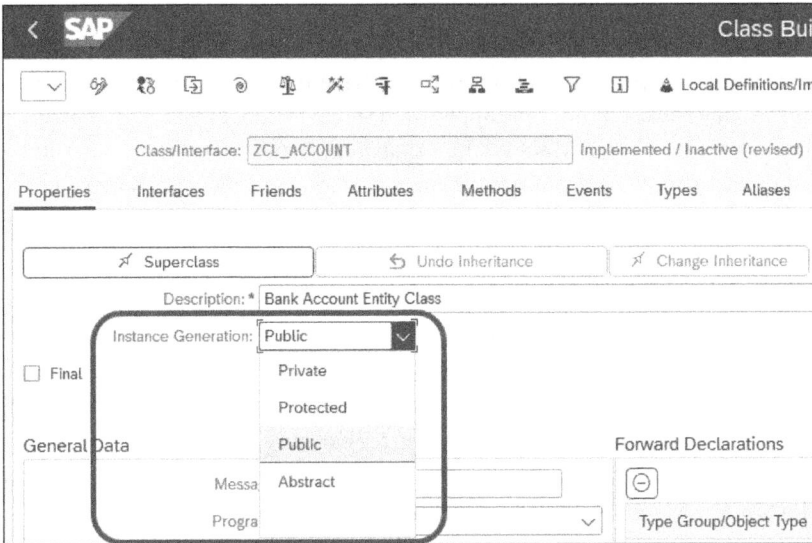

Figure 4.6 Setting the Instantiation Context for Global Classes

4.3.2   Implementing the Singleton Pattern

As we mentioned earlier, there may be times when you want to restrict the number of object instances that are created at runtime. For example, if you were to design a proxy class to broker database connections, you might want to design the class in such a way that a finite number of connections are maintained in a connection pool. Here, rather than creating a separate object instance per connection request, you would create a pool of objects and only provide users with access to one of the available connection objects from the pool.

In the most extreme case, you might only want a single object instance to exist at any one time within the system. This single object instance is referred to as a *singleton*. As a result, the software design pattern that describes how such objects are created is called the *singleton pattern*.

To demonstrate how the singleton pattern works, let's consider an example. Imagine that you need to build a number generator class that can be used to obtain a unique sequential number within some arbitrary range. To guarantee that the next number in the range is unique, from an implementation perspective, you must ensure the following:

- Only one instance of the number generator object can exist within the system at a given time.
- Every number request must be routed through that distinct object instance.
- The state of that single object must remain consistent. In other words, we must keep track of the next available number in the sequence.

> **Note**
>
> Bear in mind that this example is contrived to illustrate the singleton concept. Since SAP systems usually consist of multiple application server instances distributed across multiple host systems, it's important to note that further synchronization measures would be required to implement this pattern in a production setting (e.g., shared memory objects and/or enqueue locks).

In Listing 4.6, we've created a class called LCL_NUMBER_GEN that provides an implementation of our number generator. Here, notice how we're using the CREATE PRIVATE addition to ensure that instances of the number generator can only be created from within the LCL_NUMBER_GEN class. This forces clients to go through a static class method called get_instance() that contains conditional logic to ensure that only one instance of LCL_NUMBER_GEN exists at a time. With that mechanism in place, we can easily implement the get_next() instance method to simply return the next number in the sequence.

```
CLASS lcl_number_gen DEFINITION CREATE PRIVATE.
  PUBLIC SECTION.
    CLASS-METHODS:
      get_instance RETURNING VALUE(ro_generator) TYPE REF TO lcl_number_gen.

    METHODS:
      get_next RETURNING VALUE(rv_number) TYPE i.

  PRIVATE SECTION.
    CLASS-DATA so_instance TYPE REF TO lcl_number_gen.

    DATA mv_current_num TYPE i.

    METHODS:
      constructor.
ENDCLASS.

CLASS lcl_number_gen IMPLEMENTATION.
  METHOD get_instance.
    IF so_instance IS NOT BOUND.
      CREATE OBJECT so_instance.
    ENDIF.

    ro_generator = so_instance.
  ENDMETHOD.

  METHOD constructor.
    me->mv_current_num = 0.
  ENDMETHOD.
```

```
METHOD get_next.
  me->mv_current_sum = me->mv_current_sum + 1.
  rv_number = me->mv_current_num.
ENDMETHOD.
ENDCLASS.
```

Listing 4.6 Implementing the Singleton Pattern

The code excerpt in Listing 4.7 demonstrates how to consume the singleton number generator class within ABAP code. Here, notice that the ABAP compiler will not allow us to create instances of the number generator via the CREATE OBJECT statement. Instead, we're forced to go through the get_instance() method to create and access the number range instance. Once we obtain a reference to the current number range object, it's business as usual from there.

```
DATA lo_generator TYPE REF TO lcl_number_gen.
DATA lv_number TYPE i.

*CREATE OBJECT lo_generator. "Syntax error...
lo_generator = lcl_generator=>get_instance( ).

DO 5 TIMES.
  lv_number = lo_generator->get_next( ).
  WRITE: / 'Number #', lv_number.
ENDDO.
```

Listing 4.7 Accessing Singleton Objects from Consumer Code

4.3.3 Working with Factory Methods

Unlike other object-oriented languages, such as Java, ABAP Objects does not provide support for overloading methods. This means that you can't define multiple versions of the same method with different parameter lists. Though this is usually not a concern when defining normal methods, it does present you with a bit of a dilemma if you happen to be working with objects that need to be created in lots of different ways.

To put this scenario into perspective, imagine that you're tasked with building a data transfer object (DTO) to store information about a person. To maximize the use of your person class, you'd like to make it easy to create instances of it using all kinds of different data sources, including user master records, HR personnel records, and business partner records.

So, how do you support all those different creation scenarios? Well, one option would be to cram the various input parameters into the signature of the constructor() method, as shown in Listing 4.8. This approach works, but it's messy and hard to follow.

```
CLASS lcl_person DEFINITION.
  PUBLIC SECTION.
    METHODS:
      constructor IMPORTING iv_user_id TYPE sy-uname OPTIONAL
                            iv_empl_no TYPE hrobjid OPTIONAL
                            iv_partner TYPE bu_partner OPTIONAL
                            ...

    ...
  PRIVATE SECTION.
    DATA mv_first_name TYPE string.
    DATA mv_last_name TYPE string.
    DATA mv_phone_num TYPE string.
    DATA mv_email TYPE string.
ENDCLASS.
```

Listing 4.8 Cluttering Up the Constructor Method Signature

A better approach calls for the creation of *factory methods*, which encapsulate the various creation contexts into a series of class methods that are easy to understand and use. This approach is demonstrated in Listing 4.9. Here, the various create_from_...() methods assume the responsibility of creating person objects in terms of users, employees, and so on. This makes the code more readable and provides you with the structure for adding additional creation scenarios down the road.

```
CLASS lcl_person DEFINITION.
  PUBLIC SECTION.
    CLASS-METHODS:
      create_from_user IMPORTING iv_user_id TYPE sy-uname
                       RETURNING VALUE(ro_person)
                          TYPE REF TO lcl_person,
      create_from_employee IMPORTING iv_empl_no TYPE hrobjid
                           RETURNING VALUE(ro_person)
                              TYPE REF TO lcl_person,
      create_from_partner IMPORTING iv_partner TYPE bu_partner
                          RETURNING VALUE(ro_person)
                             TYPE REF TO lcl_person,

      ...
    METHODS:
      constructor,
      set_first_name IMPORTING iv_first_name TYPE string,
      get_first_name RETURNING VALUE(rv_first_name) TYPE string,
      ...
  ...
ENDCLASS.
```

Listing 4.9 Working with Factory Methods

4.4   Garbage Collection

Once you've finished using an object in your programs, you need to make sure that you restore its resources to the system. In early object-oriented languages, it was the programmer's responsibility to make sure that objects were properly destroyed. Fortunately, the runtime environments of modern languages like ABAP Objects come equipped with a special memory management service called a *garbage collector* that takes care of these housekeeping duties behind the scenes.

The garbage collector's job is to scan through memory and delete orphaned object instances that no longer have references associated with them. Much of the time, these references are destroyed automatically whenever an object reference variable passes out of scope (i.e., when a method, function, or subroutine terminates). However, if you're done with an object early on in a long-running routine, it's not a bad idea to explicitly remove the reference using the CLEAR statement (see Figure 4.7).

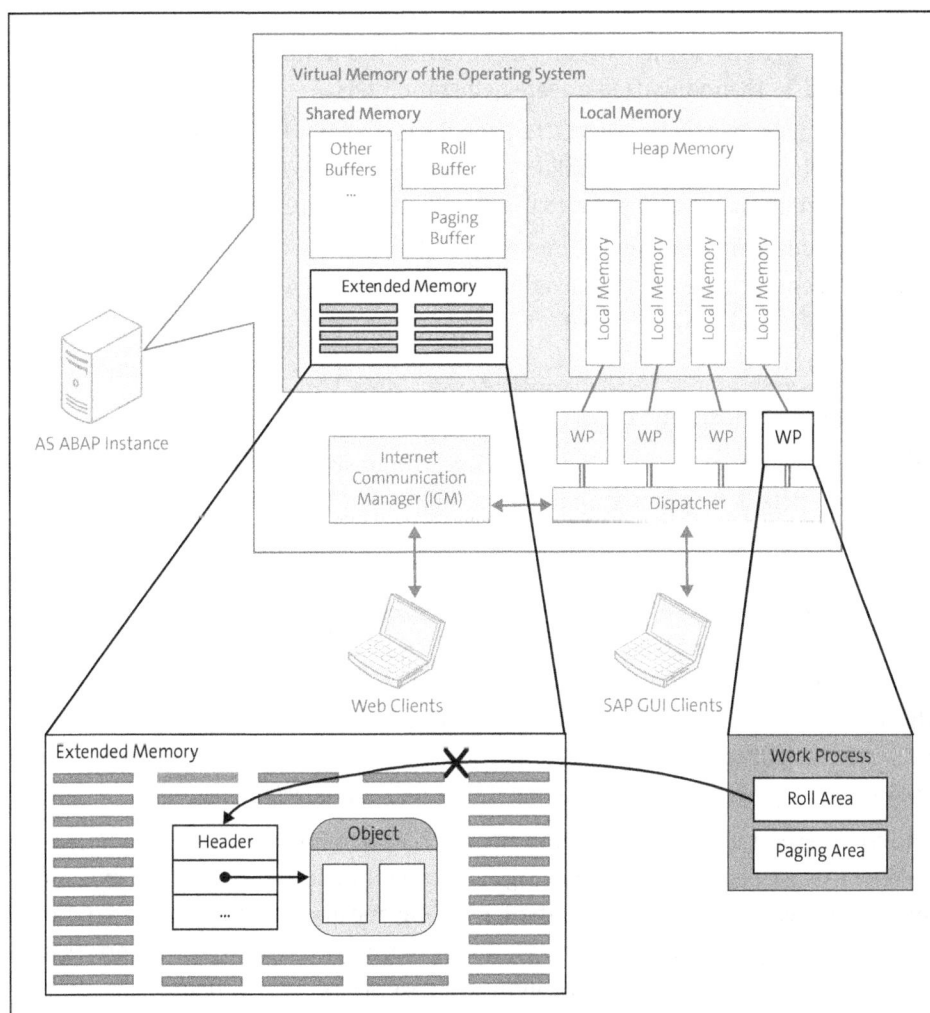

Figure 4.7 Deleting References to Objects Using the CLEAR Statement

4.5   Tuning Performance

The advanced memory management features of the ABAP runtime environment provide a safe environment for creating and destroying objects. However, it's important to remember that these features do not provide a safeguard against poor design decisions that consume excessive amounts of memory. In this section, we'll provide some basic tips that you can use to avoid these performance traps.

4.5.1   Design Considerations

Even if you don't anticipate performance problems for a given class, it's always a good idea to modularize the initialization logic of the class so that you can implement performance tuning measures later without disturbing core functionality. The following list contains some basic modularization tips that you should consider when developing your classes:

- Keep the logic inside the constructor() method to a minimum by delegating initialization tasks to modularized private helper methods.
- If you're using your class as a DTO, provide yourself with a public reset() method that can be used to clear the values of a class's instance attributes (thus, freeing up the object instances they point to for garbage collection).
- Avoid adding too many parameters to the constructor() method's interface. Instead, encapsulate the initialization process inside of a series of creational methods as described in Section 4.3.3.

4.5.2   Lazy Initialization

Sometimes, you may be working with large composite objects that contain lots of lower-level details that aren't frequently used. To put this scenario into perspective, consider a SalesOrder class that contains a list of SalesOrderItem objects that also in turn contain a list of ScheduleLine objects. Depending on the usage scenario, you may or may not need to have all the line item-level or schedule line-level detail in context to do your job. For example, if you're using the SalesOrder class to create a sales order summary report, you probably only need to report on the header-level data. Bringing the item-level data in only slows the report down and consumes a lot more memory. In this case, it makes sense to delay the initialization of the lower-level details until they're actually needed. This technique is referred to as *lazy initialization*.

The code excerpt in Listing 4.10 demonstrates how lazy initialization works. Here, you can see that we're not actually fetching sales order line item information until a consumer invokes the get_items() method. By using encapsulation techniques to funnel item lookup requests through the get_items() method, we can defer the performance hit associated with fetching the line items until we actually need them.

```
CLASS lcl_sales_order DEFINITION.
  PUBLIC SECTION.
    TYPES: ty_item_tab TYPE STANDARD TABLE OF
                         REF TO lcl_order_item.
    METHODS:
      get_items RETURNING VALUE(rt_items) TYPE ty_item_tab.

  PRIVATE SECTION.
    DATA ms_header TYPE ...
    DATA mt_items TYPE ty_item_tab.
ENDCLASS.

CLASS lcl_sales_order IMPLEMENTATION.
  METHOD get_items.
    IF lines( mt_items ) EQ 0.
      "Build the line items table on demand...
    ENDIF.

    rt_items = me->mt_items.
  ENDMETHOD.
ENDCLASS.
```

Listing 4.10 Implementing Lazy Initialization

The lazy initialization technique, though powerful, can lead to undesirable side effects if you're not careful. Typically, such side effects are the result of private methods accessing instance attributes directly instead of going through the appropriate getter methods. In this case, unexpected results might occur because of missing data.

Like most performance tuning measures, you must weigh the benefits against the risks. For example, for objects whose data is fetched from a relational database, you *might* be better off taking the up-front performance hit as opposed to sprinkling in lots of ad hoc SQL queries in getter methods—it really depends on your usage scenario.

4.5.3   Reusing Objects

The easiest way to avoid the performance hits associated with creating and destroying objects is to simply avoid this process altogether by recycling objects. Some typical candidates for recycling include temporary objects created inside loops, as well as objects created in utility methods. You may discover that you could've created an object in a higher scope that could be reused in the loop or method calls. In other cases, you might be working with a lightweight object in a loop that simply needs to be reinitialized based on the loop index. Rather than creating a new object each time, you might be able to call a reset() method to reuse the object.

4.5.4   Making Use of Class Attributes

As you design your classes, you should think about whether each object instance will require their own local copy of an attribute. If a local copy of an object is not required for an object instance, defining the attribute as a class attribute can help to avoid the creation of a lot of redundant data objects.

4.6   UML Tutorial: State Machine Diagrams

The sequence diagrams introduced in Chapter 3 showed the behavior of multiple objects interacting in a particular use case. Another type of behavioral diagram in the Unified Modeling Language (UML) is the *state machine diagram*. State machine diagrams are useful for showing the behavior of a single object throughout its lifetime.

Figure 4.8 shows a state machine diagram for a class that could be used to represent a batch job created using the built-in job scheduler tool. When a new job object is created, it's initialized in the Scheduled status. This is depicted in the diagram by an *initial pseudostate* node that points to the Scheduled state box.

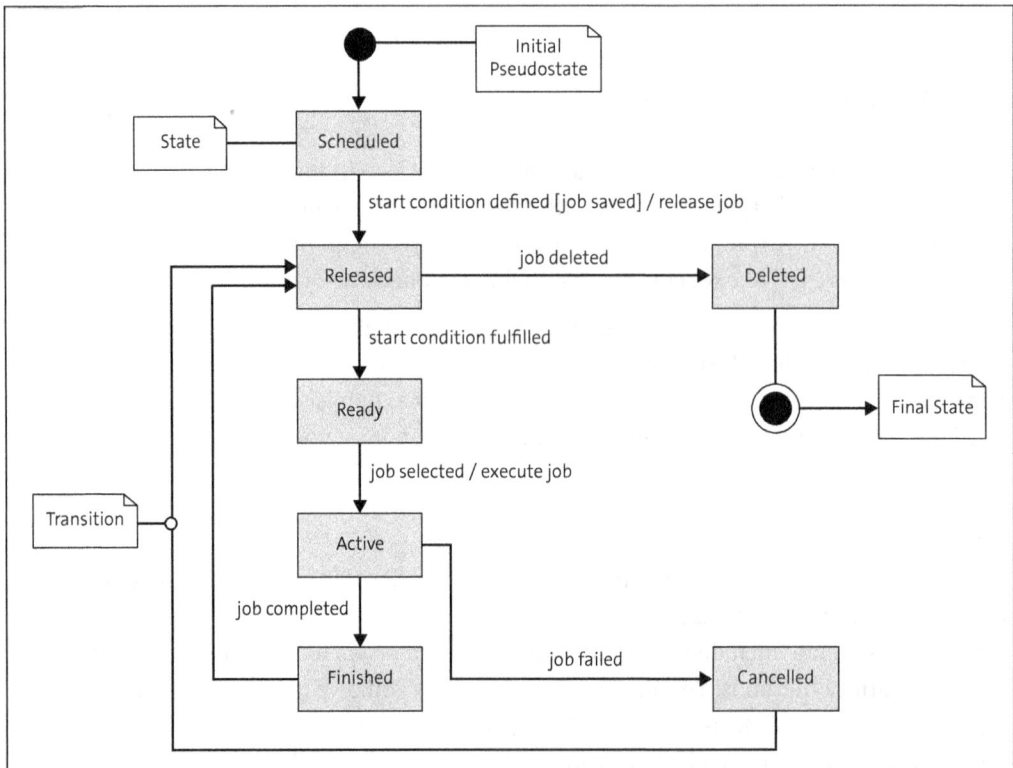

Figure 4.8 UML State Diagram for an SAP Batch Job

Each of the possible statuses of a job are shown using rounded boxes called *states*. Changes in state are represented with directed *transition* arrows. Transitions can optionally be labeled with special transition label strings using the syntax shown in Listing 4.11.

```
event(s) [guard conditions]/activity
```

Listing 4.11 Syntax Diagram for Defining Transition Labels

The `event(s)` portion of the transition label is used to describe the event (or events) that triggers a change in state for the object. If guard conditions are included in the transition label (within the square brackets), then those conditions must be true for the transition to occur. The `activity` option can be used to specify behavior that takes place during the transition. As an example, let's consider the transition between the Scheduled and Released statuses in Figure 4.8. In this case, the triggering event occurs whenever a user defines a start condition for the job in Transaction SM36. However, for the job to be released in the system, it must first be saved.

If a job is deleted in the system, then the state machine (i.e., the object) will reach its *final state*. This is shown via the arrow that points to the circular node with a dot in it (see Figure 4.8).

Like many of the diagrams that you'll see throughout this book, the state machine diagram fulfills a distinct purpose. In this chapter, we investigated some of the various ways that objects are created in the system. Most of the time, the lifecycle of an object is straightforward. However, for objects with complex lifecycles, state machine diagrams can be useful to show how an object will interact with the environment around it.

4.7   Summary

In this chapter, you learned various techniques to ensure objects are always properly initialized before they are used in a program. When you combine these methods with the encapsulation techniques described in Chapter 3, you will be able to build reliable and robust class libraries that function in a deterministic manner. In the next chapter, we'll investigate ways for reusing these classes in other contexts.

Chapter 5
Inheritance and Composition

One of the more compelling features of object-oriented programming (OOP) is its support for inheritance. In this chapter, we'll introduce this concept and look at other ways of achieving code reuse with classes.

As we noted in Chapter 3, one of the primary reasons it's so difficult to achieve code reuse with software development is because there are usually implicit assumptions made along the way that inextricably link the code to specific usage contexts. While encapsulation and implementation hiding techniques help prevent you from coding yourself into a corner, there are going to be times when you come to the realization that a particular class simply can't be stretched any further.

At that point, you've got a major dilemma on your hands. Do you try to enhance/ rework the classes to accommodate the new requirements? Or do you cut your losses and try to salvage as much of the code as possible by copying and pasting it into a new set of classes? Realistically, neither approach is ideal, because:

- Enhancing or reworking existing classes to handle the requirements threatens the integrity of preexisting programs, since it's quite possible that your enhancements could compromise existing functionality. Indeed, even the most gifted of code surgeons typically have a hard time pulling off a seamless transition.

- The copy-and-paste approach might be less risky initially, but ultimately it increases the cost of long-term maintenance efforts since the introduction of redundant code makes the overall code footprint bigger. Over time, you may find yourself having to apply the same set of changes to lots of classes that unofficially share this same code line.

In this chapter, we'll explore how the object-oriented concept of *inheritance* gives you a third option for adapting to changes. You'll find that by inheriting from a class, you can effectively make a copy of that class without disturbing the source class or introducing redundant code. As we explore this concept, you'll also learn about another technique called *composition* that provides an alternative way to reuse classes in situations where inheritance either isn't possible or doesn't make sense.

5.1 Generalization and Specialization

One of the most difficult parts of the object-oriented design process is trying to identify all the classes that are needed to model a problem domain. During the object-oriented analysis process, you might take several passes through the requirements before settling on which classes are needed, what the relationships between those classes should look like, and how instances of those classes will interact with one another at runtime. Depending on the problem domain, this analysis process can become so challenging that even the most experienced object-oriented developers rarely get it all right the first time. This begs the question: What happens if you get the cut of your class model wrong?

When you think about it, there are lots of potential mistakes you can make as you formulate your class model. For example, after several rounds of analysis, you might discover that you've failed to identify several key entities, or perhaps you've identified the entities but defined them too generically.

To put this phenomenon into perspective, imagine that you've been tasked with designing a human resources (HR) system. Early on, you naturally identify a need for an Employee entity class, among others. However, as you dig deeper, more requirements come out that describe specific functionalities relevant for certain types of employees. At this point, you could try to incorporate a lot of conditional logic into the Employee class to deal with the specialized cases, but this runs contrary to the idea that you don't want to clutter up your classes with too many responsibilities. On the other hand, abandoning the Employee class altogether in favor of lots of specialized classes (e.g., HourlyEmployee) leads to the kind of code redundancy issues you want to avoid with OOP. Fortunately, object-oriented languages like ABAP Objects provide a better and more natural way of dealing with these kinds of specialization problems: *inheritance*.

5.1.1 Inheritance Defined

Using the concept of inheritance, you can *extend* a class in such a way that you can reuse what's already developed (and hopefully tested) while at the same time expanding the class model to better fit specialized cases. The newly created class is called a *subclass* of the original class; the original class is referred to as the *superclass* of the newly created class. As the term "inheritance" suggests, subclasses *inherit* components from their superclass. These relationships allow you to build out hierarchical inheritance trees with superclasses as parent nodes and subclasses as child nodes (see Figure 5.1). In Chapter 6, you'll learn how members of this inheritance tree can be used interchangeably, providing for some very compelling generic programming options.

The Generic OBJECT Type in ABAP Objects

Though not depicted in Figure 5.1, the root of any inheritance tree in ABAP Objects is the predefined (empty) OBJECT class. So, even though you may not have realized it, every custom class that you've created thus far has implicitly inherited from this standard base class.

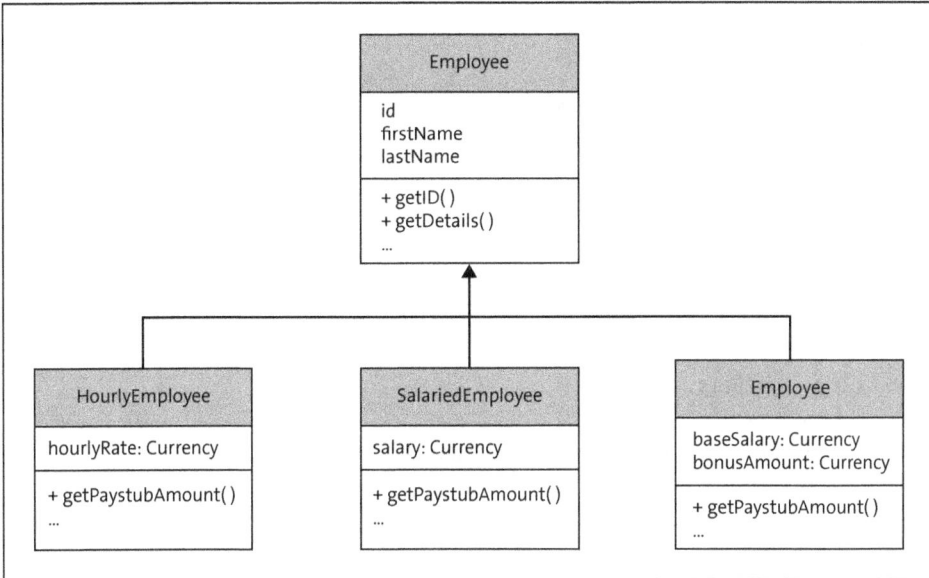

Figure 5.1 A Sample Class Hierarchy

5.1.2 Defining Inheritance Relationships in ABAP Objects

To understand how inheritance works in ABAP Objects, let's build out a portion of the Employee class hierarchy depicted in Figure 5.1. As you can see in Listing 5.1, we've defined three classes: LCL_EMPLOYEE, LCL_HOURLY_EMPLOYEE, and LCL_SALARIED_EMPLOYEE. The LCL_EMPLOYEE class is the base class in the hierarchy, while the LCL_HOURLY_EMPLOYEE and LCL_SALARIED_EMPLOYEE classes are subclasses that inherit the basic features of the LCL_EMPLOYEE superclass.

```
CLASS lcl_employee DEFINITION.
  PUBLIC SECTION.
    DATA: mv_id TYPE pernr_d READ-ONLY, "Public for demo only!!!
          mv_first_name TYPE ad_namefir,
          mv_last_name TYPE ad_namelas.

    METHODS:
      constructor IMPORTING iv_id TYPE pernr_d
                            iv_first_name TYPE ad_namefir
                            iv_last_name TYPE ad_namelas,

      get_id RETURNING VALUE(rv_id) TYPE pernr_d,

      get_details RETURNING VALUE(rv_details) TYPE string.
ENDCLASS.
```

```
CLASS lcl_employee IMPLEMENTATION.
  METHOD constructor.
    me->mv_id = iv_id.
    me->mv_first_name = iv_first_name.
    me->mv_last_name = iv_last_name.
  ENDMETHOD.

  METHOD get_id.
    rv_id = me->mv_id.
  ENDMETHOD.

  METHOD get_details.
    rv_details =
      |Employee #{ mv_id }: { mv_first_name } { mv_last_name }|.
  ENDMETHOD.
ENDCLASS.

CLASS lcl_hourly_employee DEFINITION
        INHERITING FROM lcl_employee.
  PUBLIC SECTION.
    CONSTANTS:
      CO_WORK_WEEK TYPE i VALUE 40.

    METHODS:
      constructor IMPORTING iv_id TYPE pernr_d
                            iv_first_name TYPE ad_namefir
                            iv_last_name TYPE ad_namelas
                            iv_hourly_rate TYPE bapicurr_d,

      get_paystub_amount RETURNING VALUE(rv_wages) TYPE bapicurr_d.

  PRIVATE SECTION.
    DATA mv_hourly_rate TYPE bapicurr_d.
ENDCLASS.

CLASS lcl_hourly_employee IMPLEMENTATION.
  METHOD constructor.
    super->constructor(
      EXPORTING
        iv_id = iv_id
        iv_first_name = iv_first_name
        iv_last_name = iv_last_name ).

    me->mv_hourly_rate = iv_hourly_rate.
  ENDMETHOD.
```

```
  METHOD get_paystub_amount.
    rv_wages = me->mv_hourly_rate * CO_WORK_WEEK.
  ENDMETHOD.
ENDCLASS.

CLASS lcl_salaried_employee DEFINITION
        INHERITING FROM lcl_employee.
  PUBLIC SECTION.
    METHODS:
      constructor IMPORTING iv_id TYPE pernr_d
                            iv_first_name TYPE ad_namefir
                            iv_last_name TYPE ad_namelas
                            iv_salary TYPE bapicurr_d,

      get_paystub_amount RETURNING VALUE(rv_wages) TYPE bapicurr_d.

  PRIVATE SECTION.
    DATA mv_salary TYPE bapicurr_d.
ENDCLASS.

CLASS lcl_salaried_employee IMPLEMENTATION.
  METHOD constructor.
    super->constructor(
      EXPORTING
        iv_id = iv_id
        iv_first_name = iv_first_name
        iv_last_name = iv_last_name ).

    me->mv_salary = iv_salary.
  ENDMETHOD.

  METHOD get_paystub_amount.
    rv_wages = me->mv_salary / 52.
  ENDMETHOD.
ENDCLASS.
```

Listing 5.1 Defining an Employee Class Hierarchy in ABAP Objects

For the most part, the LCL_EMPLOYEE base class in Listing 5.1 looks just like any local class you've considered up to this point. Where things get interesting is with the definition of the LCL_HOURLY_EMPLOYEE and LCL_SALARIED_EMPLOYEE subclasses. Here, notice how we're using the INHERITING FROM addition of the CLASS DEFINITION statement to define the inheritance relationship to LCL_EMPLOYEE. With this simple inclusion, we've defined new subclasses that come pre-equipped with all the instance attributes/methods of LCL_EMPLOYEE. This means, for example, that you could call the get_details() method on an LCL_HOURLY_EMPLOYEE instance and the code defined in the LCL_EMPLOYEE class's implementation would fire automagically. Similarly, instances of LCL_SALARIED_EMPLOYEE inherit/have access to public instance attributes such as mv_id, mv_first_name, and mv_last_name.

> **Note**
>
> The new subclasses may come preequipped with all the instance attributes and methods of LCL_EMPLOYEE, but we should qualify this by saying that the subclasses only inherit components that the LCL_EMPLOYEE class makes visible to them. We'll expand on this concept further in Section 5.2.

After the INHERITING FROM addition, the definition of the LCL_HOURLY_EMPLOYEE and LCL_SALARIED_EMPLOYEE subclasses proceeds as per usual. However, note that any new components you create are unique to their respective subclasses; LCL_EMPLOYEE isn't retrofitted or altered by any of the changes you make at the subclass level.

Defining Inheritance Relationships in the Class Builder

Technically speaking, the inheritance definition syntax is the same for local or global classes. However, if you prefer to work in the form-based view of the Class Builder, then you must configure the inheritance relationship manually. If you're building a new subclass from scratch, you can do so via the **Create Class** wizard screens as shown in Figure 5.2 and Figure 5.3, respectively.

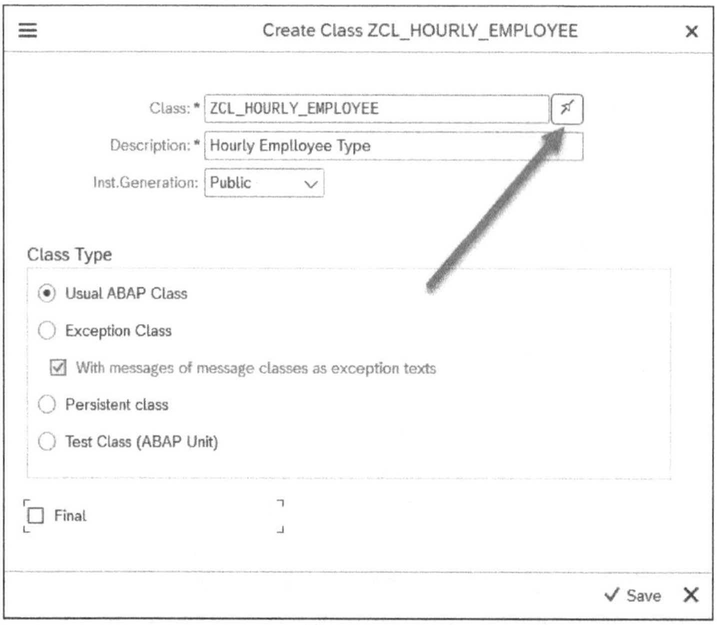

Figure 5.2 Specifying an Inheritance Relationship in the Class Builder (Part 1)

The **Create Inheritance** icon to the right of the **Class** field shown in Figure 5.2 reveals the **Superclass** field shown in Figure 5.3. Once you plug in the appropriate superclass and click the **Save** button, the Class Builder will take care of building out the appropriate syntax.

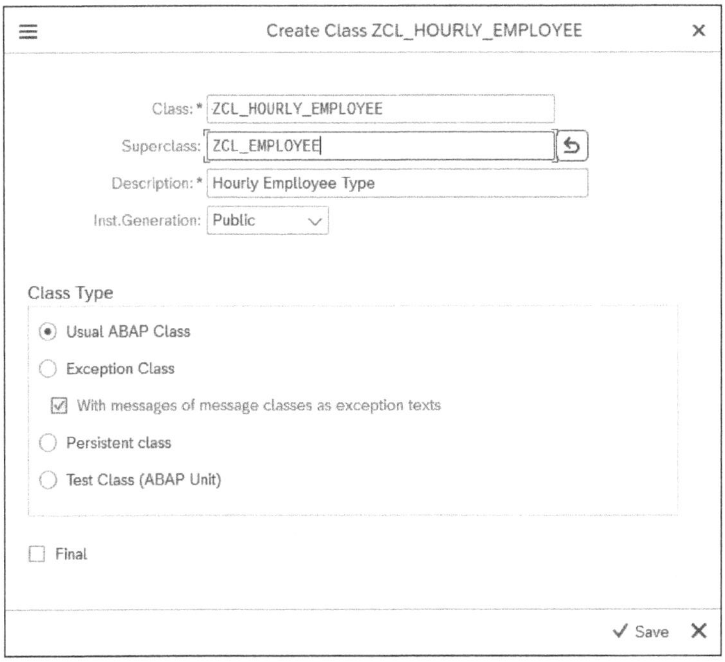

Figure 5.3 Specifying an Inheritance Relationship in the Class Builder (Part 2)

If the inheritance relationship isn't specified up front during the class creation process, then you can always maintain it after the fact on the **Properties** tab of the Class Builder as shown in Figure 5.4. On this tab, there are three buttons that you can use to adjust the inheritance relationship for the class:

- **Superclass**
  Use this button to specify a superclass for a given subclass when there isn't one defined yet. Once you select this button, you can fill in the new superclass in the correspondingly named **Superclass** field, as shown in Figure 5.4.

- **Undo Inheritance**
  Use this button to remove an inheritance relationship from a subclass. Note that this will remove all inherited components from the class, so be careful when performing this step on classes that may have methods in place that depend on these components.

- **Change Inheritance**
  As the name suggests, you can use this button to change the inheritance relationship from one superclass to another. As you might expect, this can get very tricky

depending on the compatibility of the two superclasses, so exercise caution when performing this change.

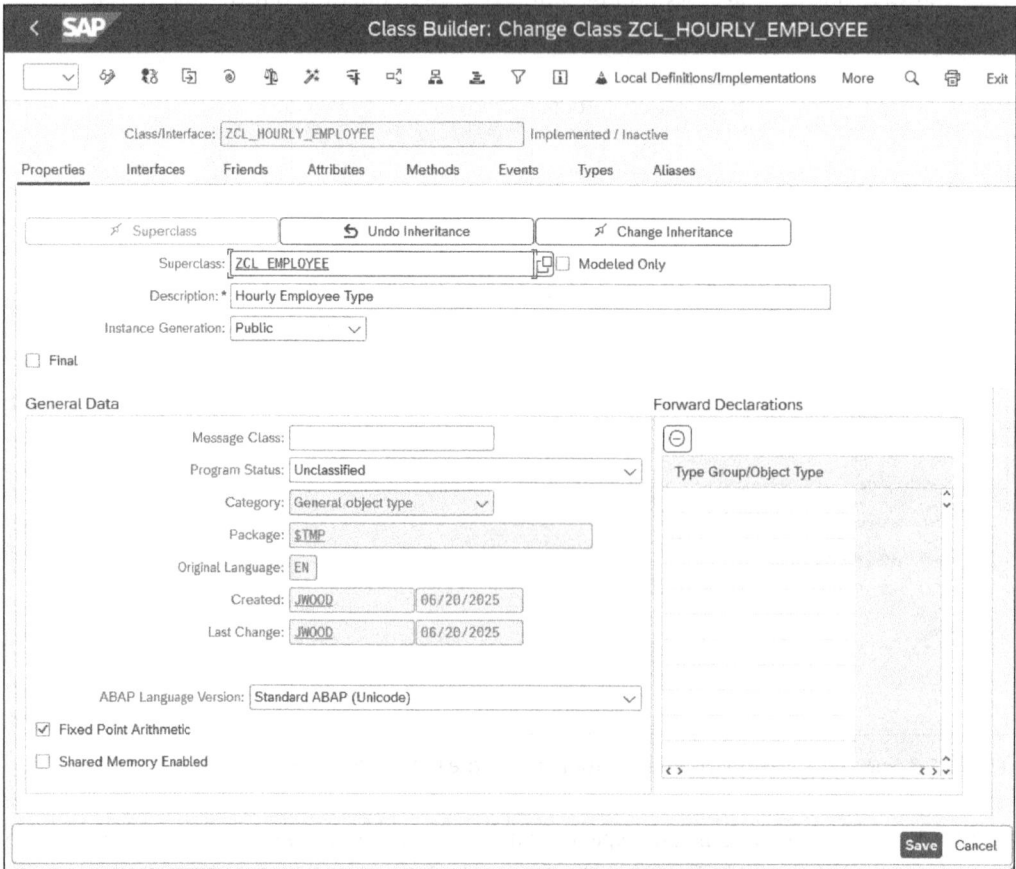

Figure 5.4 Specifying an Inheritance Relationship in the Class Builder (Part 3)

5.1.3   Working with Subclasses

The code excerpt in Listing 5.2 demonstrates how you can work with subclass instances. As you can see, there's no special syntax required to interface with the subclass instances: you instantiate them using the CREATE OBJECT statement and access instance attributes/methods as per usual. Behind the scenes, the ABAP runtime environment is smart enough to interpret incoming requests and route them appropriately. This means that calls to get_paystub_amount() are routed to the subclasses while calls to get_details() are routed to the superclass. To the end user, this sleight of hand is completely transparent.

```
DATA lo_hourly_emp TYPE REF TO lcl_hourly_employee.
DATA lo_salary_emp TYPE REF TO lcl_salaried_employee.
```

```
CREATE OBJECT lo_hourly_emp
  EXPORTING
    iv_id = '12345678'
    iv_first_name = 'Andersen'
    iv_last_name = 'Wood'
    iv_hourly_rate = '80.00'.

CREATE OBJECT lo_salary_emp
  EXPORTING
    iv_id = '23456789'
    iv_first_name = 'Paige'
    iv_last_name  = 'Wood'
    iv_salary = '150000'.

IF lo_hourly_emp->get_paystub_amount( ) GT
    lo_salary_emp->get_paystub_amount( ).
  WRITE: / lo_hourly_emp->get_details( ), `makes more money.`.
ELSE.
  WRITE: / lo_salary_emp->get_details( ), `makes more money.`.
ENDIF.
```

Listing 5.2 Working with Subclass Instances

5.1.4   Inheritance as a Living Relationship

Before we move on to more involved inheritance concepts, we should reemphasize the fact that inheritance is more than just a fancy way of copying classes into new classes. Inheritance defines a natural relationship that will likely evolve over time. To appreciate the nature of this relationship, imagine that you're asked to start keeping track of employee addresses using our fictitious Employee class model.

Depending on when you're asked to apply this change, your Employee class hierarchy could have expanded to include many type-specific subclasses. However, even if that's the case, the subclasses are not cut off from the root Employee class. So, if you need to make a fundamental change to the way that you model employees, you can simply add an instance attribute containing the address information to the Employee root class and each of the subclasses will inherit this attribute automatically.

This phenomenon is sketched out in the code excerpt in Listing 5.3. Here, we've introduced a new class called LCL_ADDRESS that models a simple address record. With this class in place, we then enhance the LCL_EMPLOYEE root class originally defined in Listing 5.1 to include an instance attribute and getter/setter methods that allow us to get our hands on this address record from an Employee object. Finally, since the getter/setter methods were defined as part of the public interface of LCL_EMPLOYEE, you can see how we can access this inherited functionality from subclasses such as LCL_HOURLY_EMPLOYEE without having to make any changes to the downstream classes themselves. Once we apply the change to the base class, the exposed functionality is available immediately downstream within the subclasses.

```abap
CLASS lcl_address DEFINITION.
  PUBLIC SECTION:
    METHODS:
      as_string RETURNING VALUE(rv_value) TYPE string,
      get_street1 RETURNING VALUE(rv_street) TYPE ad_street,
      set_street1 IMPORTING iv_street TYPE csequence,
      get_street2 RETURNING VALUE(rv_street) TYPE ad_strspp1,
      set_street2 IMPORTING iv_street TYPE csequence,
      get_city RETURNING VALUE(rv_city) TYPE ad_city1,
      set_city IMPORTING iv_city TYPE csequence,

      ...
  PRIVATE SECTION.
    DATA: mv_street1 TYPE ad_street,
          mv_street2 TYPE ad_strspp1,
          mv_city TYPE ad_city1,
          mv_region TYPE regio,
          mv_country TYPE land1.
ENDCLASS.

CLASS lcl_address IMPLEMENTATION.
  METHOD as_string.
    ...
  ENDMETHOD.
  ...
ENDCLASS.

CLASS lcl_employee DEFINITION.
  PUBLIC SECTION.
    METHODS:
      ...
      get_address RETURNING VALUE(ro_address)
                  TYPE REF TO lcl_address,
      set_address IMPORTING io_address TYPE REF TO lcl_address.

  PRIVATE SECTION.
    DATA mo_address TYPE REF TO lcl_address.
ENDCLASS.

CLASS lcl_employee IMPLEMENTATION.
  ...
```

```
  METHOD get_address.
    ro_address = me->mo_address.
  ENDMETHOD.

  METHOD set_address.
    me->mo_address = io_address.
  ENDMETHOD.
  ...
ENDCLASS.

DATA lo_hourly_emp TYPE REF TO lcl_hourly_employee.
DATA lo_address TYPE REF TO lcl_address.

CREATE OBJECT lo_address.
...
CREATE OBJECT lo_hourly_emp
  EXPORTING ...

lo_hourly_emp->set_address( lo_address ). "Inherited method
WRITE: / lo_hourly_emp->get_address( )->as_string( ).
```

Listing 5.3 Understanding the Effects of Base-Level Class Refactoring

5.2   Inheriting Components

Much like inheritance in the real world (where you inherit property from someone upon their death), inheritance relationships in the OOP space are tightly regulated. The creators of parent classes get to choose which components they wish to pass down to subclasses and which components they want to keep under wraps. This is usually for the subclass's own good, because in OOP, inheriting too much of a good thing can often become a bad thing.

In this section, we'll take an up-close look at the rules that govern inheritance relationships. Once you wrap your head around these basic concepts, you should find the inheritance definition process to be relatively straightforward and intuitive.

5.2.1   Designing the Inheritance Interface

Up to this point in the book, our discussions on component visibility have been focused on designing a class's public and private interface from the perspective of an external client (i.e., another class or function module). However, inheritance adds a new dimension into the mix, since you now also need to consider how to define the interface between a superclass and its subclasses. For example, you might run into situations where you want to provide access to selected components to subclasses without having to expose the components to the outside world via the superclass's public interface.

To put this into perspective, let's revisit the LCL_EMPLOYEE class hierarchy that we introduced in Listing 5.1. As you may recall, we included the base-level instance attributes (e.g., mv_id, mv_first_name and mv_last_name) for the employee within the public interface of the LCL_EMPLOYEE class. Why? Well, partly for demonstration purposes, but also because we wanted to provide access to these instance attributes within subclasses such as LCL_HOURLY_EMPLOYEE. Sure, we could have provided getter/setter methods to provide such access but even then, we could argue that this exposes more details than we want to share with the outside world via the LCL_EMPLOYEE class's public interface.

To address problems like this, most modern object-oriented languages introduce a third visibility section: the *protected section*. Components defined within this section are visible only to the superclass and its subclasses. To the outside world, it's as if these components were defined within the private section of the class.

In ABAP Objects, the protected section is defined in much the same way that public/private sections are defined. There's a PROTECTED SECTION statement and all the components defined within that section are assigned the protected level scope. This is demonstrated in the code excerpt in Listing 5.4. In this scenario, notice how we've moved the instance attributes for LCL_EMPLOYEE to the protected section. With this change, we've effectively cut off access to these components to the LCL_EMPOYEE class and its subclasses. Now, the only way that external clients can get their hands on these attributes is if we purposefully expose them via public methods.

```
CLASS lcl_employee DEFINITION.
  PUBLIC SECTION.
    METHODS:
      constructor IMPORTING iv_id TYPE pernr_d
                            iv_first_name TYPE ad_namefir
                            iv_last_name TYPE ad_namelas,

      get_id RETURNING VALUE(rv_id) TYPE pernr_d,

      get_details RETURNING VALUE(rv_details) TYPE string.

  PROTECTED SECTION.
    DATA: mv_id TYPE pernr_d READ-ONLY,
          mv_first_name TYPE ad_namefir,
          mv_last_name TYPE ad_namelas.
ENDCLASS.
```

Listing 5.4 Moving Inherited Components to the Protected Section

Collectively, the components that you include in the protected and public sections of a class define its *inheritance interface*. When defining this interface, it's important not to get too carried away with adding components to the protected section on the off chance that you might want to expose those components to subclasses. Once again, you should apply the encapsulation concept of *least privilege* when designing your inheritance interface.

Understanding the Concept of Least Privilege

The concept of least privilege implies that if a subclass doesn't really need to access a component, then it shouldn't be granted access to it. For example, imagine that you want to make some fundamental changes to the implementation of a base-level class that sits at the root of an inheritance tree. If the components that you want to change are defined in the protected visibility section of the superclass, then it's quite possible that the changes cannot be carried out without affecting some or all the subclasses that may be using these components.

With that being said, the general rule of thumb is to define sensitive components in the private visibility section unless you have a compelling reason to do otherwise. If a subclass needs to be granted access to these components, then access can be provided in the form of getter and setter methods that are defined in the PROTECTED SECTION of the class. This little bit of additional work ensures that a base class remains fully encapsulated.

5.2.2 Visibility of Instance Components in Subclasses

Subclasses inherit the instance components of *all* the superclasses defined in their inheritance tree. However, not all these components are *visible* at the subclass level.

A useful way of understanding how these visibility rules work is to imagine that you have a special instance attribute pointing to an instance of the superclass inside of your subclass. You can use this reference attribute to access public components of the superclass, but access to private components is restricted just as it would be for any normal object reference variable.

As it turns out, this imaginary object reference metaphor is not too far off from what's implemented in subclasses behind the scenes. Subclasses contain a special *pseudo reference* variable called super that contains a reference to an instance of an object of the superclass's type. This reference is used to access components of a superclass inside a subclass. The primary difference between the super pseudo reference variable and a normal reference variable is that the super pseudo reference can also be used to access components defined in the protected section of the superclass it points to.

The use of the super pseudo reference variable is optional (as was the case with the me self-reference variable we discussed in Chapter 2), but it can be used in situations where explicit reference to superclass components is needed. Normally, you'd simply access the components of the superclass directly, but it's important to remember that the compiler is implicitly plugging in the super pseudo reference behind the scenes to properly address these components. If you operate with this mindset, the visibility rules for accessing superclass components should become second nature to you.

Inheritance Namespace Concepts

Public and protected components of classes in an inheritance tree all belong to the same internal namespace. This implies that you cannot create a component in a subclass using the same name that was used to define a component in a superclass. There's no such restriction on naming private components, however. For example, if you define a private component called comp in a superclass, you can reuse this same name to define components in subclasses without restriction.

5.2.3   Visibility of Class Components in Subclasses

In addition to inheriting all the instance-level components, subclasses also inherit all the class components of their superclasses. As was the case with instance components, though, only the components that are defined in the public or protected visibility sections of a superclass are visible at the subclass level.

In terms of inheritance, class attributes are not associated with a single class but rather with the overall inheritance tree. The change in scope makes it possible to address these class components by binding the class component selector operator with any of the classes in the inheritance tree. This can be confusing, as class components are defined in terms of a given class and often don't have a lot of meaning outside of their defining class's context. To avoid this kind of confusion, always address class components by applying the class component selector to the defining class's name (e.g., lcl_superclass=>component). That way, your intentions are always clear.

5.2.4   Redefining Methods

The implementation of inherited methods often needs to be changed at the subclass level to support more specialized functionality. To put this concept into perspective, imagine that you decide to refactor the Employee class hierarchy by moving the definition of the get_paystub_amount() method up to the LCL_EMPLOYEE base class level. You might provide a bare-bones implementation in the base class, but your real objective is to associate this behavior with all employee types and not redundantly define the same method in each subclass.

Of course, this doesn't change the fact that the calculation of the paystub amount differs between the specific employee types. Clearly, the implementation will be type specific but, when you think about it, the signature of the method remains static in any case. So, you *define* the method in the LCL_EMPLOYEE base class and then *redefine* (or override) the implementation of the method in each of the type-specific subclasses.

The code excerpt in Listing 5.5 shows how you can redefine methods like this using ABAP Objects syntax. Notice how we've refactored the LCL_EMPLOYEE base class to define the method get_paystub_amount(). This definition fully specifies the method signature which will not change. Then, at the LCL_HOURLY_EMPLOYEE subclass level, we're using the REDEFI-NITION addition to declare our intention to redefine/re-implement the get_paystub_amount() method to include type-specific calculation logic. Finally, within the implementation section of the LCL_HOURLY_EMPLOYEE subclass, it's business as usual, as you're free to redefine the logic of the get_paystub_amount() method in any way you see fit.

```
CLASS lcl_employee DEFINITION.
  PUBLIC SECTION.
    METHODS:
      ...
      get_paystub_amount RETURNING VALUE(rv_wages)
                                   TYPE bapicurr_d.
  ...
ENDCLASS.

CLASS lcl_employee IMPLEMENTATION.
  ...
  METHOD get_paystub_amount.
    "Empty for now...
  ENDMETHOD.
ENDCLASS.

CLASS lcl_hourly_employee DEFINITION
        INHERITING FROM lcl_employee.
  PUBLIC SECTION.
    ...
    get_paystub_amount REDEFINITION.
ENDCLASS.

CLASS lcl_hourly_employee IMPLEMENTATION.
  ...
  METHOD get_paystub_amount.
    rv_wages = me->mv_hourly_rate * CO_WORK_WEEK.
  ENDMETHOD.
ENDCLASS.
```

Listing 5.5 Redefining Methods in Subclasses

You can achieve the same thing for global classes maintained in the form-based view of the Class Builder by selecting the target method and clicking on the **Redefine Method** button, as shown in Figure 5.5.

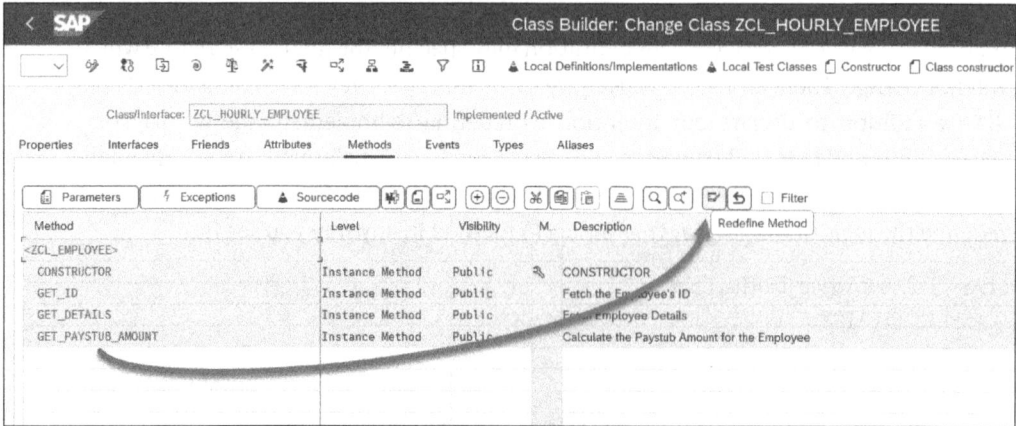

Figure 5.5 Redefining Methods in the Class Builder

Misleading Terminology

Even though the REDEFINITION addition sounds like it might change the definition of a method, it has absolutely no effect on the definition of a method's signature. This means that you can't perform tasks such as adding or removing parameters to the method or changing its visibility section assignment. In Chapter 6, you'll learn that there's a very important reason why this interface must remain intact, but for now, the main takeaway is that you can only use the REDEFINITION addition to change a method's implementation in a subclass.

Reusing the Superclass Implementation

When you redefine methods using the REDEFINITION addition, you're effectively choosing to abandon the superclass's implementation in favor of your own custom one. However, there may be times when you merely want to tweak the implementation of the superclass, not replace it altogether. In these situations, you can use the super pseudo reference within the redefined method to invoke the superclass implementation and then adjust the output as needed.

This technique is demonstrated in the code excerpt in Listing 5.6. Here, we've defined a simple estimating class and a subclass that redefines its get_estimate() method. In the subclass implementation, we're delegating most of the heavy lifting to the superclass and then adjusting the estimate by 10% after the fact. If the get_estimate() method defined importing parameters, we could have also influenced the behavior by adjusting the parameter values on the fly before handing them off to the superclass implementation.

```
CLASS lcl_estimator DEFINITION.
  PUBLIC SECTION.
    METHODS:
      get_estimate RETURNING VALUE(rv_estimate) TYPE f.
ENDCLASS.

CLASS lcl_estimator IMPLEMENTATION.
  METHOD get_estimate.
    rv_estimate = ...
  ENDMETHOD.
ENDCLASS.

CLASS lcl_conservative_estimator DEFINITION
        INHERITING FROM lcl_estimator.
  PUBLIC SECTION.
    METHODS:
    get_estimate REDEFINITION.
ENDCLASS.

CLASS lcl_conservative_estimator IMPLEMENTATION.
  METHOD get_estimate.
    rv_estimate = super->get_estimate( ).
    rv_estimate = rv_estimate + ( rv_estimate * '0.10' ).
  ENDMETHOD.
ENDCLASS.
```

Listing 5.6 Invoking the Superclass's Implementation of a Redefined Method

5.2.5 Instance Constructors

Unlike regular instance components such as attributes or methods, constructors *are not* inherited. If you think about it, this makes sense, since each class only knows how to initialize objects of its own type.

Even though constructor methods exist independently in each class within an inheritance hierarchy, the signature of the constructor() method in subclasses must remain relatively stable. While it's perfectly acceptable to add new type-specific parameters to the signature of the method, you'll want to leave the preexisting parameter interface intact. This is because constructor methods in subclasses are required to call the constructor of their superclass within the implementation of the constructor() method.

You saw an example of this with the LCL_EMPLOYEE class hierarchy defined in Listing 5.1, where you saw how type-specific subclasses such as LCL_HOURLY_EMPLOYEE expand the interface of the constructor() method to include additional parameters as needed.

The rest of the parameters are passed onto the superclass's constructor via the call to super->constructor() to facilitate the initialization of the base-level employee instance. Then, after that initialization is complete, the subclass's constructor method logic kicks in to complete the initialization of the type-specific employee instance.

5.2.6   Class Constructors

Each subclass is also allowed to define its own unique class constructor. This constructor is called right before the class is addressed in a program for the first time. However, before it's executed, the ABAP runtime environment will work its way up the inheritance tree to make sure that the class constructor has been called for each superclass in the inheritance hierarchy. These class constructor calls are guaranteed to occur in the proper order.

For example, imagine that you have a class hierarchy with four classes A, B, C, and D. When a program tries to access class D for the first time, the runtime environment will first check to see if the class constructors have been called for classes A, B, and C. If the class constructor has already been called for class A, but not for B and C, then the order of class constructor calls will be B, C, and D. This ensures that the class attributes of a superclass are always properly initialized before a subclass is loaded into context.

5.3   The Abstract and Final Keywords

As you've observed, inheritance adds a dimension to your class design that forces you to take a hard look at *where* you should define certain components within a class hierarchy. As you come to these decisions, it's important to be able to lock the hierarchy down so that these design choices are honored in subclasses. In this section, we'll look at a couple of keywords that make this possible in ABAP Objects.

5.3.1   Abstract Classes and Methods

In Section 5.2.4, you saw the potential value in moving the get_paystub_amount() up the Employee class hierarchy to the LCL_EMPLOYEE root class. Though we'll explore the details of this in Chapter 6, this refactoring job does identify an important point: The higher you make your way up the inheritance tree, the more generic things become. Indeed, when you get to the LCL_EMPLOYEE root class, things are so generic that there's not really a reasonable implementation to provide for the get_paystub_amount() method. That's why we ended up leaving the implementation empty in Listing 5.5.

While this approach might seem harmless enough, it can be quite dangerous in practice. For example, imagine that you neglected to redefine the get_paystub_amount() method in type-specific subclasses such as LCL_HOURLY_EMPLOYEE. If this were to happen, all calls to this method on type-specific employee instances at runtime would be routed to the implementation in the LCL_EMPLOYEE superclass, which does nothing.

Rather than leave all this to chance, ABAP Objects allows you to identify these gaps in a class definition using the ABSTRACT keyword. With this keyword, you can explicitly delegate undefined features to subclasses.

To understand how this works, let's once again refactor the LCL_EMPLOYEE class from Listing 5.5 by redefining the get_paystub_amount() method as abstract. As you can see in Listing 5.7, this minor change was carried out by strategically plugging in the ABSTRACT keyword in two places within the class definition:

- First, we included the ABSTRACT keyword in the overall CLASS DEFINITION statement. This is necessary for any class that will contain one or more abstract methods. You'll learn more about what this means in just a moment.

- Then, we added the ABSTRACT keyword to the definition of the get_paystub_amount() method.

Aside from these two syntactical changes, the only other notable change is to the implementation section of the LCL_EMPLOYEE class. Here, notice that we've no longer included an empty implementation of the get_paystub_amount() method. This is not an accidental omission in the source code. When you declare a method as abstract, you can no longer provide an implementation for it in the class that defines it. Instead, the implementation must come from subclasses such as LCL_HOURLY_EMPLOYEE which, as you can see in Listing 5.7, redefine the method as per usual.

```
CLASS lcl_employee DEFINITION ABSTRACT.
  PUBLIC SECTION.
    METHODS:
      ...
      get_paystub_amount ABSTRACT RETURNING VALUE(rv_wages)
                                   TYPE bap1curr_d.
  ...
ENDCLASS.

CLASS lcl_employee IMPLEMENTATION.
  ...
ENDCLASS.

CLASS lcl_hourly_employee DEFINITION
      INHERITING FROM lcl_employee.
  PUBLIC SECTION.
    ...
    get_paystub_amount REDEFINITION.
ENDCLASS.
```

```
CLASS lcl_hourly_employee IMPLEMENTATION.
  ...
  METHOD get_paystub_amount.
    rv_wages = me->mv_hourly_rate * CO_WORK_WEEK.
  ENDMETHOD.
ENDCLASS.
```

Listing 5.7 Defining Abstract Classes and Methods

For global classes maintained in the form-based view of the Class Builder, you can define abstract methods by performing the following steps:

1. First, select the target method on the **Methods** tab of the form-based view, as shown in Figure 5.6, and click on the **Goto Properties** icon.

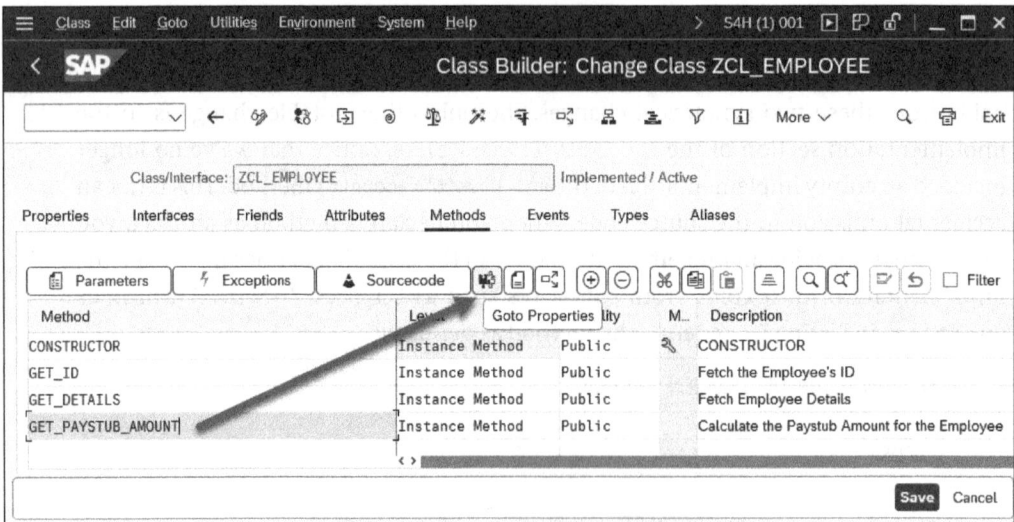

Figure 5.6 Defining an Abstract Method in the Class Builder (Part 1)

2. Then, in the **Change Method** dialog box, click on the **Abstract** checkbox to declare the method as abstract. This will open up the **Information** dialog box shown in Figure 5.7 advising that the implementation of the method will be deleted.

3. Finally, after confirming the information message, click on the **Change** button to complete the assignment. At this point, the Class Builder will implicitly change the instantiation context to *abstract*, indicating that the class itself is now abstract. You can view and/or set this context directly on the **Properties** tab, as shown in Figure 5.8.

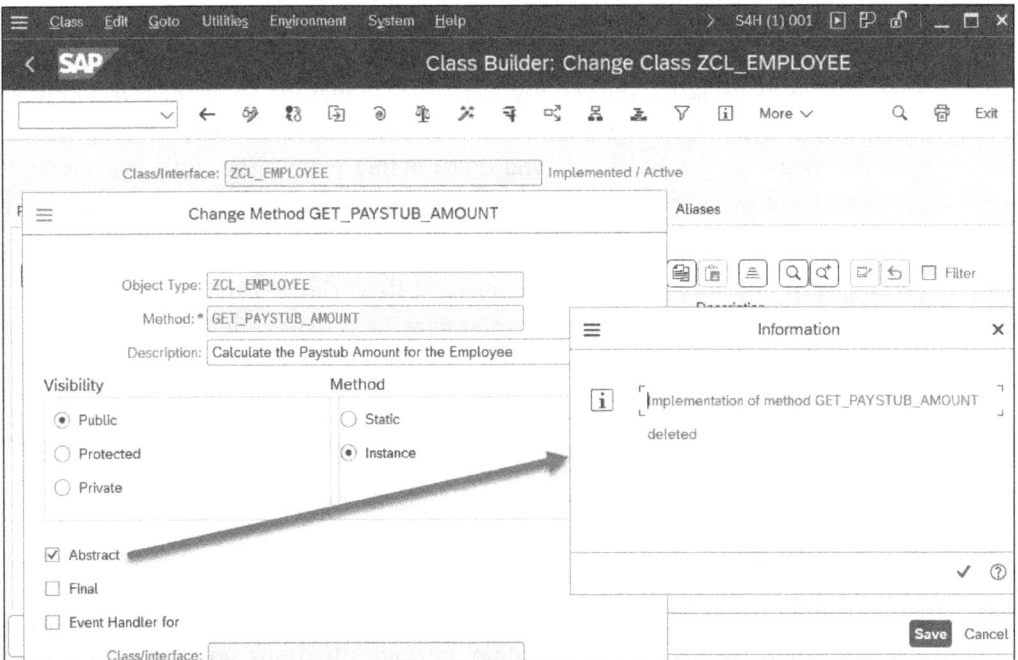

Figure 5.7 Defining an Abstract Method in the Class Builder (Part 2)

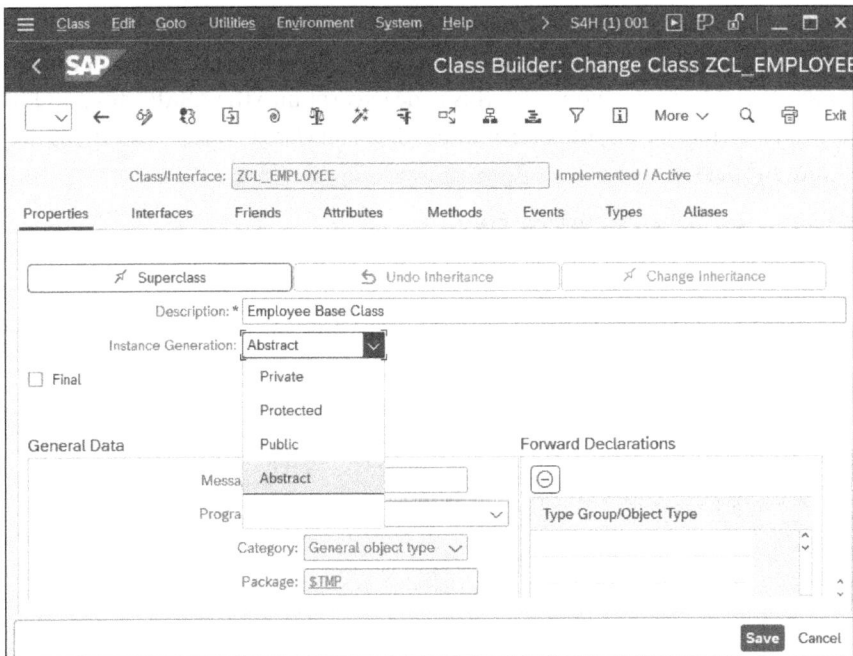

Figure 5.8 Defining an Abstract Method in the Class Builder (Part 3)

While these abstract declarations might seem rather cosmetic, the use of the ABSTRACT keyword has some significant effects on the way these classes are utilized at runtime. For example, by declaring the get_paystub_amount() method as abstract in Listing 5.7, you're implying that the LCL_EMPLOYEE class is incomplete from an implementation perspective. While LCL_EMPLOYEE can (and does) define many other fully implemented methods, there are certain elements of its design that are abstract in nature. Therefore, by extension, LCL_EMPLOYEE is an *abstract class*.

Because their implementation is incomplete, abstract classes like LCL_EMPLOYEE cannot be instantiated on their own. Instead, their purpose is to provide a common *template* that makes it easier to implement specialized subclasses. For instance, by stubbing out core behaviors such as the get_paystub_amount() method in the abstract LCL_EMPLOYEE superclass, you've provided a more accurate template for what an employee class should look like—regardless of type.

5.3.2   Final Classes

As you build out your class inheritance trees, you may sometimes reach a point where a class shouldn't be extended any further. In these situations, you can use the FINAL keyword to lock the class down so that it can no longer be inherited.

The code excerpt in Listing 5.8 shows how the FINAL keyword is being used to finalize a class called LCL_PASSWORD_UTILS. Here, we simply add the FINAL keyword to the CLASS DEFINITION statement. Once this designator is set, it's no longer possible to inherit from LCL_PASSWORD_UTILS. Indeed, if you paste this code excerpt into the ABAP editor and try to activate it, you'll receive a syntax error indicating that class LCL_PASSWORD_UTILS may not have any subclasses.

```
CLASS lcl_password_utils DEFINITION FINAL.
  PUBLIC SECTION.
    METHODS:
      is_valid_password RETURNING VALUE(rv_valid) TYPE abap_bool,
      ...
ENDCLASS.

CLASS lcl_password_utils IMPLEMENTATION.
  METHOD is_valid_password.
    ...
  ENDMETHOD.
ENDCLASS.

CLASS lcl_malware DEFINITION  "<== Syntax error
      INHERITING FROM lcl_password_utils.
  PUBLIC SECTION.
```

```
METHODS:
    is_valid_password REDEFINITION.
ENDCLASS.

CLASS lcl_malware IMPLEMENTATION.
  METHOD is_valid_password.
    "Try to bypass core behavior through inheritance...
    rv_valid = abap_true.
  ENDMETHOD.
ENDCLASS.
```

Listing 5.8 Marking a Class as Final

You can achieve the same effect for global classes maintained in the form-based view of the Class Builder by selecting the **Final** checkbox on the **Properties** tab as shown in Figure 5.9.

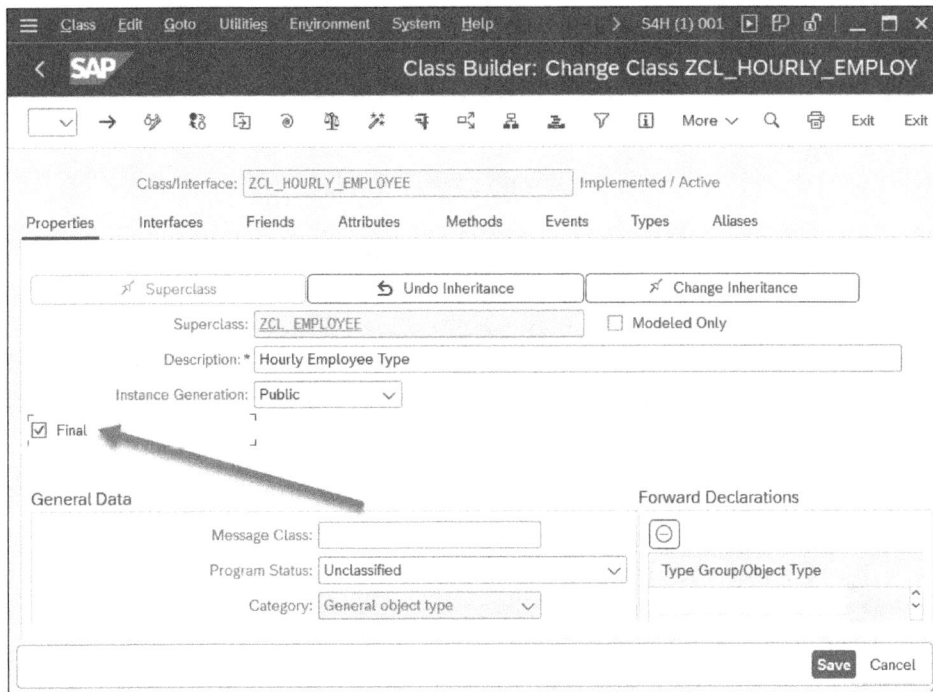

Figure 5.9 Setting the Final Indicator on a Global Class in the Class Builder

Though you don't see the **Final** indicator used as much with custom classes developed at customer sites, it's very common to see SAP and their partners apply this flag to prevent customers from making further changes to their standard-delivered content.

5.3.3   Final Methods

Sometimes, we may find ourselves in situations where we want to lock down certain methods of a class for inheritance without locking down the entire class. To accommodate these kinds of scenarios, ABAP Objects also allows us to apply the FINAL keyword to individual method definitions.

This approach is demonstrated in Listing 5.9. Here, you can see how we've re-opened the LCL_PASSWORD_UTILS class for inheritance but kept the is_valid_password() method locked down.

```
CLASS lcl_password_utils DEFINITION.
  PUBLIC SECTION.
    METHODS:
      is_valid_password FINAL RETURNING VALUE(rv_valid)
                              TYPE abap_bool,
      ...
ENDCLASS.

CLASS lcl_password_utils IMPLEMENTATION.
  METHOD is_valid_password.
    ...
  ENDMETHOD.
ENDCLASS.

CLASS lcl_malware DEFINITION   "<== This is now allowed
      INHERITING FROM lcl_password_utils.
  PUBLIC SECTION.
    METHODS:
      is_valid_password REDEFINITION. "<== But this isn't
ENDCLASS.

CLASS lcl_malware IMPLEMENTATION.
  METHOD is_valid_password.
    "Try to bypass core behavior through inheritance...
    rv_valid = abap_true.
  ENDMETHOD.
ENDCLASS.
```

Listing 5.9 Marking Individual Methods as Final

You can achieve the same effect for methods maintained in the form-based view of the Class Builder by selecting the target method, clicking on the **Goto Properties** icon, and checking the **Final** checkbox as shown in Figure 5.10. Be sure to click the **Change** button to confirm your changes.

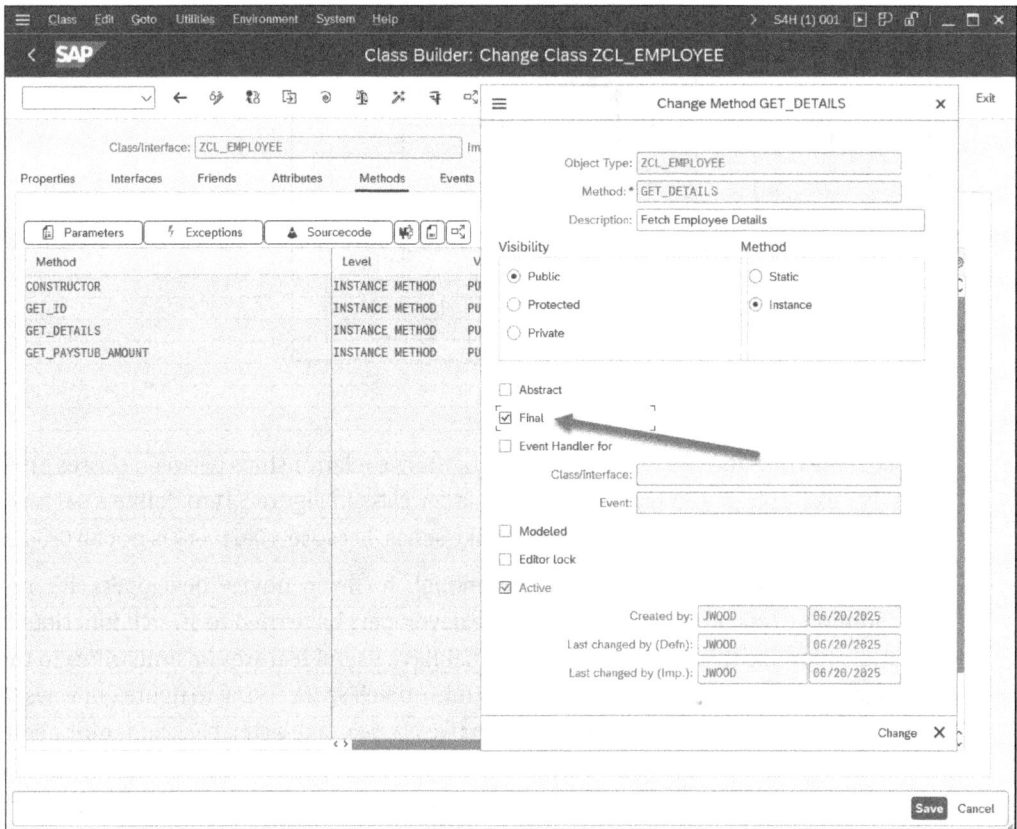

Figure 5.10 Setting the Final Indicator for a Method Using the Form-Based View of the Class Builder

5.4   Inheritance Versus Composition

With all the hype surrounding inheritance, novice object-oriented developers sometimes assume that if they're not using inheritance extensively that they must be doing something wrong. It's important to realize that while inheritance is powerful, it's not always the best solution for reusing code from existing classes. Indeed, one of the worst mistakes you can make is to try to stretch classes to fit into some sort of contrived inheritance relationship.

Whenever you're thinking of defining a new class in terms of some preexisting class, the first question you should ask yourself is whether or not the relationship between the subclass and superclass fits into the *is-a* relationship mold. To illustrate this, let's consider an inheritance tree for various types of orders (see Figure 5.11). At each level of the tree, you should be able to apply the is-a relationship between a subclass and its superclass, and it should make sense. For example, a SalesOrder *is an* Order, a CashOrder *is a* SalesOrder, and so on.

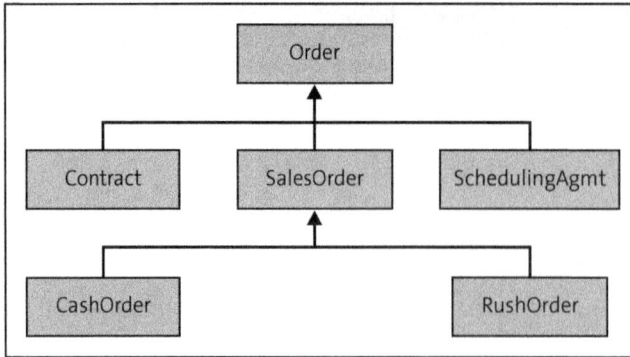

Figure 5.11 Inheritance Tree for Order Types

Most of the time, the is-a test makes the inheritance relationships between classes obvious. For example, if you try to extend the Order class in Figure 5.11 to define a Delivery subclass, the is-a relationship wouldn't make sense, because a Delivery is not an Order.

While the is-a test should be intuitive enough for even novice developers, it's not uncommon to encounter situations where developers have tried to stretch inheritance relationships like this to leverage classes that have useful features or similarities to the ones they are trying to implement. If you find yourself stuck trying to figure out ways to define an inheritance relationship between two classes, take a step back and think about the relationship between the classes from a logical perspective. If you think about it, a Delivery is not an Order, but an Order does have one or more Deliveries associated with it. This *has-a* association is commonly referred to as a *composition* relationship.

The term *composition* is used in object-oriented circles to describe the reuse of existing functionality in classes by integrating objects of those classes as attributes in your new class. You can use these attributes in the same way that you use ordinary attributes based on elementary types and structures.

The code excerpt in Listing 5.10 demonstrates how the composition technique is used to model the has-a relationship between a delivery class called LCL_DELIVERY and an order class called LCL_ORDER. As you can see, we're not really doing anything special here syntax-wise. Instead, we're just incorporating functionality from the LCL_DELIVERY class into LCL_ORDER so that we don't have to reinvent the wheel.

```
CLASS lcl_delivery DEFINITION.
  PUBLIC SECTION.
    METHODS:
      constructor,
      get_delivery_date RETURNING value(rv_date)
                                  TYPE sydatum.

  PRIVATE SECTION.
    DATA: mv_delivery_date TYPE sydatum.
```

```abap
ENDCLASS.

CLASS lcl_delivery IMPLEMENTATION.
   METHOD constructor.
      mv_delivery_date = sy-datum.
   ENDMETHOD.

   METHOD get_delivery_date.
      rv_date = mv_delivery_date.
   ENDMETHOD.
ENDCLASS.

CLASS lcl_order DEFINITION.
   PUBLIC SECTION.
      METHODS:
         constructor IMPORTING iv_id TYPE i,
         release,
         track.

   PRIVATE SECTION.
      DATA: mv_id       TYPE i,
            mo_delivery TYPE REF TO lcl_delivery.
ENDCLASS.

CLASS lcl_order IMPLEMENTATION.
   METHOD constructor.
      mv_id = iv_id.
   ENDMETHOD.

   METHOD release.
      "Create an outbound delivery for the order...
      CREATE OBJECT mo_delivery.
      ...
   ENDMETHOD.

   METHOD track.
     DATA lv_message TYPE string.
     lv_message =
       |Order #{ me->mv_id } was shipped on | &&
       |{ mo_delivery->get_delivery_date DATE = user}|.
     WRITE: / lv_message.
   ENDMETHOD.
ENDCLASS.
```

Listing 5.10 Reusing Classes Using the Composition Technique

In many respects, composition is rather like building with blocks: you're stacking classes together to build subassemblies that are more powerful than the sum of their parts. Sometimes, the easiest way to achieve code reuse is to simply reuse the classes directly. Indeed, unless the inheritance relationship is obvious, you should generally favor the use of composition over inheritance in day-to-day development because it provides you with more flexibility. In Chapter 6, we'll shed further light on this and show you how inheritance brings along some unwanted baggage that can clutter up your design if you're not careful.

5.5   Working with ABAP Refactoring Tools

As you've seen, inheritance provides a natural way of extending classes to adapt to changing functional requirements. Of course, while this all sounds good on paper, there will be times when you discover that modeling an inheritance relationship requires more than just creating a subclass or two. Indeed, depending on when you discover these relationships, you might end up having to modify other classes in the inheritance hierarchy to make the relationships work.

You've seen a couple of instances of this throughout this chapter as the LCL_EMPLOYEE class hierarchy has evolved. For example, in Section 5.2.4, we decided to move the get_ paystub_amount() out of type-specific subclasses such as LCL_HOURLY_EMPLOYEE and into the LCL_EMPLOYEE root class, since this is an operation that exists for any employee type. While this may seem like a simple cut-and-paste code modification, structural changes like this are normally described using a more official term in object-oriented circles: *refactoring*.

What Is Refactoring?

In his book *Refactoring: Improving the Design of Existing Code* (Addison-Wesley Professional, 1999), Martin Fowler describes refactoring as a process whereby selective code modifications are applied to improve the underlying structure of a system without affecting its external behavior. While many would argue that you should never touch code that's not "broken," proponents of refactoring refute this claim by pointing out that small structural improvements greatly extend the lifespan of the code.

In his book *Refactoring to Patterns* (Addison-Wesley Professional, 2004), Joshua Kerievsky puts it this way: "By continuously improving the design of code, we make it easier and easier to work with. This is in sharp contrast to what typically happens: little refactoring and a great deal of attention paid to expediently adding new features. If you get into the hygienic habit of refactoring continuously, you'll find that it is easier to extend and maintain code." If you've ever stumbled across shared ABAP modules such as BAdI implementation classes or function modules with lots of cross-cutting functions cobbled together, then you can appreciate just how messy code can become if it isn't cleaned up periodically.

We won't attempt to sell you on the merits of refactoring in this book. Fowler's *Refactoring: Improving the Design of Existing Code* describes a series of *refactorings* (or patterns) that can guide you in making good design decisions whenever you need to alter the structure of your classes. Though you can perform these refactorings manually using cut-and-paste style code changes, the Class Builder offers a much better alternative: the *Refactoring Assistant*.

You can use the Refactoring Assistant, which is based in the ABAP Workbench, to automatically perform some of the most common refactorings. Besides saving you some keystrokes, the automation of this process ensures that you don't accidentally make a mistake by missing a manual step or two along the way.

You can start the Refactoring Assistant from within the form-based view of the Class Builder by selecting **Utilities • Refactoring • Refactoring Assistant** in the top-level menu bar. This will open the **Refactoring Help** dialog box shown in Figure 5.12.

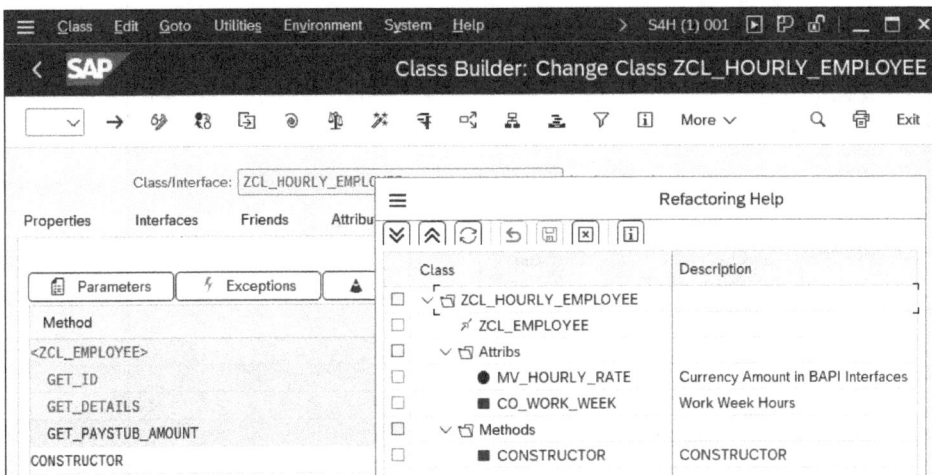

Figure 5.12 Working with the Refactoring Assistant

Within the Refactoring Assistant, you can implement various move method refactorings by simply dragging-and-dropping components from one class onto another. For example, Figure 5.13 shows how to implement the move method refactoring described in Section 5.2.4 to move the get_paystub_amount() method up to the ZCL_EMPLOYEE base class level. After you drop the method onto ZCL_EMPLOYEE, click the **Save** button, and the Refactoring Assistant will take care of making all the necessary changes. From there, all that's left to do is activate the changes in the respective classes.

In addition to the class-specific refactorings provided by the Refactoring Assistant, ABAP development tools provides automated support for many useful refactoring patterns, including renaming variables and methods as well as extracting methods. Many of these patterns are provided by the contextual Quick Fix features described in Chapter 2, Section 2.5.3. Others are included in the contextual source code features accessible via the **Source** menu shown in Figure 5.14.

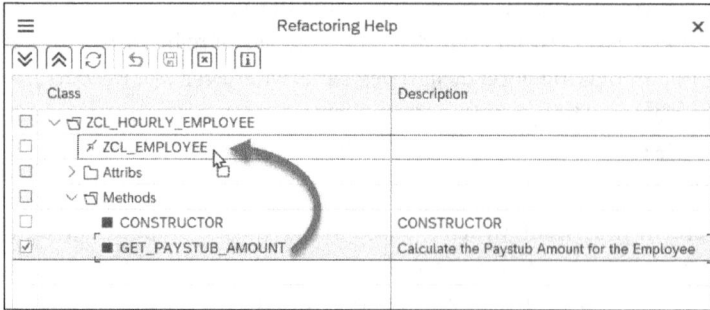

Figure 5.13 Performing a Move Method Refactoring Using the Refactoring Assistant

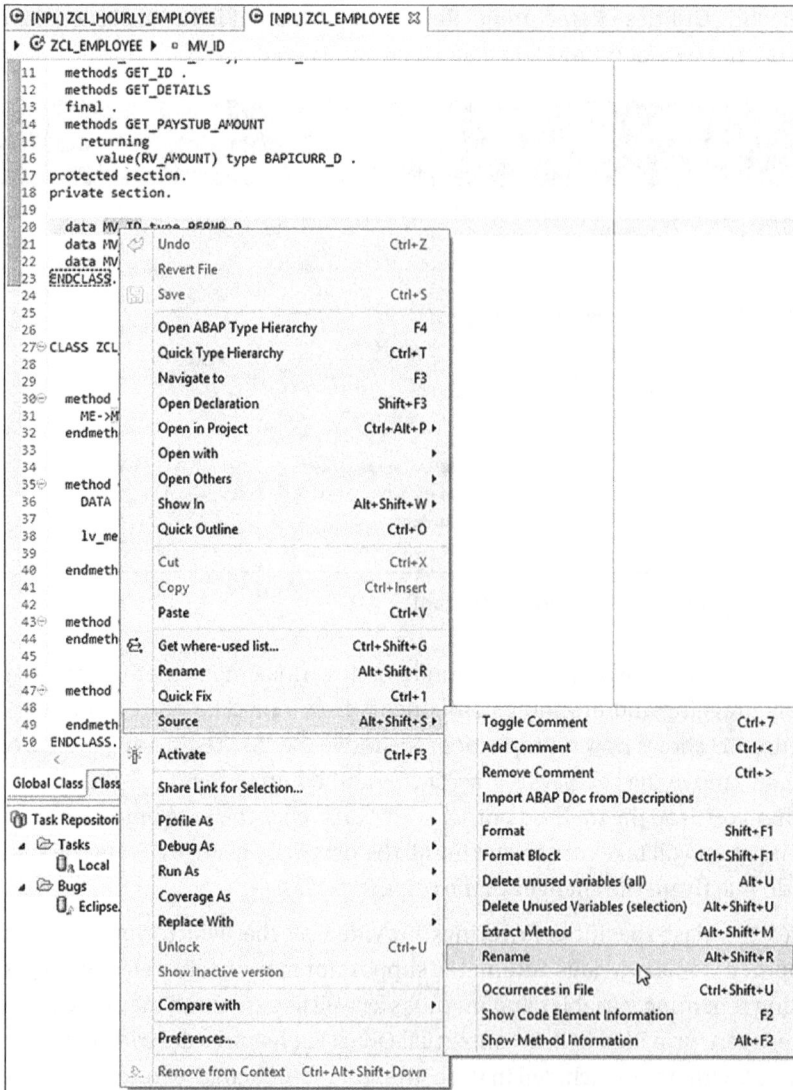

Figure 5.14 Accessing Refactoring Patterns in ABAP Development Tools

Over time, it's likely that we'll see the Eclipse-based refactoring feature list grow to eclipse (no pun intended) that of the ABAP-based Refactoring Assistant. Eclipse has long included many useful refactoring tools for other supported languages such as Java, so it's only a matter of time before these same tools make their way into the ABAP world.

5.6   UML Tutorial: Advanced Class Diagrams, Part I

In Chapter 1, we introduced some of the basic elements of a class diagram, showing you how to model rudimentary classes along with their attributes and behaviors. In this chapter and in Chapter 6, we'll expand our discussion of class diagrams to incorporate some of the more advanced concepts that we've covered in the past several chapters.

5.6.1   Generalizations

Most of the time, discussions on inheritance tend to focus on specializations at the subclass level. However, if you look up the inheritance tree, you see that superclasses become more generalized as you make our way upwards towards the top of the tree. Perhaps this is why the creators of the Unified Modeling Language (UML) decided to describe the notation used to depict inheritance relationships between classes in a class diagram as a *generalization* relationship.

Figure 5.15 shows a basic class diagram that depicts a superclass called Account along with two subclasses (CheckingAccount and SavingsAccount). Notice that each subclass has a connector drawn upward towards their superclass. The triangle at the top of the association identifies the relationship between the two classes as a generalization.

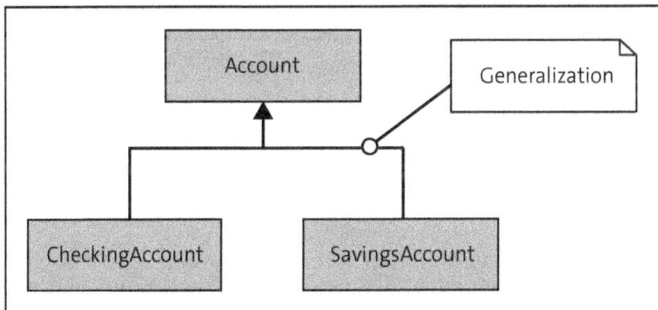

Figure 5.15 UML Class Diagram Notation for Generalizations

5.6.2   Dependencies and Composition

In Section 5.4, we described the concept of composition in terms of a has-a relationship between two classes. Up until now, the only way we could represent this kind of relationship is with a UML association. However, an association depicts a fairly loose relationship between two classes. In the case of composition, the terms of the relationship between the classes are much more pronounced. To model these relationships, UML provides us with *dependency associations.*

Dependency associations are depicted using directed lines with an arrow pointing towards the class that the source class depends on. Figure 5.16 shows how dependencies are depicted using the notation. Here, you can follow the direction of the association to determine that the Order class is dependent on the Delivery class.

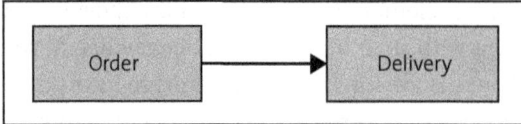

Figure 5.16 Defining a Dependency Relationship Between Classes

UML also provides a specific notation for depicting composition relationships. In Figure 5.17, we've used this notation to show that an instance of class Address can be embedded inside either class Customer or class Vendor, but not both. This notation also implies that any instances of class Address will be deleted whenever the instance of the composing Customer or Vendor class is deleted.

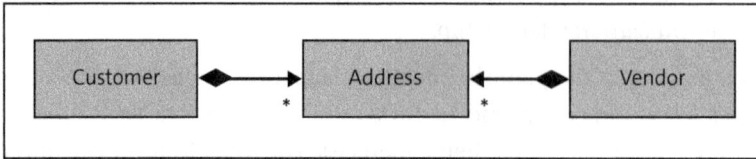

Figure 5.17 Defining Composition Relationships in Class Diagrams

Because the UML interpretation of composition relationships is more detailed than the common view of composition applied in practice, most developers prefer to model these relationships using dependencies. This leaves more flexibility to the implementer to implement the actual relationships from a code perspective.

5.6.3 Abstract Classes and Methods

Figure 5.18 shows the UML notation for depicting abstract classes and methods. As you can see, the only requirement here is to italicize the class or method name to indicate that the class or method is to be defined as abstract. However, since the italics are sometimes hard to read, you'll sometimes see developers tag abstract classes using the non-normative <> keyword as shown in Figure 5.19.

5

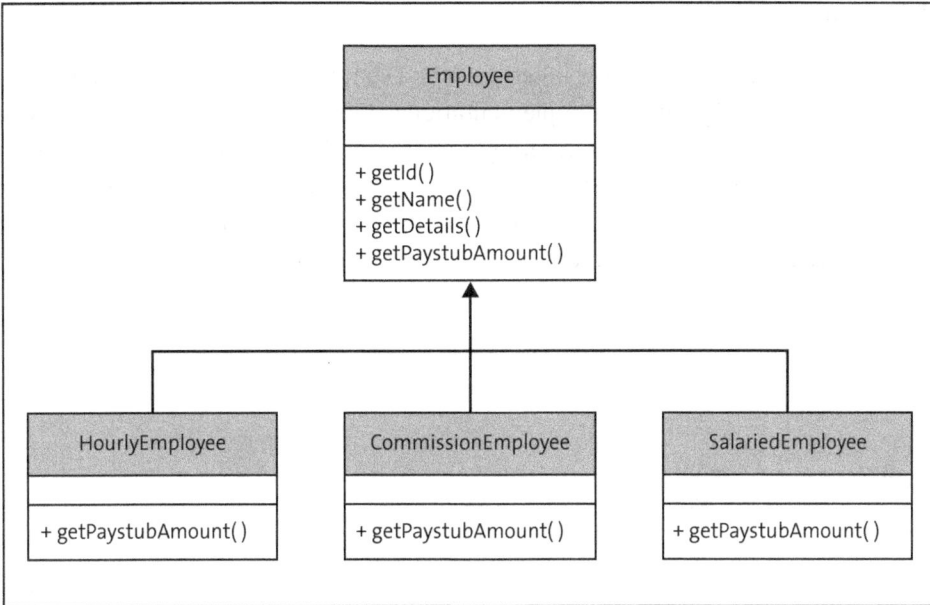

Figure 5.18 Defining Abstract Classes and Methods

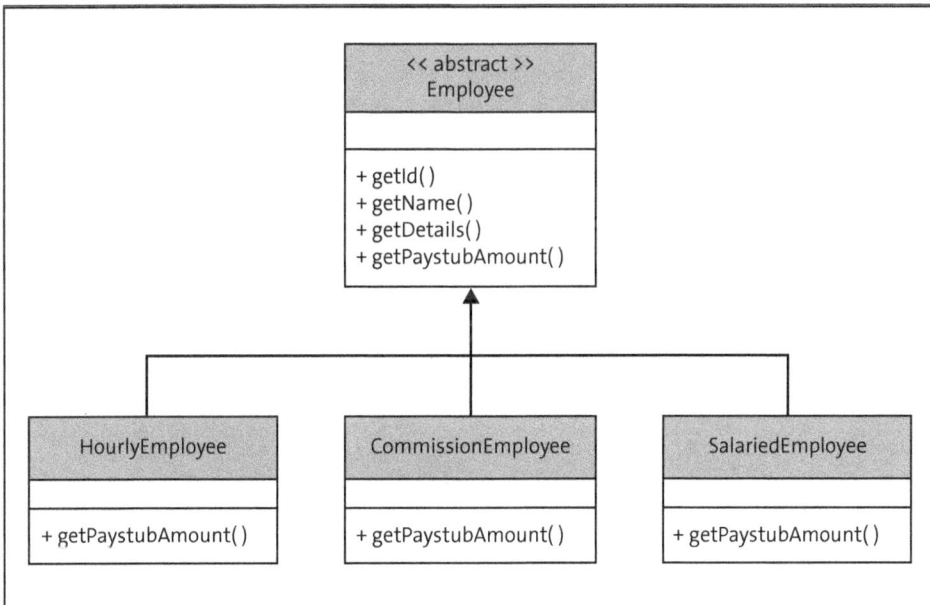

Figure 5.19 Non-Normative Form for Defining Abstract Classes

5.7   Summary

In this chapter, you learned how inheritance and composition techniques can be used to quickly and safely reuse the implementations of existing classes. We concentrated our focus in this chapter on inheriting a class's implementation. However, there's another dimension of the inheritance relationship that we have not yet considered. In the next chapter, we'll explain how you can use *type inheritance* to further exploit inheritance relationships and make your designs even more flexible.

Chapter 6
Polymorphism

The term polymorphism literally means "many forms." From an object-oriented perspective, polymorphism works in concert with inheritance to allow various types within an inheritance tree to be used interchangeably. In this chapter, you'll learn how to harness this power to create highly flexible program designs using ABAP Objects.

In the previous chapter, you learned how to define inheritance relationships between related classes. You'll recall that the basic litmus test you used to identify these relationships was to ask yourself whether or not a given class was a more specific type of a particular superclass. For example, a Dog is a specific type of Mammal. So, instead of creating a standalone Dog class, it makes sense to define the Dog class as a *subclass* of Mammal so that it can inherit selected features from Mammal.

When you look at inheritance in this light, it's easy to see its obvious benefits in terms of code reuse. However, as it turns out, there's another (and arguably more important) dimension of the *is-a* relationship that we haven't yet considered.

Since the classes in an inheritance tree share a common public interface, it's technically possible for a given subclass to respond to any request (i.e., method call) directed at its superclass. This aspect of the inheritance relationship is referred to as *interface inheritance*. When you combine interface inheritance with subclasses' ability to redefine or override the implementation of their inherited methods, you end up in a situation where clients no longer need to worry about the types of objects they're interfacing with: they simply issue requests (method calls) and let the objects figure out how to process the requests in type-specific ways. In other words, types within an inheritance tree become interchangeable from a client perspective.

In this chapter, you'll see how we can exploit this functionality to simplify your object-oriented designs and improve flexibility. Along the way, we'll also introduce you to another powerful tool in the object-oriented developer's toolkit: *interfaces*.

6.1 Object Reference Assignments Revisited

In Chapter 2, you learned how to use the assignment operator (=) to perform assignments between object reference variables. As you may recall, object reference assignments copy the pointer stored in the source object reference variable into the target object reference variable. After the assignment is complete, both object reference variables point to the same object instance.

Though we glossed over the details in Chapter 2, there's an important rule that you must follow when making object reference assignments: the assignments must be made using *compatible types*. For example, making an assignment between an object reference variable pointing to a Material class and an object reference variable pointing to a Customer class doesn't make sense because these two types are not compatible with one another. This is demonstrated in the code excerpt in Listing 6.1.

```
DATA lo_material TYPE REF TO lcl_material.
DATA lo_customer TYPE REF TO lcl_customer.

CREATE OBJECT lo_material.

lo_customer = lo_material.  "<== Syntax error
```

Listing 6.1 Type Compatibility Issues with Object Reference Assignments

Strictly speaking, two variables are only compatible if they share the same type. Despite this, developers routinely make assignments between variables having incompatible types (e.g., between built-in types such as an integer and a floating-point number). Such types are said to be *convertible* in the sense that there exists some kind of conversion rule that tells the ABAP runtime environment how to convert the contents of the source variable into a format compatible with the target variable.

In the case of object reference assignments, conversions don't make sense because an object reference stores a pointer to an object and not the object itself. Therefore, for an object reference variable to point at an object that has a different type, it must be *enhanced* with additional type information that provides visibility to the components of the actual object that it points to. In this section, you'll learn how to achieve this and perform object reference assignments between families of related types. Understanding how these assignments work is a prerequisite for learning how to implement generic designs using polymorphism.

6.1.1   Static and Dynamic Types

Up until now, the object reference assignment examples that we've used have been between object reference variables that share the same *static type*. The static type of an object reference variable is the class type (or interface type, as you'll learn in Section 6.3) used to define the object reference variable. For example, the lo_oref object reference variable depicted in Listing 6.2 has the class type LCL_SOME_CLASS.

```
DATA lo_oref TYPE REF TO lcl_some_class.
```

Listing 6.2 Determining the Static Type of an Object Reference Variable

You may often want to perform assignments between object reference variables that do not share the same static type. For example, since instances of a superclass and its subclasses are intended to be interchangeable, it should be possible to perform an assignment between object reference variables whose static types are part of the same inheritance tree. The code excerpt in Listing 6.3 provides an example of the kind of assignment you might want to perform between related types.

```
CLASS lcl_parent DEFINITION.
   PUBLIC SECTION.
      METHODS: a,
               b.
ENDCLASS.

CLASS lcl_parent IMPLEMENTATION.
   METHOD a.
      WRITE: / 'In method a.'.
   ENDMETHOD.

   METHOD b.
      WRITE: / 'In method b.'.
   ENDMETHOD.
ENDCLASS.

CLASS lcl_child DEFINITION
               INHERITING FROM lcl_parent.
   PUBLIC SECTION.
      METHODS: c.
ENDCLASS.

CLASS lcl_child IMPLEMENTATION.
   METHOD c.
      WRITE: / 'In method c.'.
   ENDMETHOD.
ENDCLASS.

DATA: lo_parent TYPE REF TO lcl_parent,
      lo_child  TYPE REF TO lcl_child.

CREATE OBJECT lo_parent.
CREATE OBJECT lo_child.
lo_parent = lo_child.
```

Listing 6.3 Performing a Cast with an Object Reference Assignment

To help you understand how an object reference assignment like the one depicted in Listing 6.3 works, let's revisit the remote control/TV metaphor we first introduced in Chapter 2. There, we noted that object reference variables are like remote controls that can be used to interface with object instances of a particular type. By default, object reference variables, like remote controls, are only able to communicate with a specific type of object.

Now, imagine that you decide to purchase a universal remote to replace the default remote that came with the TV. In this case, even though the *static type* of the universal remote is more generic than the one provided by the manufacturer, it's still *compatible* with the public interface provided by the TV (i.e., it supports common operations such as turning the TV on and adjusting volume). However, before you can use the universal remote with the TV, it must be *programmed* with information about the actual TV model it's interfacing with. Similarly, object reference variables that are assigned (or reassigned) to point to objects that have a different static type must be programmed (or reprogrammed) with dynamic type information at runtime.

The *dynamic type* of an object reference variable is the class type of the actual object instance pointed to by the reference variable. In the example given in Listing 6.3, the statement CREATE OBJECT lo_parent instantiates an object of type LCL_PARENT and assigns a pointer to that object to the lo_parent object reference variable. At this point, the static and dynamic type of the lo_parent reference variable is the same. However, when you perform the assignment statement lo_parent = lo_child, the dynamic type of the lo_parent reference is changed by the ABAP runtime environment to point to the LCL_CHILD class type. This information is crucial for the ABAP runtime environment to be able to interact with the compatible components of the LCL_CHILD object instance that's now being pointed to by the lo_parent reference variable.

It's worth mentioning that you cannot arbitrarily set the dynamic type of an object reference to just any class type. In other words, these kinds of assignments don't make sense without some kind of an inheritance relationship between the source and target object reference variables. In Section 6.1.2, we'll explore the rules that determine how these assignments work.

6.1.2   Casting

If the static type of the source and target object reference variables is not the same in an assignment operation, a special operation called a *cast* must occur for the assignment to work. A cast operation is allowed whenever the static type of the target object reference is the same as or more general than the dynamic type of the source object reference. There are two different types of cast operations, which we'll discuss in the following sections: a *narrowing cast* and a *widening cast*.

Narrowing Casts

A narrowing cast occurs in an object reference assignment statement whenever the static type of a target object reference variable is more generic than the static type of the source object reference variable. Since you're moving upwards in the inheritance tree, these narrowing casts are also sometimes referred to as *up casts*.

The assignment statement from Listing 6.3 provided an example of a narrowing cast between the reference variables lo_parent and lo_child. This type of assignment is a narrowing cast because the class type LCL_PARENT is more general than LCL_CHILD, effectively *narrowing* the scope of the components that can be accessed in the LCL_CHILD object to those defined in the LCL_PARENT superclass.

This reduction in scope prevents the target object reference variable (i.e., LO_PARENT) from accessing components that are not defined in its static type definition. For example, the method call that is commented out in Listing 6.4 would cause a syntax error since method c() is not defined in class LCL_PARENT. Of course, this reduction in scope does not imply that the object itself is changed or truncated in some way. The LCL_CHILD object instance in Listing 6.4 is still a full-fledged object of type LCL_CHILD; it's just that the lo_parent object reference variable doesn't have visibility to the components defined in LCL_CHILD.

```
DATA: lo_parent TYPE REF TO lcl_parent,
      lo_child  TYPE REF TO lcl_child.
CREATE OBJECT lo_parent.
CREATE OBJECT lo_child.
lo_parent = lo_child.
lo_parent->c( ).   "Syntax Error!
```

Listing 6.4 Attempting to Call a Method That's Out of Scope

In addition to normal object reference assignments, it's also possible to perform narrowing casts using inline syntax. For example, if you know you want to perform an up cast during the object instantiation process from the get-go, then you can carry out the narrowing cast using the TYPE addition of the CREATE OBJECT statement. Listing 6.5 demonstrates this syntax is demonstrated. Here, the CREATE OBJECT statement creates an object of type LCL_CHILD and then performs a narrowing cast as it assigns a pointer to the object back to the lo_parent object reference.

```
DATA lo_parent TYPE REF TO lcl_parent.
CREATE OBJECT lo_parent TYPE lcl_child.
```

Listing 6.5 Performing a Narrowing Cast During Object Creation

You also have the option of performing up casts using the CAST operator. The code excerpt in Listing 6.6 demonstrates how you can use this syntax to implement the same casting scenario that we demonstrated in Listing 6.4. Here, you simply plug in the static type of the source object reference and the CAST operator takes care of the rest.

```
DATA(lo_child) = NEW lcl_child( ).
DATA(lo_parent) = CAST lcl_parent( lo_child ).
```

Listing 6.6 Performing an Up Cast Using the CAST Operator

Widening Casts

In cases where the static type of the target object reference variable is *more* specific than the static type of the source object reference variable, a *widening cast* must be applied for the assignment statement to pass muster with the ABAP compiler. Widening casts allow you to take control of the assignment process by telling the compiler that you know what you're doing when you perform an assignment between object references with different static types.

Of course, this delegation doesn't mean that a validity check never takes place; it just means that the check is deferred until runtime when the dynamic type of the source object reference is known. As we stated in Section 6.1.1, the static type of the target object reference must be the same as or more general than the dynamic type of the source object reference—otherwise, an exception will occur. In Chapter 8, we'll share how you can gracefully recover from these types of exceptions. Nevertheless, we strongly recommend that you use widening casts carefully, as they can be confusing as well as hazardous to your code.

Because of the dangerous nature of widening casts, the ABAP language specification requires that you formally declare your intent to perform a widening cast by using a special assignment operator called the *casting operator*. There are actually *two* casting operators that you can use for performing widening casts: the legacy casting operator (?=) and the aforementioned CAST operator.

The code excerpt in Listing 6.7 shows you how to perform a widening cast using the legacy casting operator. This contrived example performs an up cast to copy an object reference of type LCL_CHILD into an object reference variable with static type LCL_PARENT. Then, you assign the object reference back to the lo_child object reference variable using a widening cast. During all of this shuffling, the LCL_CHILD object created via the CREATE OBJECT remained intact, so after the widening cast, you can access the object via the lo_child object reference variable as per usual.

```
DATA: lo_parent TYPE REF TO lcl_parent,
      lo_child TYPE REF TO lcl_child.

CREATE OBJECT lo_child.
lo_parent = lo_child.
* lo_child = lo_parent.   "<== Syntax Error
lo_child ?= lo_parent.
```

Listing 6.7 Performing Widening Casts Using the Legacy Casting Operator

The code excerpt in Listing 6.8 shows how you can achieve the same effect using the CAST operator. Here, notice how the usage remains the same whether we're performing an up cast or a down cast. This approach improves readability by keeping the syntax consistent for any type of cast operation.

```
DATA(lo_child) = NEW lcl_child( ).
DATA(lo_parent) = CAST lcl_parent( lo_child ).
lo_child = CAST lcl_child( lo_parent ).
```

Listing 6.8 Performing a Down Cast Using the CAST Operator

6.2 Dynamic Method Call Binding

Now that you have a better idea of how to use casting operations to perform assignments between related class types, you're ready to start examining how polymorphism really works. It all starts with an object-oriented language's ability to support dynamic method call binding.

To understand how dynamic method call binding works, consider this example. The report program ZPOLYTEST in Listing 6.9 contains an abstract class called LCL_ANIMAL, a pair of subclasses called LCL_CAT and LCL_DOG, and a test driver class called LCL_SEE_AND_SAY. The LCL_SEE_AND_SAY class is modeled loosely after the "See-n-Say®" educational toys manufactured by Mattel, Inc. Here, we want to build a simulation in which the LCL_SEE_AND_SAY class can play the sounds made by many different animals.

```
REPORT zpolytest.
CLASS lcl_animal DEFINITION ABSTRACT.
  PUBLIC SECTION.
    METHODS:
      get_type ABSTRACT RETURNING VALUE(rv_type) TYPE string,
      speak ABSTRACT RETURNING VALUE(rv_message) TYPE string.
ENDCLASS.

CLASS lcl_cat DEFINITION
        INHERITING FROM lcl_animal.
  PUBLIC SECTION.
    METHODS: get_type REDEFINITION,
            speak REDEFINITION.
ENDCLASS.

CLASS lcl_cat IMPLEMENTATION.
  METHOD get_type.
    rv_type = 'Cat'.
  ENDMETHOD.
```

6

```abap
  METHOD speak.
    rv_message = 'Meow'.
  ENDMETHOD.
ENDCLASS.

CLASS lcl_dog DEFINITION
        INHERITING FROM lcl_animal.
  PUBLIC SECTION.
    METHODS: get_type REDEFINITION,
             speak REDEFINITION.
ENDCLASS.

CLASS lcl_dog IMPLEMENTATION.
  METHOD get_type.
    rv_type = 'Dog'.
  ENDMETHOD.

  METHOD speak.
    rv_message = 'Bark'.
  ENDMETHOD.
ENDCLASS.

CLASS lcl_see_and_say DEFINITION.
  PUBLIC SECTION.
    CLASS-METHODS:
      play IMPORTING io_animal
                TYPE REF TO lcl_animal.
ENDCLASS.

CLASS lcl_see_and_say IMPLEMENTATION.
  METHOD play.
    DATA(lv_message) =
      |The { io_animal->get_type( ) } | &&
      |says "{ io_animal->speak( ) }".|.
    WRITE: / lv_message.
  ENDMETHOD.
ENDCLASS.

START-OF-SELECTION.
  DATA(lo_cat) = NEW lcl_cat( ).
  DATA(lo_dog) = NEW lcl_dog( ).
```

```
lcl_see_and_say=>play( lo_cat ).
lcl_see_and_say=>play( lo_dog ).
```

Listing 6.9 Dynamic Binding with Method Calls

If you look at the sample code in Listing 6.9, you can see how the various classes are implemented and the inheritance relationships are formed between the abstract LCL_ANIMAL base class and the animal-specific subclasses. Where things start to get interesting is in the definition of the LCL_SEE_AND_SEE class. If you look closely at the signature of method play(), you can see that it receives an importing parameter of type LCL_ANIMAL. However, in the START-OF-SELECTION event of program ZPOLYTEST, notice how we're calling this method with object reference parameters of type LCL_CAT and LCL_DOG. In this case, the ABAP runtime environment performs an implicit narrowing cast during the assignment of the importing parameter io_animal.

Within the play() method, the logic is purposefully generic. From the perspective of the LCL_SEE_AND_SAY class, it doesn't matter what the dynamic type of the io_animal parameter is because any LCL_ANIMAL object is guaranteed to provide implementation for methods get_type() and speak(). The play() method calls these methods on the io_animal parameter and the requests are routed to the type-specific subclasses. This subtle feature makes it possible for the code inside the play() method to be implemented 100% generically.

The driving force behind all this is dynamic method call binding. The dynamic type information associated with an object reference variable allows the ABAP runtime environment to dynamically bind a method call with the implementation defined in the static type of the object instance pointed to by the object reference variable. This is crucial because the io_animal parameter points to an abstract type that could never be instantiated on its own. The implementation must come from a concrete subclass such as LCL_CAT or LCL_DOG.

Dynamic binding provides for tremendous flexibility in designs. In the simple example from Listing 6.9, we only considered an implementation for a cat and a dog. In the future though, we may decide to implement subclasses for lots of other types of animals such as a horse, a cow, or a pig. However, since the LCL_SEE_AND_SAY class works with the generic LCL_ANIMAL type, we can integrate these new types into the "See-n-Say" service quite seamlessly. Such designs are said to be *extensible* in the sense that you can easily introduce new functionality by simply creating a new subclass and plugging it in at runtime.

6.3   Interfaces

Throughout this book, we've used the term *interface* to describe the various interaction points between classes and their clients. For example, a method's signature defines an interface that's used by clients wishing to call that method. From an object-oriented perspective, you can think of an interface as a type of *protocol* that defines rules for communicating with objects.

This protocol analogy should be familiar since you interact with many types of protocols every day. For instance, the *HyperText Transfer Protocol* (HTTP) defines the rules that web clients (i.e., browsers like Mozilla Firefox or Google Chrome) and web servers must adhere to in order to reliably publish and retrieve content on the internet. These rules make it possible for web browsers to request web pages from many different types of web server implementations (e.g., Microsoft Internet Information Services [IIS] or Apache) without having to worry about how these servers are implemented. Similarly, you've seen how you can use polymorphism to dynamically bind many different types of implementations to a single interface.

As the theory behind object-oriented programming (OOP) evolved, many language designers began to see the value in being able to define an interface independently from any particular class. Such interfaces do not have an implementation associated with them and therefore cannot be instantiated on their own. While you may not realize it at first, you can do a lot of interesting things with interfaces. In this section, we'll look at how you can use interfaces to expand a class's scope into multiple dimensions.

6.3.1   Interface Inheritance Versus Implementation Inheritance

Some object-oriented languages support a *multiple inheritance* model, allowing you to define several inheritance relationships within a given class. As you may have guessed by now, ABAP Objects only supports a *single inheritance* model. This is a design decision employed by many modern object-oriented languages in an effort to avoid the ambiguity that can arise with complex inheritance hierarchies.

To illustrate some of the potential problems associated with a multiple inheritance model, let's consider an example. The class diagram in Figure 6.1 depicts a diamond-shaped class hierarchy. In this case, let's imagine that classes B and C have both redefined method someMethod() from class A. If class D does not redefine method some-Method(), from which implementation does it inherit: B or C? This problem is known as the *diamond problem*.

A single-inheritance model avoids these kinds of vagaries since a subclass always inherits from a single superclass. However, interfaces enhance this model by providing a way to extend the *type* of a class without having to bring along the implementation baggage associated with multiple inheritance. In Section 6.3.4, you'll see how the implementation of an interface allows a class to be used polymorphically wherever a reference of that interface type is used.

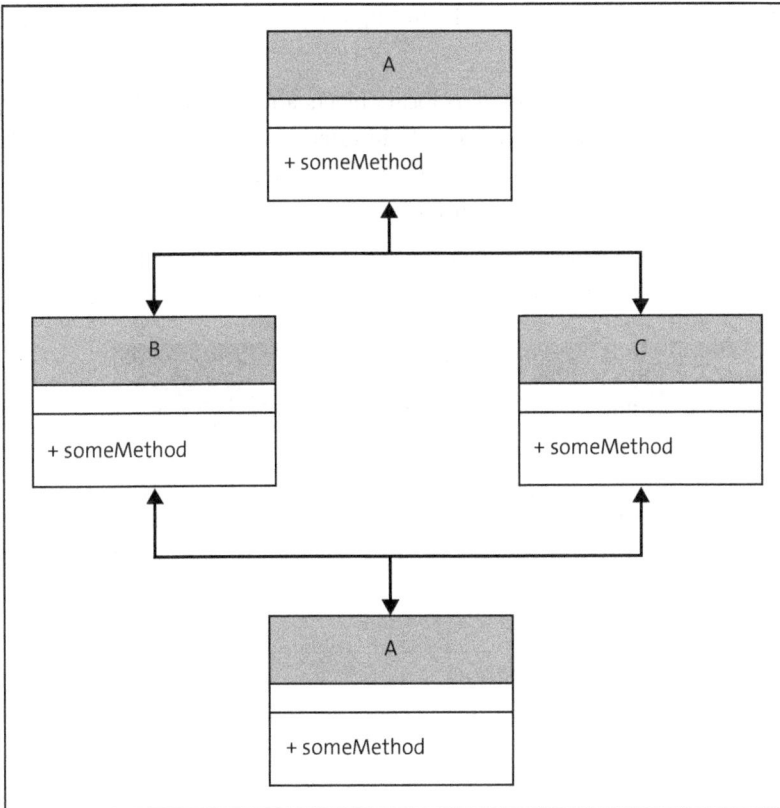

Figure 6.1 Example of the Diamond Problem with Multiple Inheritance

6.3.2   Defining Interfaces

The syntax required to define an interface is very similar to the syntax used in the declaration part of a class definition. Listing 6.10 shows an example of how to define a local interface called LIF_IFACE. Notice that none of the interface components have been defined within a visibility section. This is because all components of an interface are implicitly defined within the public visibility section. This makes sense, given that the purpose of an interface is to expand the public interface of implementing classes.

```
INTERFACE lif_iface.
  DATA a TYPE string.
  METHODS: m.
  EVENTS: e.
ENDINTERFACE.
```

Listing 6.10 Syntax for Defining a Local Interface

Most of the time, interfaces are used to add additional methods to the public interface of a class. However, you can technically define all the same types of components that you can define for classes in an interface, including attributes, methods, events, and types.

Just like classes, interfaces can be defined as local objects and global ABAP Repository objects. In the latter case, you can create and edit interfaces using the Class Builder. To illustrate how this works, let's create a new global interface called ZIF_COMPARABLE that can be used to specify an arbitrary ordering for implementing classes, following these steps:

1. From within the ABAP Workbench, select the appropriate package, right-click on it, and select the **Create · Class Library · Interface** from the menu (see Figure 6.2).

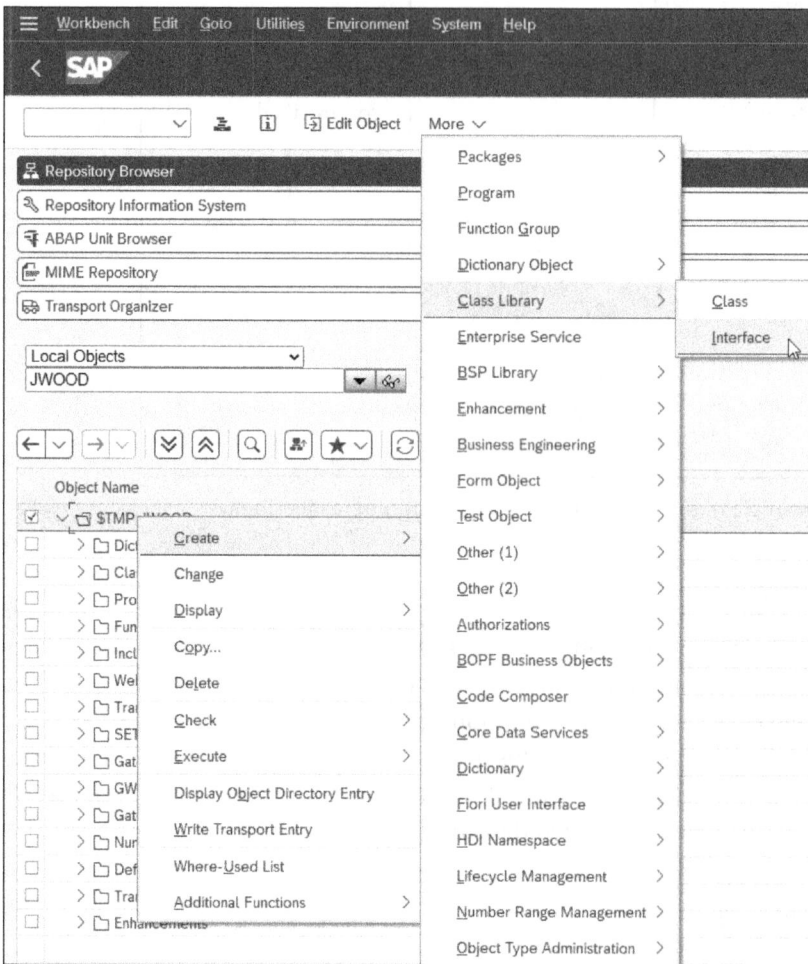

Figure 6.2 Creating a Global Interface (Part 1)

2. This opens up the **Create Interface** dialog box shown in Figure 6.3. Enter a name for the **Interface** and a brief **Description**. The name of the interface should be defined according to the convention {namespace}IF_{meaningful_name}. In the example, we use ZIF_COMPARABLE, which is defined in the customer Z namespace.

Figure 6.3 Creating a Global Interface (Part 2)

3. After you fill out the details in the **Create Interface** dialog box shown in Figure 6.3, click on the **Save** button to create the interface. You'll be prompted with the **Create Object Directory Entry** dialog box shown in Figure 6.4.

Figure 6.4 Creating a Global Interface (Part 3)

4. Finally, after confirming the repository and transport details, the interface will be created and opened in the Class Builder as shown in Figure 6.5. Just as with classes, you can edit interfaces in the form-based view or the source code-based view.

Figure 6.5 Editing a Global Interface in the Class Builder

Within the ZIF_COMPARABLE interface, we've defined a single method called compare_to() that can be used to compare two objects to determine their relative ordering. Figure 6.6 shows the signature of this method. As you can see, it defines two parameters:

- IO_OBJECT
  This importing parameter represents the object that's being compared with the host object (i.e., the object that drives the comparison). Here, notice that we've defined the type of this parameter using the generic OBJECT type. This will allow us to compare *any* type of object using this interface.

- RV_RESULT
  This returning parameter specifies the ordering of the two objects as an integer value. The resulting value will be a negative integer, zero, or a positive integer, depending on whether the host object is less than, equal to, or greater than the object passed via the io_object parameter.

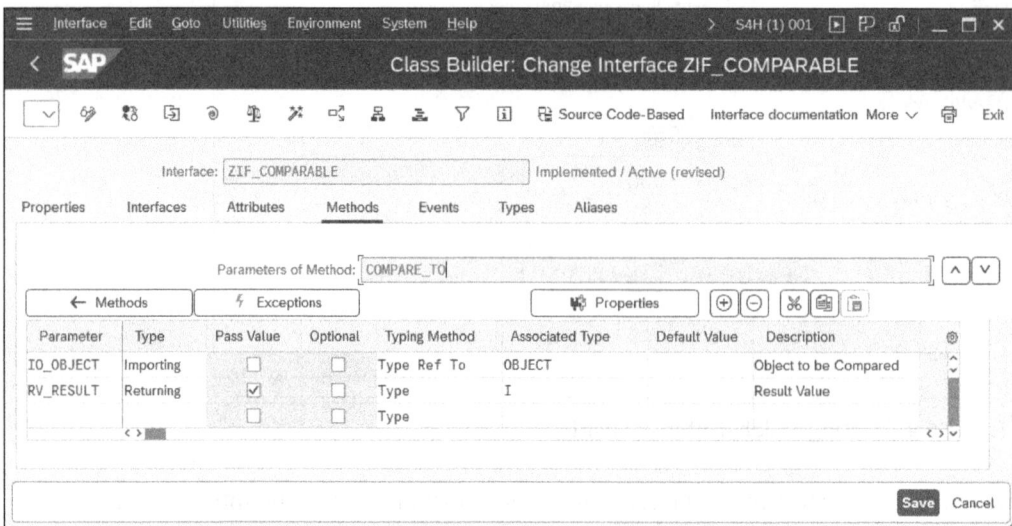

Figure 6.6 Building Out the ZIF_COMPARABLE Interface

Listing 6.11 shows the source code generated for the ZIF_COMPARABLE interface. Since interfaces don't contain any sort of implementation, that's all there is to our ZIF_COMPARABLE interface. In Section 6.3.3 and Section 6.3.4, you'll see how this interface is used to build a container class that collects a series of related objects together and provides value-add operations such as sorting.

```
interface ZIF_COMPARABLE
  public .

  methods COMPARE_TO
    importing
      !IO_OBJECT type ref to OBJECT
```

```
    returning
      value(RV_RESULT) type I .
endinterface.
```

Listing 6.11 Source Code for Interface ZIF_COMPARABLE

6.3.3 Implementing Interfaces

Interfaces are not all that interesting—until you start implementing them in classes. Within a class definition, you can implement an interface using the INTERFACES keyword. You specify one or more interfaces that you want to implement in a comma-separated list. Logically, this has a similar effect to adding the INHERITING FROM addition to the CLASS DEFINITION statement: the components from the selected interfaces will be inherited in the implementing class.

To demonstrate how interface implementation works, let's consider a brief example. Listing 6.12 shows how a local class called LCL_IMPLEMENTER is implementing an interface called LIF_IFACE. As you can see, implementing the LIF_IFACE interface in LCL_IMPLE-MENTER requires that you provide implementation for the methods defined in LIF_IFACE: m1() and m2().

```
INTERFACE lif_iface.
  METHODS: m1,
           m2.
ENDINTERFACE.

CLASS lcl_implementer DEFINITION.
  PUBLIC SECTION.
    INTERFACES: lif_iface.
ENDCLASS.

CLASS lcl_implementer IMPLEMENTATION.
  METHOD lif_iface~m1.

    ...
  ENDMETHOD.

  METHOD lif_iface~m2.

    ...
  ENDMETHOD.
ENDCLASS.
```

Listing 6.12 Implementing an Interface in a Class Definition

If you look closely at the implementation of the LCL_IMPLEMENTER class, you can see that the inherited methods are qualified with a prefix of the source interface (LIF_IFACE) and the tilde character (~). The ~ between the interface name and the interface component name is called the *interface component selector* operator. This qualification makes the inheritance relationships clear in cases where a class might implement multiple interfaces that happen to have components that share the same name.

When working with the form-based view of the Class Builder, you can implement interfaces by simply plugging in the interface name on the **Interfaces** tab, as shown in Figure 6.7. Behind the scenes, the Class Builder will automatically add the inherited components to the class definition and stub out the implementation section.

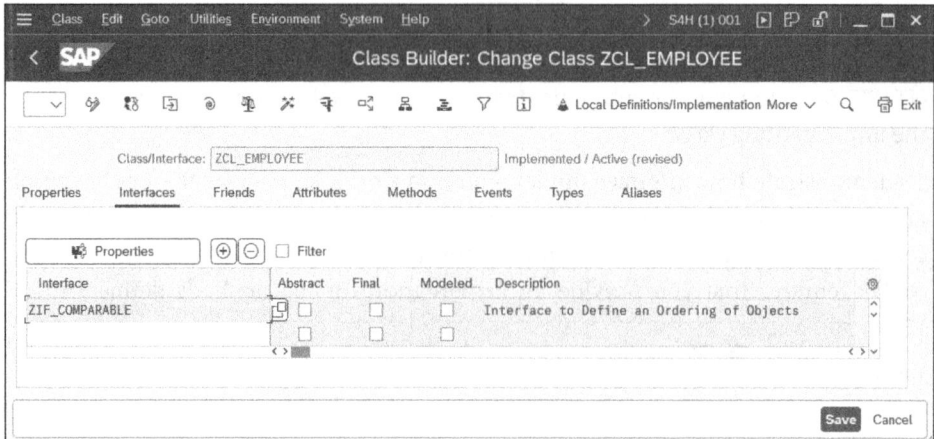

Figure 6.7 Implementing an Interface Using the Form-Based View of the Class Builder

Now that you have a better understanding of the nuts-and-bolts of implementing interfaces, let's turn our attention back to the implementation of our ZIF_COMPARABLE interface which we've been working on in this section. As you can see in Figure 6.7, we've implemented this interface in the ZCL_EMPLOYEE entity object that we developed in Chapter 5. The intent here is to provide a means of comparing employee objects so that we can sort them.

The code excerpt in Listing 6.13 provides a demonstration of how you might implement the compare_to() method for your ZCL_EMPLOYEE class. Here, we chose to define the sort ordering by the employee's ID value, but we could have just as easily defined the order another way, such as by the employee's last name. As you can see in the code, the actual comparison is straightforward. The only trick to making this work is implementing the widening cast on the generic io_object parameter. By performing a downcast on the object, you can address its components (e.g., the get_id() method) in the evaluation logic.

```
method ZIF_COMPARABLE~COMPARE_TO.
  "Perform a widening cast on the comparison object
  "so that we can access its components during the
  "comparison process:
```

```
  DATA(lo_employee) = CAST zcl_employee( io_object ).

  "Compare the two employees based on their ID number:
  IF me->get_id( ) > lo_employee->get_id( ).
    rv_result = 1.
  ELSEIF me->get_id( ) < lo_employee->get_id( ).
    rv_result = -1.
  ELSE.
    rv_result = 0.
  ENDIF.
endmethod.
```

Listing 6.13 Implementing the compare_to() Method

Ultimately, implementing an interface is rather like inheriting from an abstract class: the interface defines the structure of the components, and the implementing class must fill in the implementation. Within the methods themselves, there's nothing special going on—just regular ABAP Objects code as per usual.

Over the years, one of the pain points of working with interfaces in ABAP Objects is that you must provide an implementation for every method defined by the interface in the implementing class. While purists would no doubt argue that this is proper, it does make our job as developers a bit tedious at times. To address this phenomenon, SAP introduced the METHODS statement, which can be used to specify whether a given method must be implemented by an implementing class. The code excerpt in Listing 6.14 shows how this addition works.

```
INTERFACE lif_big_interface.
  METHODS:
    important_method DEFAULT FAIL,
    optional_method DEFAULT IGNORE.
  ...
ENDINTERFACE.
```

Listing 6.14 Working with the DEFAULT Addition

In the code excerpt in Listing 6.14 we have an interface called LIF_BIG_INTERFACE that defines a pair of methods called important_method() and optional_method(), respectively. In the case of important_method(), we want to ensure that any implementing class does provide an implementation for this method. With the DEFAULT FAIL addition, the behavior works as it would under normal circumstances.

In the case of optional_method(), the DEFAULT IGNORE addition makes it possible for implementing classes to skip over the implementation of this method if desired. At runtime, if a polymorphic call is made to this method, you won't see a runtime error.

Instead, the ABAP runtime environment will silently skip over the method as if the call never took place.

The use of the DEFAULT addition is optional and supports complete backwards compatibility with previously defined interfaces. Whether you choose to use it to define your interfaces is up to you, but it can simplify the implementation process in some cases.

6.3.4   Working with Interfaces

At this point, we don't have a lot to show for all the hard work we put into defining and implementing the ZIF_COMPARABLE interface. However, this doesn't mean that interfaces are a waste of time—you just have to learn how to harness their capabilities in our designs. In this section, we'll look at a practical example which uses the ZIF_COMPARABLE interface to create a generic collection class that aggregates and organizes a set of related object instances.

At first glance, you might think that creating a custom collection class is a waste of time, since ABAP already defines a built-in container for object instances with internal tables. You can create collections of object instances using syntax similar to that of Listing 6.15. However, while internal tables like lt_employees provide a convenient mechanism for storing and looping through object references, there are certain things they can't do. For example, this internal table has no idea how to sort the employee records it contains. A custom collection class, on the other hand, can be enhanced to include this kind of expanded functionality. This is where the ZIF_COMPARABLE interface begins to show its value.

```
DATA lt_employees TYPE STANDARD TABLE OF REF TO zcl_employee.
```

Listing 6.15 Defining an Internal Table with an Object-Based Line Type

If we build our collection class to accept objects that implement the ZIF_COMPARABLE interface, then we have everything we need to build out a generic sort function within the collection. However, before we get into the nuts and bolts of this, let's first look at the basic design of our custom collection class (see Listing 6.16). For our first pass at this, we'll create a new class called ZCL_COLLECTION that defines three methods whose names are pretty self-explanatory: add_element() adds an element to the collection, remove_element() removes an element from the collection, and sort() performs an in-place sorting on the elements. As you can see in the method signatures, as well as the mt_elements internal table attribute that stores the elements internally, all element references are defined in terms of the generic OBJECT type. This design choice allows you to store *any* type of object within your collection.

```
CLASS zcl_collection DEFINITION.
  PUBLIC SECTION.
    TYPE-POOLS abap.
```

```
  METHODS:
    add_element IMPORTING io_element TYPE REF TO object,
    remove_element IMPORTING io_element TYPE REF TO object
                   RETURNING VALUE(rv_result) TYPE abap_bool,
    sort.

  PRIVATE SECTION.
    DATA mt_elements TYPE STANDARD TABLE OF REF TO object.
ENDCLASS.
```

Listing 6.16 Defining the Shell of the ZCL_COLLECTION Class

With this basic framework in place, we can now turn our attention to the sort() method's implementation. Listing 6.17 shows an implementation based on the *insertion sort* algorithm. This simple (though somewhat inefficient) algorithm operates similar to the way you might sort a hand of playing cards. Assuming the cards are in random order and lying face down on the table, you pick up cards one by one and arrange them in your hand based on the value of the card. When applied to the sorting of collection elements, the insertion sort algorithm gets the ordering from the compare_to() method of the ZIF_COMPARABLE interface.

```
method SORT.
  DATA: lo_key     TYPE REF TO object,
        lo_element TYPE REF TO object,
        lo_compare TYPE REF TO zif_comparable,
        lo_temp    TYPE REF TO object,
        lv_i       TYPE i,
        lv_j       TYPE i VALUE 2,
        lv_index   TYPE i.

  "Sort the collection elements using the Insertion Sort
  "algorithm:
  LOOP AT me->mt_elements INTO lo_key FROM 2.
    lv_i = lv_j - 1.
    READ TABLE me->mt_elements INDEX lv_i INTO lo_element.
    lo_compare ?= lo_element.

    WHILE lv_i GT 0 AND
          lo_compare->compare_to( lo_key ) EQ
            zif_comparable=>co_greater_than.
      READ TABLE me->mt_elements INDEX lv_i INTO lo_temp.
      lv_index = lv_i + 1.
      MODIFY me->mt_elements FROM lo_temp INDEX lv_index.
```

```
    lv_i = lv_i - 1.
    READ TABLE me->mt_elements INDEX lv_i INTO lo_element.
    lo_compare ?= lo_element.
  ENDWHILE.

  lv_index = lv_i + 1.
  MODIFY me->mt_elements FROM lo_key INDEX lv_index.
  lv_j = lv_j + 1.
  ENDLOOP.
endmethod.
```

Listing 6.17 Implementing a Generic Sort Algorithm in the ZCL_COLLECTION Class

If you look closely at the implementation of the sort() method in Listing 6.17, you can see that we never refer to any particular type of element class within the sorting logic. Instead, we use a widening cast on the element objects to reference them as ZIF_COMPA-RABLE instances. The target of this widening cast is an *interface reference variable* called lo_compare whose static type is defined as ZIF_COMPARABLE. This interface reference variable allows us to address the object in exactly the same way we use object reference variables to address object components: via the familiar instance selector operator (->). This is evidenced by the call to the compare_to() method that drives the sorting logic.

Since the logic in the sort() method is focused on the implementation of the ZIF_COM-PARABLE interface, the only requirement for incorporating element classes into the collection is that the element class implements the ZIF_COMPARABLE interface. How that element class chooses to implement the comparison logic is completely irrelevant to the ZCL_COLLECTION class; it just assumes the element classes know what they're doing.

> **Note**
>
> The ZIF_COMPARABLE implementation validation is enforced in the add_element() method using features that we haven't yet covered in this book. You can find detailed information about the approach we're taking here in the ZCL_COLLECTION class within the book's source code bundle (available at *www.sap-press.com/6093*).

Though you could achieve similar results with regular class-based inheritance, such a design approach would limit the usefulness of the collection because it restricts you from incorporating element classes that already have inheritance relationships in place. For example, since the ZCL_HOURLY_EMPLOYEE class introduced in Chapter 5 already inherits from ZCL_EMPLOYEE, it can't also inherit from an abstract class such as ZCL_COM-PARABLE. However, since you can implement as many interfaces as you want in an ABAP Objects class, it's a trivial matter to incorporate the ZIF_COMPARABLE interface to extend a class to make it *comparable*.

6.3.5   Nesting Interfaces

So far, we've only considered simple, elementary interfaces. However, it's possible to *nest* interfaces inside of a compound or *nested interface*. Interfaces embedded inside of a nested interface are called *component interfaces*. Listing 6.18 shows an example of the syntax used to nest the component interface LIF_COMPONENT inside of the nested interface LIF_NESTED. As you can see, interfaces are themselves nested using the familiar INTERFACES statement.

```
INTERFACE lif_component.
  METHODS: c1,
           c2.
ENDINTERFACE.


INTERFACE lif_nested.
  INTERFACES: lif_component.
  METHODS: n1,
           n2.
ENDINTERFACE.
```

Listing 6.18 A Nested Interface Example

You can achieve the same effect for ABAP Repository-based interfaces in the form-based view of the Class Builder by plugging in the component interface in the **Includes** column of the **Interfaces** tab, as shown in Figure 6.8.

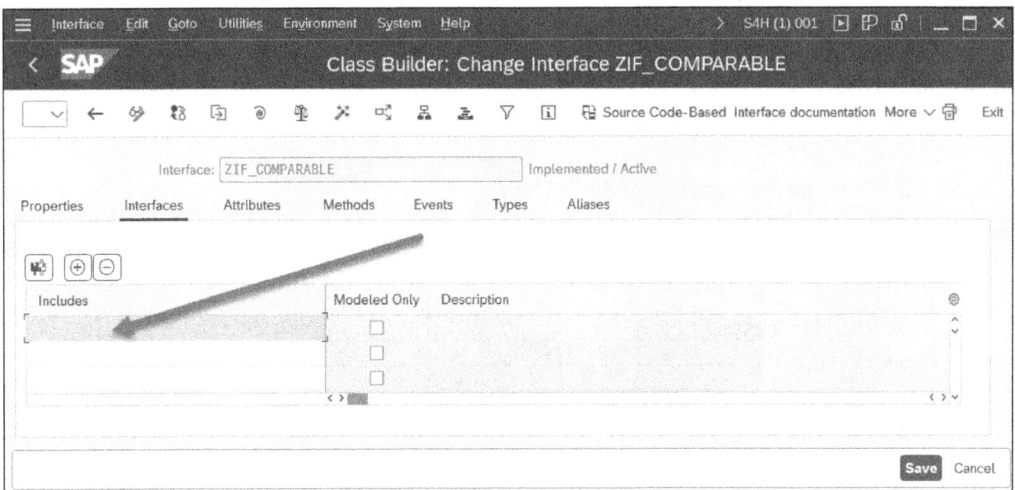

Figure 6.8 Nesting Interfaces in the Form-Based View of the Class Builder

Each of the component interfaces within a nested interface exists at the same level. If a given component interface happens to be nested more than once, there will only be a single set of the components defined within that component interface inside the nested interface.

By default, components of component interfaces are not directly visible within the nested interface. To make these components visible via the public interface, you must define *alias names* for the components. Listing 6.19 shows how the ALIASES statement is used in the LIF_NESTED interface definition to define aliases for a pair of methods inherited from the LIF_COMPONENT component interface. With this promotion, the class LCL_NESTED_IMPL is now able to address and implement the nested methods provided via the LIF_COMPONENT interface, even though it only formally implements the overarching LIF_NESTED interface.

```
INTERFACE lif_component.
  METHODS: c1,
           c2.
ENDINTERFACE.


INTERFACE lif_nested.
  INTERFACES: lif_component.
  ALIASES: c1 FOR lif_component~c1,
           c2 FOR lif_component~c2.
  METHODS: n1,
           n2.
ENDINTERFACE.



CLASS lcl_nested_impl DEFINITION.
  PUBLIC SECTION.
    INTERFACES: lif_nested.
ENDCLASS.



CLASS lcl_nested_impl IMPLEMENTATION.
  METHOD lif_nested~n1.
  ENDMETHOD.

  METHOD lif_nested~n2.
  ENDMETHOD.

  METHOD lif_nested~c1.
  ENDMETHOD.

  METHOD lif_nested~c2.
  ENDMETHOD.
ENDCLASS.
```

Listing 6.19 Working with Alias Names

Much like you observed in Section 6.1.2, you can also perform casts in assignments between interface reference variables. For example, Listing 6.20 shows an example of how you can perform a narrowing cast between interface reference variables defined using the static types LIF_COMPONENT and LIF_NESTED introduced in Listing 6.18 and Listing 6.19. In this case, the narrowing cast is allowed because LIF_COMPONENT is a component interface of LIF_NESTED.

```
DATA: lo_component TYPE REF TO lif_component,
      lo_nested    TYPE REF TO lif_nested.
CREATE OBJECT lo_nested TYPE lcl_nested_impl.
lo_component = lo_nested.
lo_component->c1( ).
```

Listing 6.20 Performing Narrowing Casts Using Interface References

6.3.6   When to Use Interfaces

Since interfaces are expressions of pure design, it can be difficult at times to figure out when and where to employ them in your own designs. Rest assured you're not alone in this; object-oriented enthusiasts have been debating this point for quite some time. On one side of the argument, you have those who advocate widespread use of interfaces. Indeed, in the introductory title *The Java Programming Language, Fourth Edition* (Addison-Wesley Professional, 2005), the authors assert that "every major class in an application should be an implementation of some interfaces that captures the contract of that class." The other side of the debate takes a softer stance, preferring to use interfaces more on an as-needed basis (e.g., to implement multiple inheritance).

While we won't attempt to persuade you one way or another in this debate, we would be remiss if we didn't at least highlight a couple of the more common design scenarios where you might want to consider the use of interfaces. Over time, we think you'll find that these design choices will become second nature, but it helps to first examine these scenarios up close, as we'll do in the following sections.

Interfaces Versus Abstract Classes

Since abstract classes are basically interfaces with a bit of reusable implementation content in tow, it's only natural for developers to gravitate toward the use of abstract classes over interfaces. From an implementation perspective, there's no doubt about it: Abstract classes are easier to work with. Nevertheless, there are times when the extra effort associated with going the interface route pays off.

Since ABAP Objects doesn't support multiple inheritance, the only way you can utilize abstract classes is by directly inheriting from them. This works well when you're clearly dealing with main types such as Employee, SalesOrder, or Delivery, but what about secondary types like the Comparable type we considered in Section 6.3? For certain class types, the *is-a* relationship might apply to multiple types, not just the main type.

To illustrate this concept further, let's look at a fairly typical example of interface usage in ABAP Objects: SAP Business Workflow. In SAP Business Workflow, the processing logic behind workflow tasks can be defined using the instance methods of ABAP Objects classes. In order for the workflow engine to communicate with these objects, there's a requirement that these workflow classes implement the SAP standard IF_WORKFLOW interface, which defines callback methods the workflow engine can use to load objects. Though SAP could have opted to define this functionality in an abstract class (e.g., CL_WORKFLOW), such a design choice would have limited developers from plugging any old ABAP Objects class into workflow scenarios since such classes might already inherit from some other base class. On the other hand, adding an implementation of the IF_WORKFLOW interface on top of an existing class is relatively easy and doesn't disturb any of the existing functionality. In general, we recommend that you model any sort of secondary type using interfaces. You can usually identify these secondary types by considering whether they represent a core concept in the object model or overlapping concepts that are secondary in nature. For example, when we implemented the ZIF_COMPARABLE interface in the ZCL_EMPLOYEE in Section 6.3.3, we didn't change the main concept of employee types; we just said that these employee types are *also* comparable. You can think of this as an *is-also* relationship.

Using Interfaces to Hide Implementation Details

As you learned in Section 6.3.4, clients can use interface reference variables to address object instances polymorphically. While this may not seem all that interesting at first, there are lots of things you can do with this capability.

One common way that developers exploit this feature is by creating *factories* that clients can use to obtain instances of objects that implement a particular interface. If you've ever worked with the iXML library for XML parsing in ABAP, you've seen an example of this design pattern in practice. The code excerpt in Listing 6.21 demonstrates how you can use the CL_IXML factory class to obtain access to the iXML library. As you can see, on the client side, you deal exclusively with the interfaces defined by the iXML library: IF_IXML, IF_IXML_DOCUMENT, and so forth. Such interfaces define the core functionality of the iXML library in abstract terms.

```
DATA lo_ixml TYPE REF TO if_ixml.
DATA lo_document TYPE REF TO if_ixml_document.
lo_ixml = cl_ixml=>create( ).
lo_document = lo_ixml->create_document( ).
...
```

Listing 6.21 The Factory Pattern Applied to the iXML Library

The reason SAP went to the trouble of creating all these interfaces was to provide an abstraction around a delicate XML parsing library that's built on top of kernel modules (i.e., modules written using C/C++ native code in lieu of ABAP). By defining the application programming interface (API) in terms of interfaces, SAP shields customers from the underlying implementation details. As a result, if SAP decided to scrap the kernel module approach and rewrite the library, they could easily do so, and simply plug in the new concrete classes at runtime via the create() factory method of class CL_IXML.

While customers might be less likely to encounter direct requirements like this in practice, think about how the use of interfaces might improve the flexibility of the APIs you decide to develop in-house. As you progress further through this book, you'll encounter several other examples of this to give you a better sense of the benefits of this kind of design pattern.

6.4   UML Tutorial: Advanced Class Diagrams, Part II

In this section, we'll complete our coverage of Unified Modeling Language (UML) class diagrams by introducing the notation for working with interfaces and their components.

6.4.1   Interfaces

The notation for defining interfaces in a UML class diagram is almost identical to the one used to define classes. The only difference is the addition of the << interface >> tag in the top name section of the interface notation (see Figure 6.9).

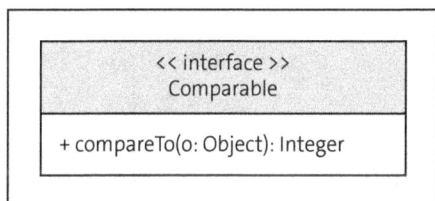

Figure 6.9 Notation for Defining Interfaces

The relationship between a nested interface and its component interfaces is shown using the same generalization notation used to depict inheritance relationships. For example, in Figure 6.10, the Nested interface inherits the components from interface Component.

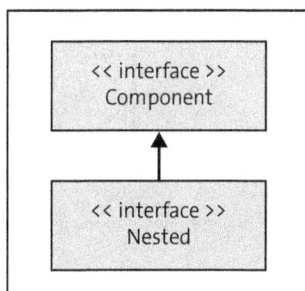

Figure 6.10 Notation for Defining Nested Interfaces

6.4.2 Providing and Required Relationships with Interfaces

Figure 6.11 shows the two kinds of relationships that a class can have with an interface. The dashed line between class Employee and interface Comparable indicates that class Employee *provides* (or implements) the Comparable interface. Notice how the notation for this relationship is similar to the one you've seen for generalization relationships. In this case, the interface Comparable represents one kind of generalization for class Employee. Implicitly, this tells you that you can substitute instances of class Employee in places where the interface Comparable is used. The dashed arrow between class Collection and interface Comparable represents a dependency, indicating that class Collection *requires* the Comparable interface in some way. As you saw in Section 6.3.4, this dependency exists in method sort(), which performs comparisons between collection elements using the compareTo() method defined in interface Comparable.

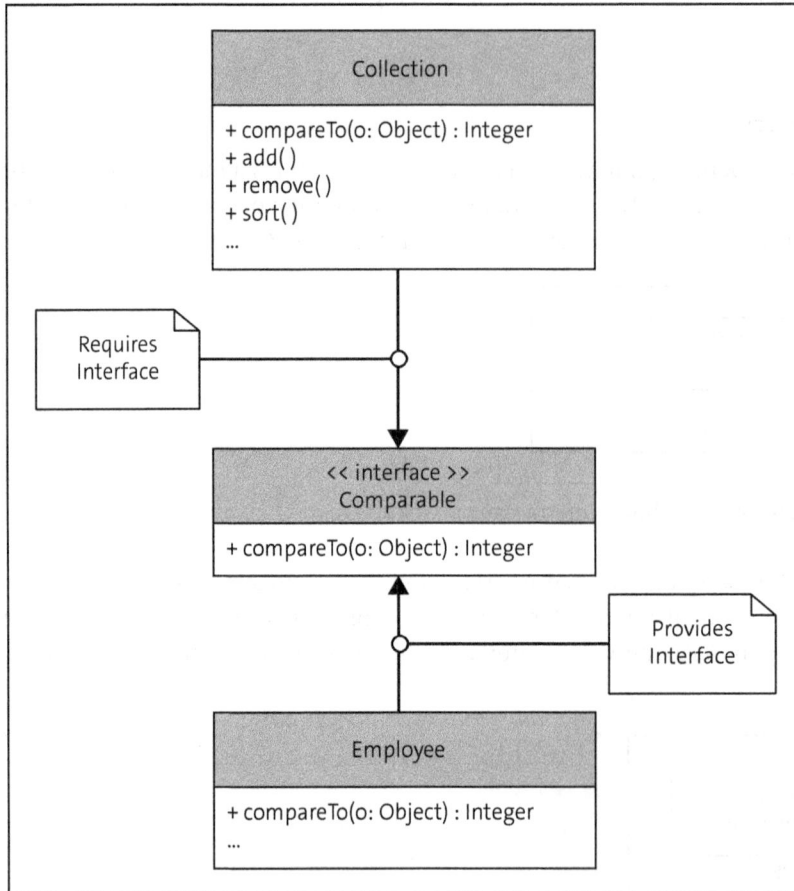

Figure 6.11 Defining Providing and Required Relationships

6.4.3 Static Attributes and Methods

Whether you're defining them at the class level or the interface level, you can identify static components within a class diagram by simply underlining them. Figure 6.12 illustrates this notation for a standard utilities class provided by SAP called CL_ABAP_CHAR_UTILITIES.

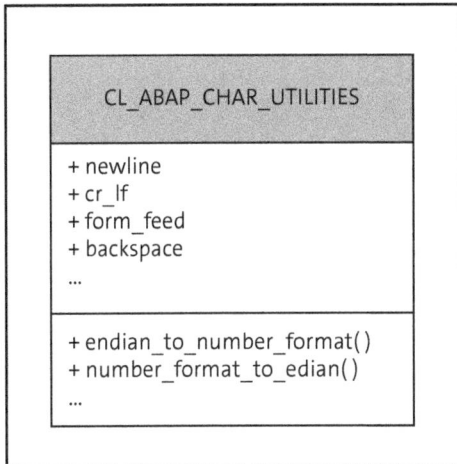

Figure 6.12 Defining Static Attributes and Methods

6.5 Summary

This chapter concludes our basic introduction to OOP. In many ways, the powerful designs that you learned to implement in this chapter represent part of the big payoff for all the hard work that goes into designing families of abstract data types. As you progress further through the book, you'll see many more examples that demonstrate how the three main pillars of OOP (encapsulation, inheritance, and polymorphism) allow you to implement designs that can stand up to any future changes. In the world of business software, such flexibility is vital to keep pace with ever-changing business requirements.

In the next chapter, we'll take these abstractions to the next level and show you how to organize your class libraries into high-level software components using packages.

Chapter 7
Component-Based Design Concepts

A component-based approach to software design breaks a system down into a series of logical components that communicate using well-defined interfaces. When designed properly, these components become reusable software assets that you can mix and match to rapidly adapt to ever-changing business processes. In this chapter, we'll look at ways to implement component-based designs in ABAP.

Now that you've learned the basic principles of object-oriented software development, we'll broaden our focus and look at ways of organizing class libraries and their related resources into reusable *software components*. This process begins with the assignment of classes to modular software units called *packages*. Packages bring structure to the ABAP development process, transforming fine-grained code libraries into more coarse-grained development components.

In this chapter, you'll learn how to create and work with packages. You'll also see how packages fit into the overall SAP component-based software logistics model. Collectively, these concepts will help you keep your software catalog organized as class libraries evolve over time.

7.1 Understanding SAP's Component Model

To understand how to effectively implement component-based software designs in an ABAP development environment, it's helpful to review the component model that SAP uses to manage their own software logistics. As you can see in Figure 7.1, SAP assembles its software products using high-level software units called software components. Software components are comprised of a series of packages that organize the development objects that make up the implementation part of the system (e.g., classes, function groups, and ABAP Dictionary objects).

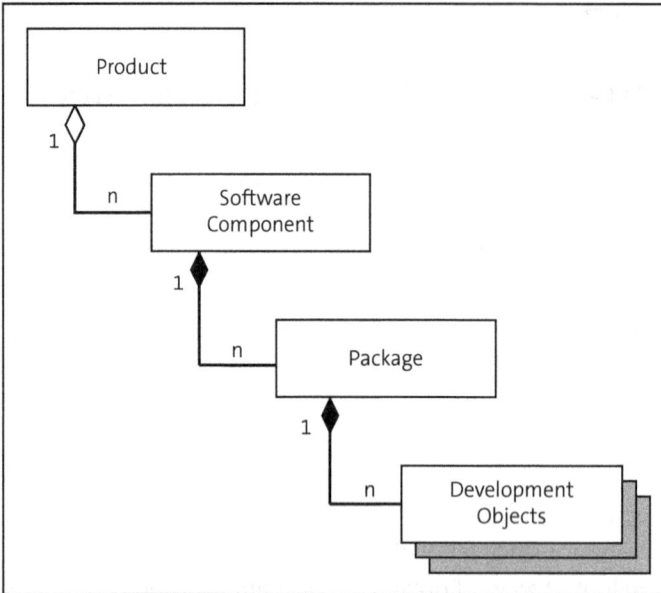

Figure 7.1 SAP's Component Model for ABAP Software Logistics

Each of the component model elements depicted in Figure 7.1 exist in *versions*. To put this into perspective, consider the architecture of one of the more well-known products in the SAP landscape: SAP S/4HANA. At the time of writing, many SAP customers have migrated to this product, with various enhancement packages and support packages installed on top of it. Underneath the hood, an SAP S/4HANA product installation consists of a series of installed software components that also exist in versions: S4CORE, SAP_HR, and so forth. You can see which software component versions are installed on any SAP Business Suite system by selecting **System • Status** in the top-level menu bar and clicking on the **Component Information** button (magnifying glass), as shown in Figure 7.2.

Software Component Versioning Concepts

If you look closely at Figure 7.2, you can see that there are multiple dimensions to a software component's version:

- Release Number
- Support Package (SP) Level
- Support Package Patch (SPP) Level

This versioning scheme is consistent with the major, minor, build, revision scheme you've probably seen in other software products. With such a large customer base, SAP needs to be able to issue periodic support packs that bundle together a series of fixes for a particular software component version. Such changes are meant to follow the break/fix model and shouldn't disturb day-to-day activities within an enterprise. Support pack patches are similar to hot fixes, where SAP issues a patch to correct a reported product defect.

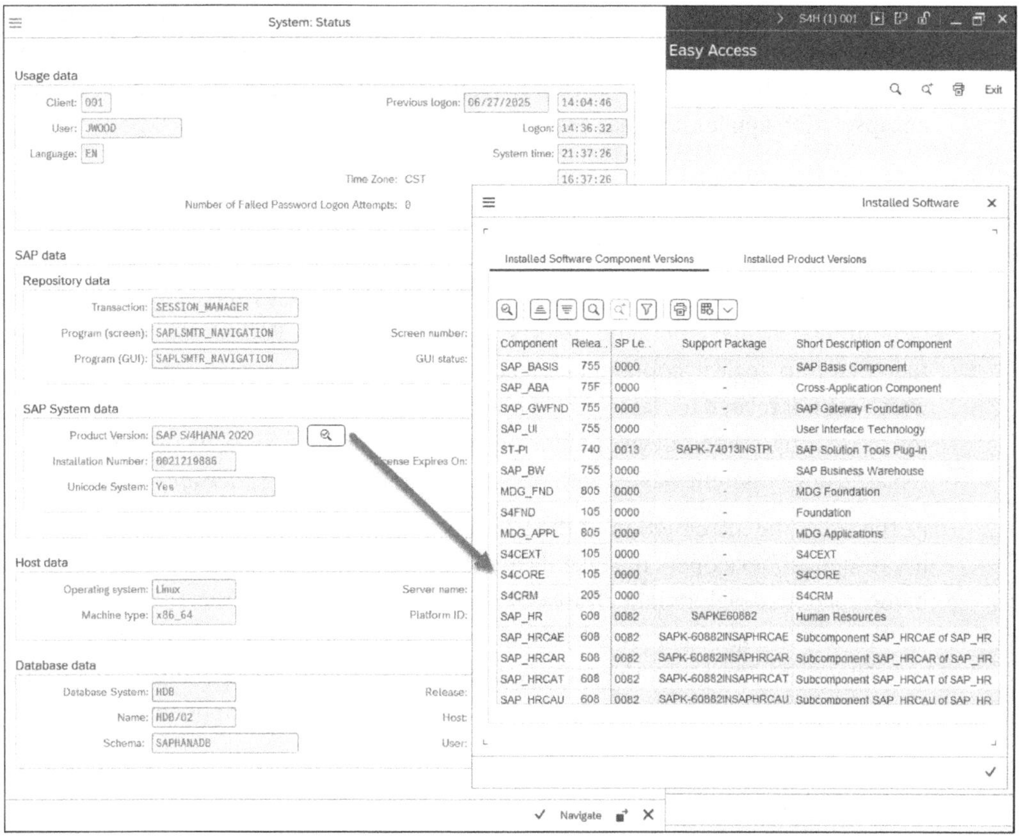

Figure 7.2 Browsing the Installed Software Components in an SAP System

If you dissect a particular software component version, you'll find a series of ABAP development objects organized into packages. Since each of these development objects exist in versions, you can think of a software component version as a snapshot of related development objects bundled together to create an installation unit. Such software components are then logically grouped together into a product version.

From a logistics perspective, adopting a layered approach to system design offers many advantages. First, it helps organize the software into logical pieces that are encapsulated and therefore easier to work with. This is particularly important for large organizations like SAP that employ development teams working on interrelated projects around the globe. Second, it promotes the reuse of common components in other systems. Within SAP Business Suite, you see many examples of this with foundational components such as the SAP_BASIS and SAP_ABA components being reused across SAP Business Suite solutions such as SAP Customer Relationship Management (SAP CRM), SAP Supply Chain Management (SAP SCM), and so on. Finally, it makes the software more extensible because defined dependencies between components make it easier to determine how to integrate new or revised components into the system.

In the sections that follow, you'll see how you can apply these concepts on a smaller scale in custom ABAP development performed at customer sites. Even though your goal may not be to produce a commercial product, the need for component-based encapsulation applies just the same.

7.2   The Package Concept

Prior to release 6.10 of the SAP Web Application Server (today called application server [AS] ABAP), all development objects within the ABAP Repository were grouped together into logical containers called *development classes*. Development classes, which are not related to classes in the object-oriented paradigm, provided a simple way for organizing related development objects by functional area. In release 6.10, SAP replaced development classes with packages and the so-called *package concept*.

At the time, the introduction of the package concept was not met with much fanfare because most developers mistakenly assumed that the term "package" was just a new name given to development classes. However, as you'll learn in this section, the package concept brings much more to the table than just a folder-like organizational structure. When used properly, the package concept allows developers to organize development objects into coarse-grained and reusable development components.

7.2.1   Why Do You Need Packages?

Throughout this book, you've seen how the object-oriented development approach lends itself toward the creation of lots of small, individualized classes, interfaces, and related artifacts. This is a far cry from the old days of ABAP, where the main units of development were large, monolithic ABAP report programs, classic Web Dynpro-based module pool programs, and the occasional subroutine pool or function group.

While such decomposition is undoubtedly a good thing, there is a natural side effect: Modern ABAP development projects produce lots of custom development objects that you must account for. It's no longer realistic to think that you can organize objects using specialized naming conventions or functional area-specific development classes or packages (e.g., ZFI for financial accounting-related artifacts and ZHR for human resources).

This phenomenon is brought into sharp relief when you consider the number of artifacts generated in a Web Dynpro ABAP application, for example. In even the simplest of application scenarios, you might produce upwards of 25 custom objects when you consider the ABAP Dictionary objects and helper classes as well as the Web Dynpro ABAP-related artifacts themselves. If you multiply that number by the number of custom applications you might build in a particular functional area (e.g., logistics), your package can fill up in a hurry. This makes it difficult for developers who are unfamiliar with the history of the development objects contained within these packages to figure out the purpose of each object after the fact. Attempts to reorganize large packages into smaller subpackages eventually break down, too, because they fail to address the main problem: The development objects within a package require boundaries.

In many ways, these logistical problems are not unlike the ones you observed in Chapter 3 when you learned about encapsulation concepts. For example, if you compare the organization of development objects into development classes and packages with the functional decomposition process used to break large, monolithic programs into subroutines and procedures, you can see that neither process establishes any sort of boundaries. Whenever you build applications and libraries, you want to group related development objects together, but you also want to be able to restrict access to objects that are part of the underlying implementation and subject to change. Packages allow you to achieve both objectives. You'll see how this works in the sections to come.

7.2.2   Introducing Packages

Within the package concept, there are three types of packages you can use to organize development objects: *structure packages*, *main packages*, and *development packages*. These package types and their interrelationships are depicted in Figure 7.3. We'll describe each of these package types in detail in the following subsections.

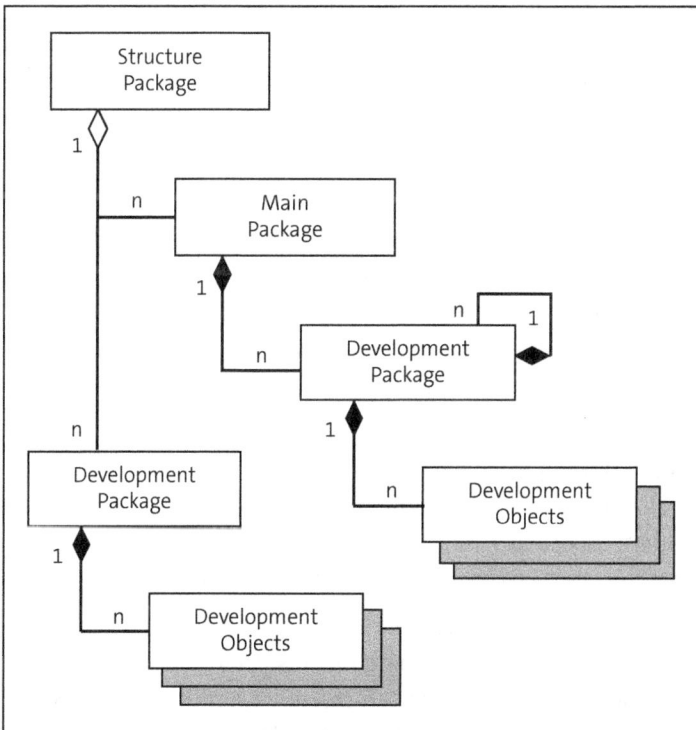

Figure 7.3 Structure of the ABAP Package Hierarchy

Structure Packages

Structure packages, as the name suggests, are used to provide *structure* around lower-level packages within the hierarchy. Within the package concept, structure packages are closely aligned with the encapsulating software component (refer to Figure 7.1). For example, the structure package ABA encapsulates all the packages and development objects contained within the SAP_ABA software component. Though this is the typical use case, structure packages can be used any time you want to organize a large-scale development effort. Customers who aren't interested in reselling software components can use structure packages to organize development objects by functional area, as the development classes of old.

Regardless of how they're used, it's important to bear in mind that structure packages are not *extensible* in the sense that you can't embed development objects directly beneath them. Instead, you can only embed other packages within a given structure package.

Main Packages

Underneath structure packages, you can further organize your development into high-level modules called *main packages*. Main packages are typically used to group development objects by *function*. Development objects embedded inside a main package are logically related in some way. Often, a main package is used to group together modules related to the development of complex applications. However, main packages, just like structure packages, cannot have development objects embedded directly beneath them (refer to Figure 7.3).

Development Packages

At the bottom of the package hierarchy, you have *development packages*. It's at this level that you can begin embedding ABAP development objects (e.g., classes, function groups, and ABAP Dictionary objects). Such assignments occur within the ABAP Workbench tools whenever new development objects are created. This is shown in Figure 7.4, where you can see that a package assignment is a required attribute in the **Create Object Directory Entry** screen.

Within the package concept, development packages are used to organize development objects that are closely related. For example, if you use ABAP Objects classes to model a business object (e.g., a material) then it makes sense to bundle these classes, their related ABAP Dictionary objects, and so on within the same development package.

Ultimately, the goal in defining development packages is to maintain a reasonable level of cohesion with the underlying development objects. There's no restriction on the number of development packages that you can create within the system, so you shouldn't be afraid to create as many as you need to keep things organized.

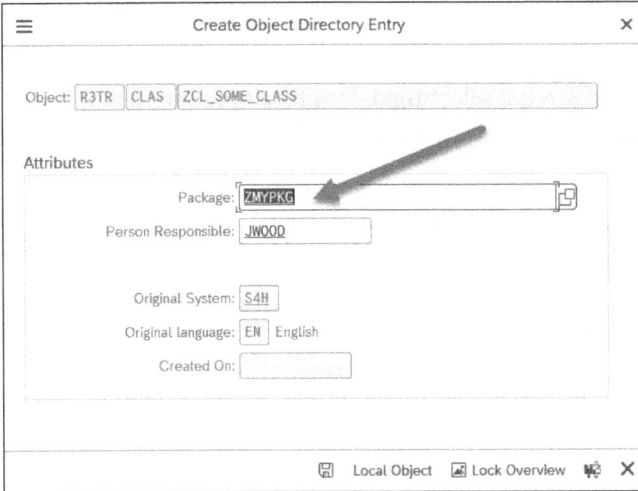

Figure 7.4 Assigning ABAP Repository Objects to Packages

These package assignments may change as the software architecture evolves, and that's okay. At the end of the day, you want to make it easy for developers to find development objects without having to consult naming convention guides. If you take the right approach to your development packages, the process of locating development objects will be intuitive.

7.2.3 Creating Packages Using the Package Builder

Packages can be maintained in one of two places: directly within the ABAP Workbench (Transaction SE80) or in the standalone Package Builder (Transaction SE21). Figure 7.5 shows the initial screen of the Package Builder with the familiar **Create**, **Change**, and **Display** functions. Once you get past the initial screen of the Package Builder, the look and feel is largely the same in both transactions. As such, most developers will prefer to work within the ABAP Workbench directly due to its handy context-sensitive features.

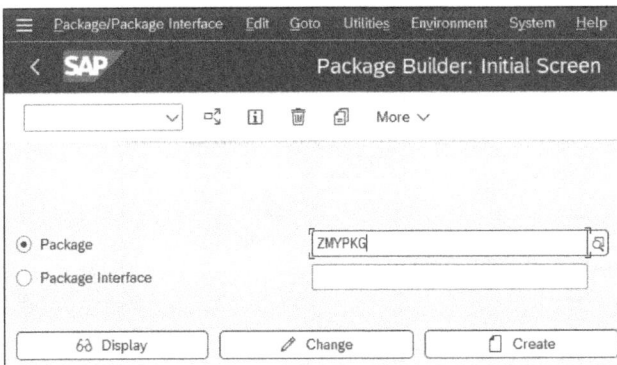

Figure 7.5 Package Builder Transaction: Initial Screen

Within the ABAP Workbench, you can create a new package by performing the following steps:

1. On the left-hand side of the screen, select the **Repository Browser** view if it's not selected already. From here, select the **Package** option in the object list as shown in Figure 7.6.

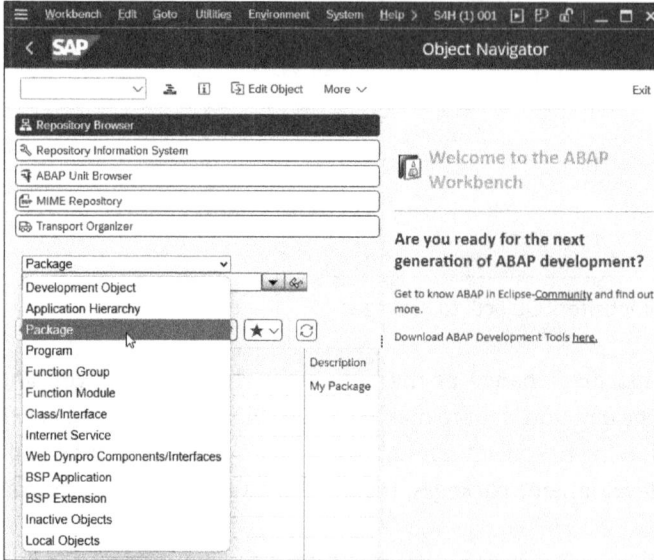

Figure 7.6 Creating a New Package (Part 1)

2. Next, in the input field directly beneath the object list box, enter the name of your new package (e.g., ZPKGDEMO) and press the Enter key. This will open the **Create Pack.** dialog box shown in Figure 7.7. Click on the **Yes** button to proceed with the creation of the new package.

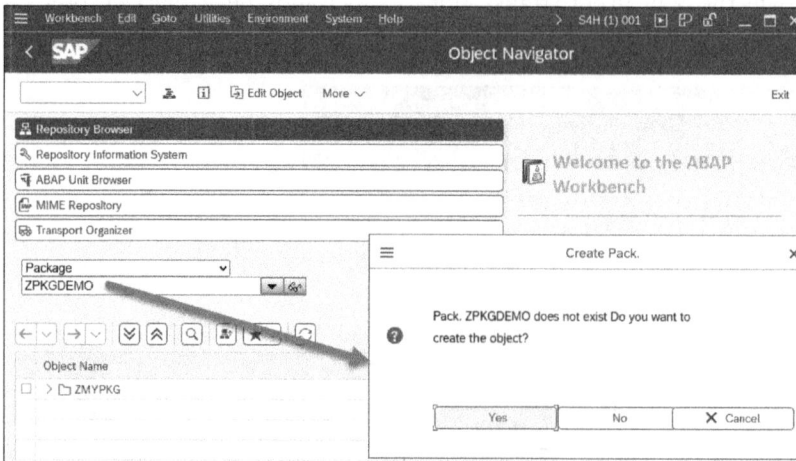

Figure 7.7 Creating a New Package (Part 2)

3. At the **Create Package** screen shown in Figure 7.8, specify the basic attributes for your new package. We'll explore the purpose of these attributes later in this section. Once the attributes are set, click on the green checkmark to continue.

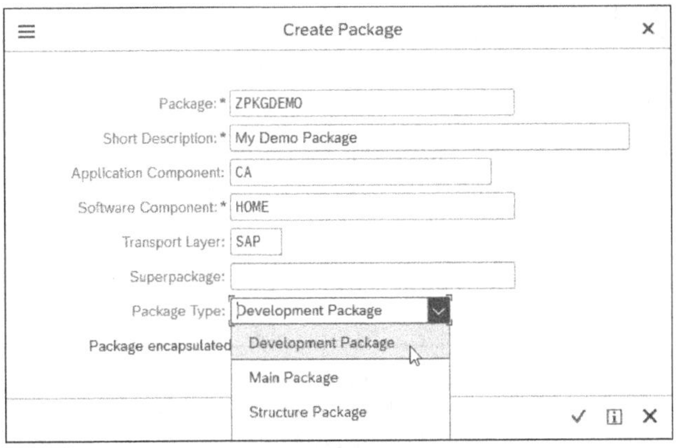

Figure 7.8 Creating a New Package (Part 3)

4. Finally, after assigning the package to a transport request, you'll end up with an empty package in the object list as shown in Figure 7.9. To edit this package, simply double-click on it, just like any other object in the ABAP Workbench.

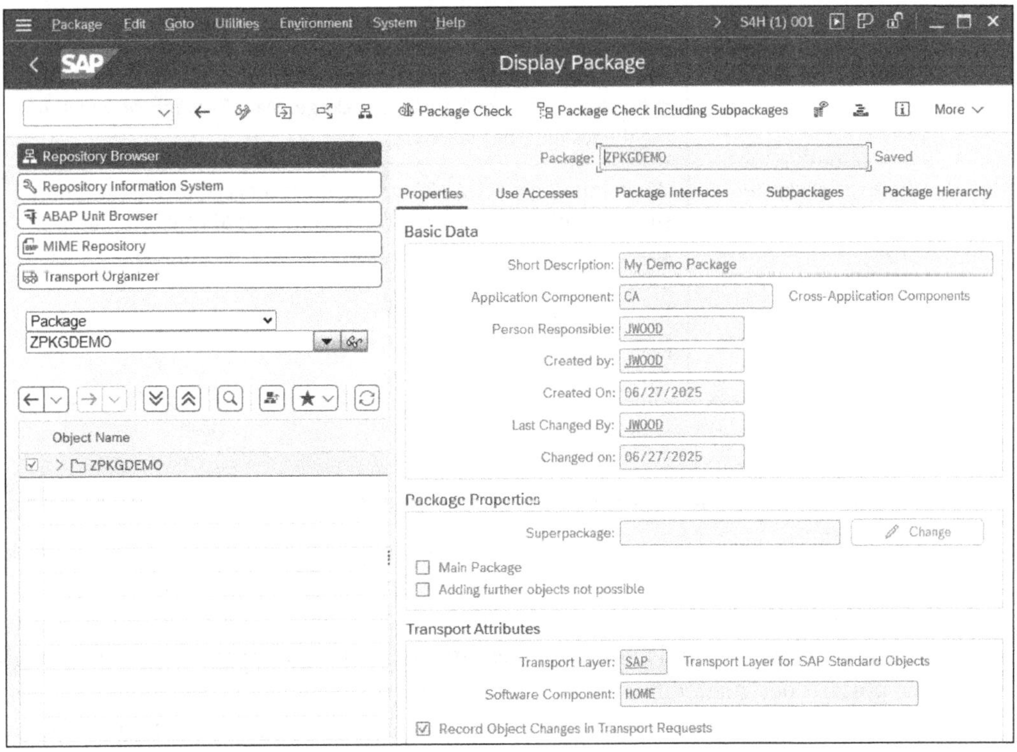

Figure 7.9 Creating a New Package (Part 4)

Now that you know how packages are created within the ABAP Workbench, let's examine some of the various attributes that make up a package definition in Table 7.1. These attributes correspond with the ones entered on the input screen shown in Figure 7.8.

Attribute Name	Description
Short Description	This attribute provides a short text description for the package and its intended use.
Application Component	This attribute aligns a package within an application component in the SAP application hierarchy. We'll explore the benefits of this assignment later in this section.
Software Component	This attribute assigns the package to a software component (refer to Figure 7.1 to visualize this relationship).
Transport Layer	This attribute links a package (and by extension, its embedded development objects) with a transport layer definition in SAP's Change and Transport System (CTS). If you're not familiar with the CTS, suffice it to say that a transport layer definition defines the transport path which guides changes through the landscape (e.g., from development to quality assurance to production).
Superpackage	This attribute embeds a given package underneath an existing one. We'll explore this concept further in Section 7.2.4. You'll also find that the ABAP Workbench will implicitly fill this attribute out for you when you create subpackages using the contextual flyout menus.
Package Type	As the name suggests, this attribute defines the package type. You can choose between **Structure Package**, **Main Package**, or **Development Package** (see Figure 7.8).
Package encapsulated	When this flag is set, ABAP development tools enforces strict encapsulation within the package. This means that ABAP clients cannot address development objects within the package unless the objects are exposed via the package's interface. In object-oriented terms, this setting introduces the concept of visibility sections within a package definition. We'll explore this idea further in Section 7.2.5 and Section 7.2.6, respectively.

Table 7.1 Attributes of Packages

Aligning Packages with the SAP Application Hierarchy

If you've worked with SAP software for a long time, then you've probably encountered the SAP Support Portal (available online at *https://support.sap.com*). Among other things, this portal provides customers with links to download SAP software products, search for SAP Notes and knowledge base articles, and report product defects. Though the portal is designed to make it easy for customers to search for relevant items, it certainly helps to understand how the SAP software catalog is organized. Otherwise, it can feel like you're searching for a needle in a haystack.

Internally, SAP organizes its software catalog into logical *application components*, which are in turn organized into a hierarchy called the *SAP application hierarchy*. Using this hierarchy, you can quickly narrow down the scope of your search to components within a particular application area. For example, when searching for SAP Notes related to purchase orders, you could plug in the MM-PUR-PO application area as shown in Figure 7.10 and Figure 7.11. Here, you're narrowing down the search to the materials management application area first, then the purchasing application area, and then specifically purchase orders. Of course, if you don't know the target area(s) offhand, you can also use wildcards (i.e., *, the asterisk character) to broaden the scope of your selection.

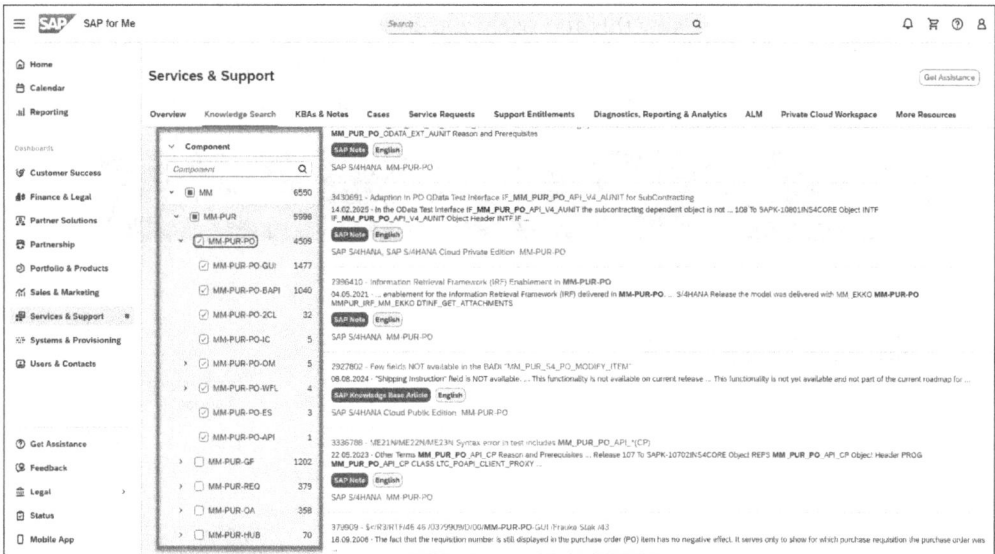

Figure 7.10 Searching for SAP Notes by Application Area (Part 1)

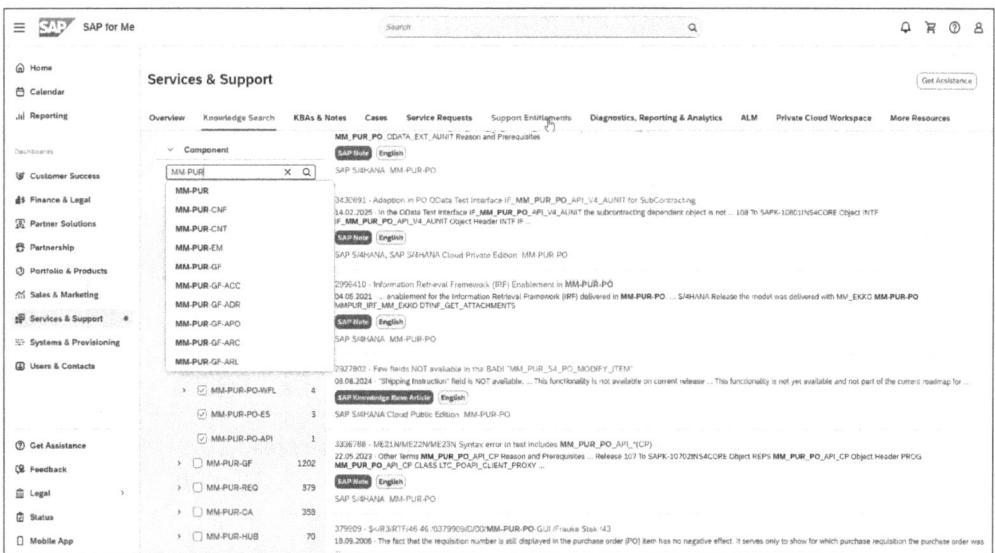

Figure 7.11 Searching for SAP Notes by Application Area (Part 2)

What does all this have to do with packages? Well, since package definitions contain an application component attribute, you can technically align them with the SAP application hierarchy as well. This allows users to search for development objects by application area in a couple of different ways:

1. Within the **Repository Browser** perspective of the ABAP Workbench, you can search for specific types of development objects (e.g., classes) within a particular application component (or set of application components, if you use the * wildcard operator on the end of the search expression). This is demonstrated in Figure 7.12. Here, you're searching for custom employee classes created within the **CA-HR** application area (or subareas). Using this approach, you don't have to guess the naming convention of the classes or which package they might reside in. Instead, you just look for classes in their logical application area and whittle down the list from there.

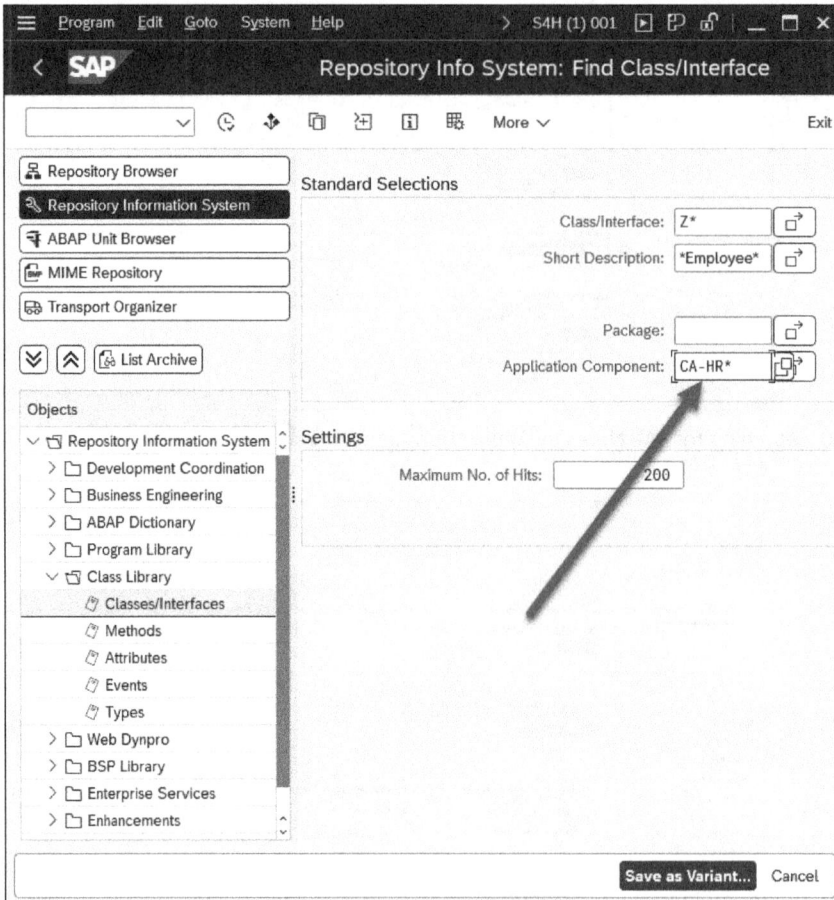

Figure 7.12 Searching for Development Objects by Application Area

2. If a more hierarchical search method is preferred, you can search for relevant development packages using the application hierarchy transaction (Transaction SE81). Here, the SAP application hierarchy is arranged in tree-like form, allowing you to drill into generic application areas and discover specific subareas. Figure 7.13 demonstrates how you can use this transaction to pinpoint custom packages within the HR area. From here, you can double-click on the package names (e.g. ZEMPLOYEE_MODEL) and search for the target development objects.

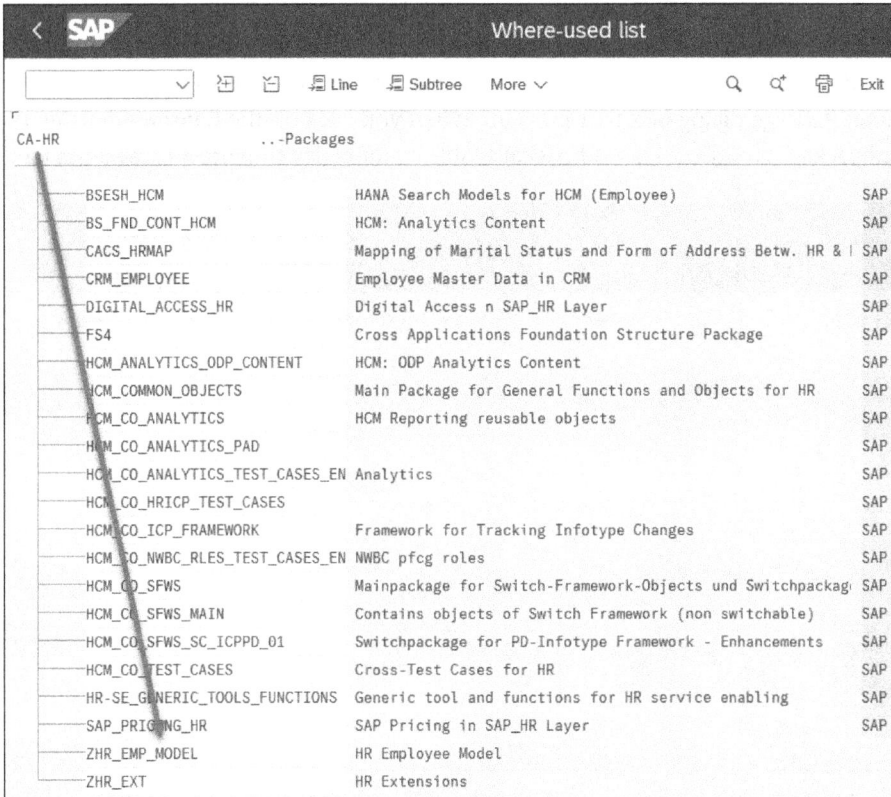

CA-HR	...-Packages	
BSESH_HCM	HANA Search Models for HCM (Employee)	SAP
BS_FND_CONT_HCM	HCM: Analytics Content	SAP
CACS_HRMAP	Mapping of Marital Status and Form of Address Betw. HR &	SAP
CRM_EMPLOYEE	Employee Master Data in CRM	SAP
DIGITAL_ACCESS_HR	Digital Access n SAP_HR Layer	SAP
FS4	Cross Applications Foundation Structure Package	SAP
HCM_ANALYTICS_ODP_CONTENT	HCM: ODP Analytics Content	SAP
HCM_COMMON_OBJECTS	Main Package for General Functions and Objects for HR	SAP
HCM_CO_ANALYTICS	HCM Reporting reusable objects	SAP
HCM_CO_ANALYTICS_PAD		SAP
HCM_CO_ANALYTICS_TEST_CASES_EN	Analytics	SAP
HCM_CO_HRICP_TEST_CASES		SAP
HCM_CO_ICP_FRAMEWORK	Framework for Tracking Infotype Changes	SAP
HCM_CO_NWBC_RLES_TEST_CASES_EN	NWBC pfcg roles	SAP
HCM_CO_SFWS	Mainpackage for Switch-Framework-Objects und Switchpackag	SAP
HCM_CO_SFWS_MAIN	Contains objects of Switch Framework (non switchable)	SAP
HCM_CO_SFWS_SC_ICPPD_01	Switchpackage for PD-Infotype Framework - Enhancements	SAP
HCM_CO_TEST_CASES	Cross-Test Cases for HR	SAP
HR-SE_GENERIC_TOOLS_FUNCTIONS	Generic tool and functions for HR service enabling	SAP
SAP_PRICING_HR	SAP Pricing in SAP_HR Layer	SAP
ZHR_EMP_MODEL	HR Employee Model	
ZHR_EXT	HR Extensions	

Figure 7.13 Searching Within the Application Hierarchy

This search capability is quite powerful and far superior to relying solely on naming conventions to identify objects. This is particularly the case for developers coming in off the street who need to orient themselves within a new customer landscape. While a developer might not be familiar with the naming conventions or object history, they should at least be able to navigate within the application hierarchy and narrow down the scope of their search to objects within a particular application area.

7.2.4   Embedding Packages

To build out package hierarchies like the one shown in Figure 7.3, you must have a way to *embed* a package within another package. This can be achieved in several different ways:

- If you know the name of the superpackage you want to embed a new package in up front, you can specify its name in the **Superpackage** attribute shown in Figure 7.8 (see Section 7.2.3 for a refresher on the package creation process).

- Otherwise, you can embed a package after it's initially created by opening the superpackage in the Package Builder and navigating to the **Subpackages** tab shown in Figure 7.14. From here, you can embed the subpackage by clicking on the **Add existing package** button and filling in the subpackage name in the **Package** field of the **Enter Package** dialog box that pops up. When you click on the **Continue** button, the subpackage is added to the subpackage table contained within the **Subpackages** tab.

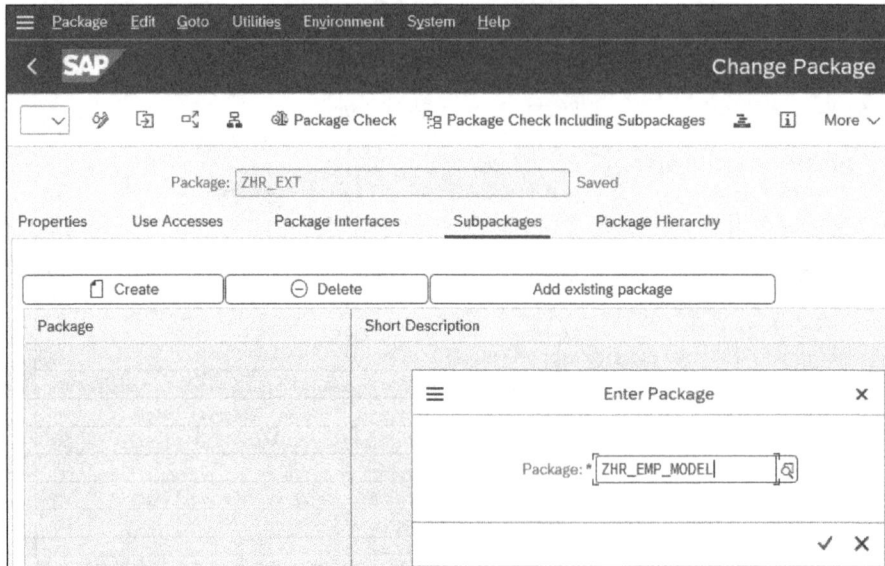

Figure 7.14 Embedding a Package in the Package Builder

- Another option is to create the subpackage directly within the **Subpackages** tab of the superpackage by clicking on the **Create** button (see Figure 7.14). This will bring up the **Create Package** dialog box we reviewed in Section 7.2.3 with the superpackage already filled in.

- Finally, you can create subpackages directly within the ABAP Workbench by selecting the superpackage in the **Repository Browser** view and then right-clicking on the package name and choosing **Create · Packages · Package** from the context menu shown in Figure 7.15. This also brings up the **Create Package** dialog box we reviewed in Section 7.2.3 with the superpackage already filled in.

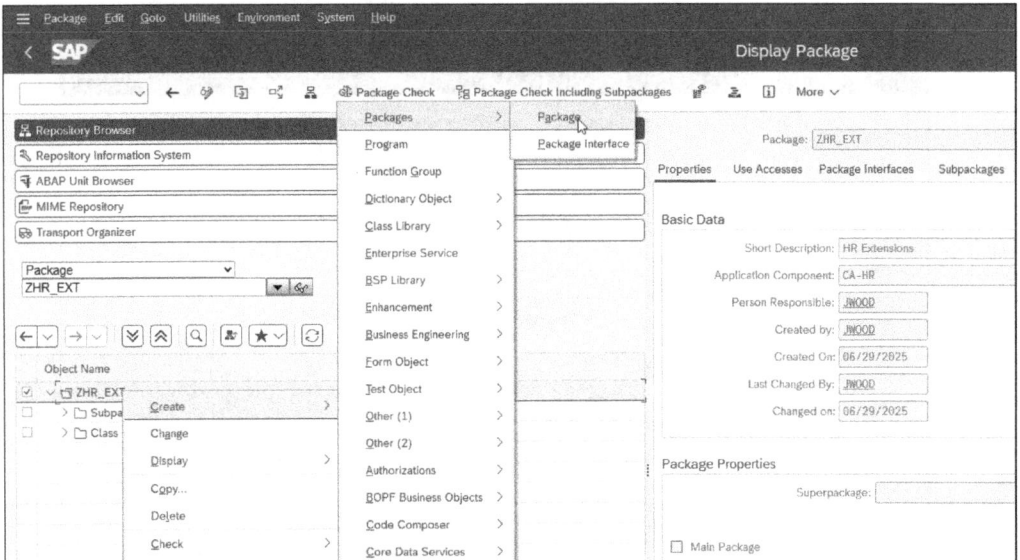

Figure 7.15 Creating a Subpackage in the ABAP Workbench

7.2.5   Defining Package Interfaces

Typically, you want to build your packages and components like black boxes and hide as much of the internal development object details as you can. This approach is partly meant to keep developers from getting their hands on internal objects that are subject to change, but it also highlights the objects that you *do* want to share with clients. By adding these selected development objects to a package's interface, you can make it easier for external clients to understand how a library works without having to comb through each of the individual development objects.

To understand how this works, it's helpful to first see how package interfaces are created within the Package Builder. Here, the steps are relatively straightforward:

1. Within the Package Builder transaction, navigate to the **Package Interfaces** tab and click on the **Create** button. This will open the **Create Package Interface** dialog box shown in Figure 7.16.

2. Within the **Package Interface** dialog box, provide a name for the **Package Interface** and a **Short Description**. Since the package interface is defined as a separate object in the ABAP Repository, it's important that you give it the appropriate namespace prefix, as well as provide a name that conveys meaning. Click on the **Continue** button to finish creating the package interface.

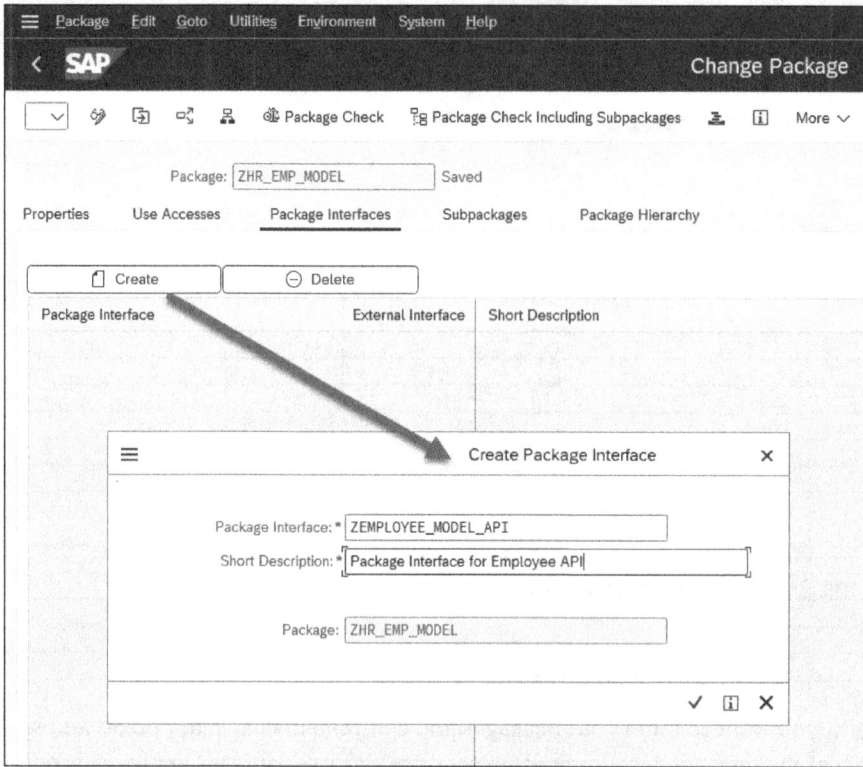

Figure 7.16 Creating a Package Interface

After the package interface is initially created, double-click on it to open the **Change Package Interface** view of the Package Builder shown in Figure 7.17. Here, you'll add and remove elements to the package interface on the **Visible Elements** tab. You can use the toolbar buttons to maintain content (e.g., the **Add Element** and **Remove Element** buttons) or use the drag-and-drop capabilities of the ABAP Workbench to drag over relevant objects from the object list, as demonstrated in Figure 7.17. Both approaches get you to the same place, though the drag-and-drop approach is generally faster and easier to work with.

In the case of nested package hierarchies, you can *promote* development objects from subpackages into the package interface of superpackages to roll related subobjects into higher-level package interfaces. The embedded objects get added to the **From Subpackages** folder shown in Figure 7.18. As is the case with regular development objects, you can use the toolbar buttons (e.g., the **Add Package Interface** button) or the drag-and-drop capabilities of the ABAP Workbench to add these subobjects to the package interface (see Figure 7.18).

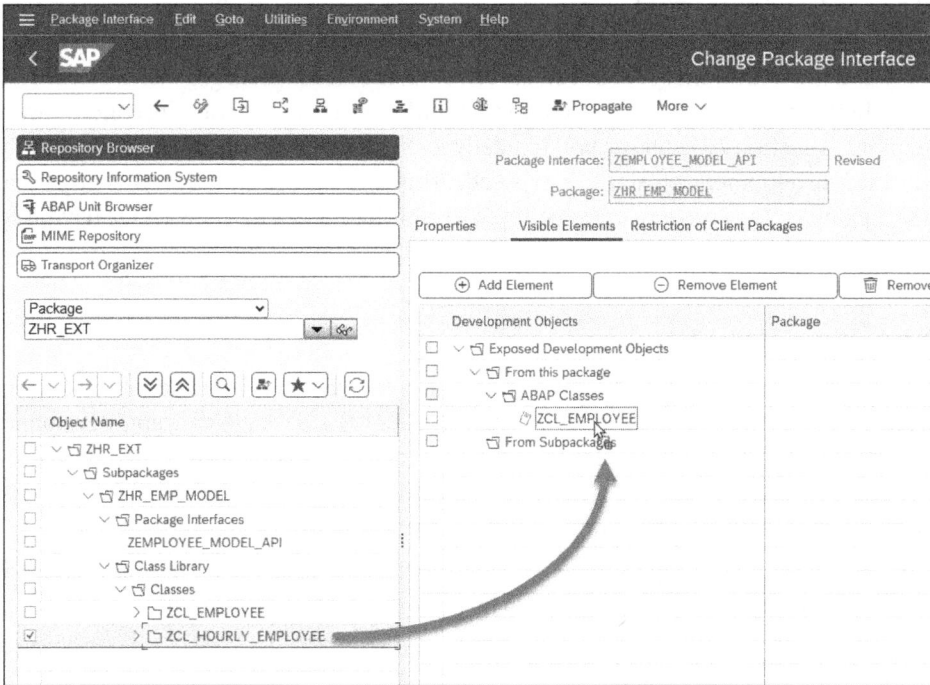

Figure 7.17 Adding Development Objects to a Package Interface

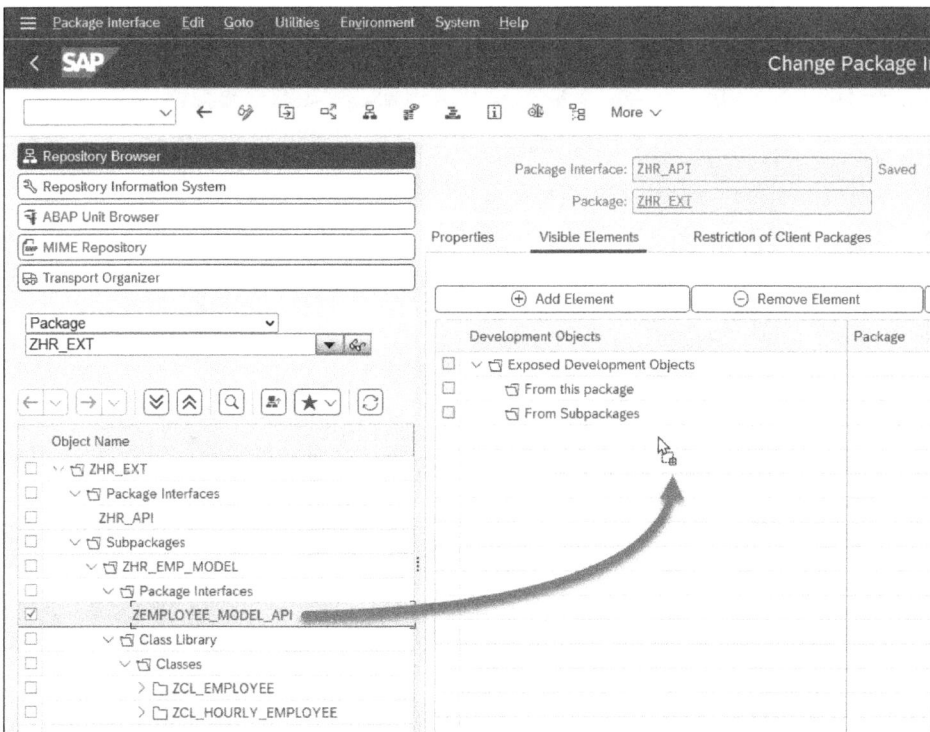

Figure 7.18 Exposing Objects of a Subpackage in the Package Interface of Its Superpackage

When adding development objects to the package interface, you'll once again want to apply the *least privilege* principle described in Chapter 5. This means that you should only add the minimum set of development objects outside packages will need to carry out their work. This approach, combined with the client restriction concept we'll explore in Section 7.2.6, allows you to neatly lock down a package and ensure that only selected development objects are exposed. From a design perspective, this gives you the flexibility to change or remove internal development objects without having to worry about what kind of problems it might cause for other programs.

7.2.6   Creating Use Accesses

Frequently, development objects in one package depend on development objects defined in another package. Prior to the release of the package concept, developers could create such dependencies at whim without any restrictions. This was highly problematic from a logistics perspective since it was next to impossible to prevent developers from using development objects they shouldn't.

With the package concept, you avoid this problem by creating explicit *use accesses* between packages. You can achieve this by performing the following steps:

1. In the Package Builder, open the package that contains the development object(s) that you intend to use with objects defined in another package.

2. To create the use access, navigate to the **Use Accesses** tab shown in Figure 7.19. Click on the **Create** button to create a new use access.

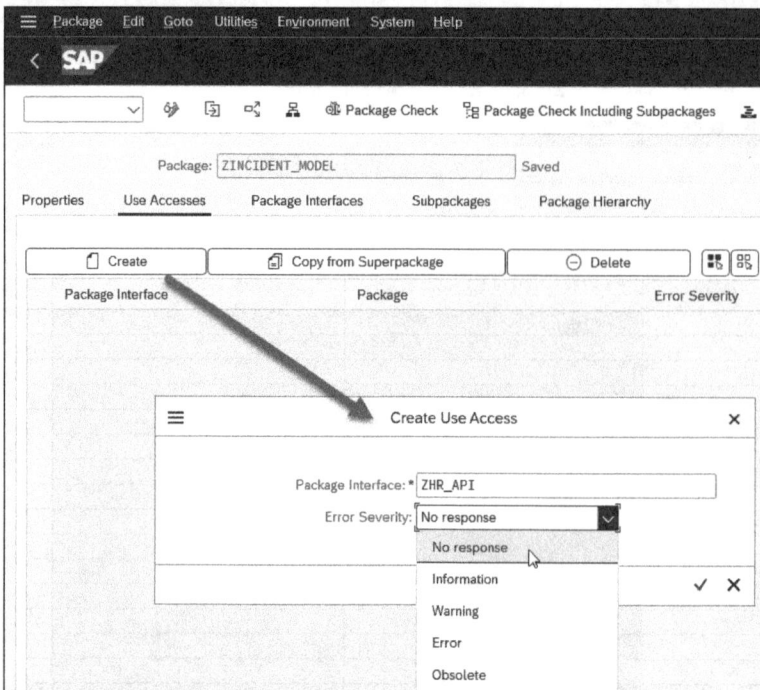

Figure 7.19 Creating Use Accesses in Packages

3. In the **Create Use Access** dialog box, specify the package interface that exposes the object(s) you want to leverage and choose an appropriate **Error Severity**. Normally, you'll select the default **No response** value here.

4. Finally, once the properties are set, click on the **Continue** button to create the use access.

Once a use access is in place, the system now has information that tells it which packages plan on leveraging publicly exposed objects in other packages. In the next section, you'll see how you can use this information to enforce best practices on ABAP development projects.

7.2.7   Performing Package Checks

By default, the package interfaces and use accesses you create serve little purpose other than to formally document a package's public interface and the way you intend for it to be used. While this is helpful up to a point, the real value of going to all this trouble is to be able to enforce best practices in the development process. Just as visibility sections lock down access to the components of ABAP Objects classes, package interfaces should restrict access to internal development objects contained within packages.

To enable this kind of functionality, there's an important setting that you need to remember to select when creating packages: the **Package encapsulated** checkbox, shown in Figure 7.20. When you select this checkbox, the contents of the host/server package are protected against unwanted use.

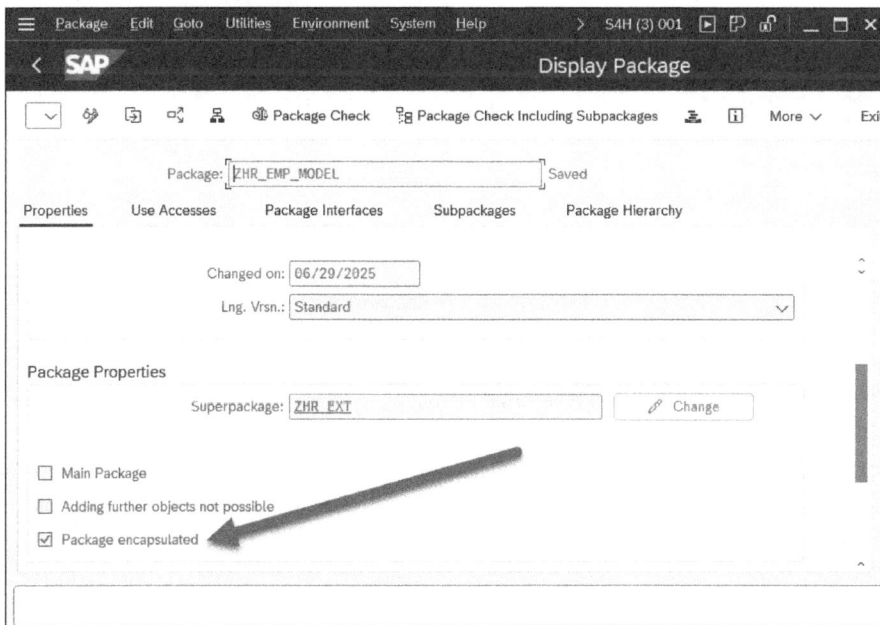

Figure 7.20 Defining Encapsulation Settings on a Package

To demonstrate how this works, imagine that you've created a package hierarchy to contain development objects related to SAP Environment, Health, and Safety Management (SAP EHS Management). Among other things, your extensions to this application will include incident-related objects that have touch points to HR employee objects (employees that might be involved in an incident record, for example). Rather than reinvent the wheel, you should leverage the ZCL_EMPLOYEE entity class we introduced in Chapter 5 in a utilities class called ZCL_PERSON_PROXY that's used to search for involved people.

With the package check turned on, though, you can't just arbitrarily access ZCL_EMPLOYEE from within your SAP EHS Management package hierarchy. That's because access to your employee package hierarchy is locked down to just include the elements you've purposefully exposed via the package interface. If you perform a package check on your ZCL_PERSON_PROXY class (by selecting **Class** • **Check** • **Package Check** from the Class Builder menu), you end up seeing an error report like the one shown in Figure 7.21. This error tells you that you must create a use access for the ZCL_EMPLOYEE class before you can use it in your SAP EHS Management package.

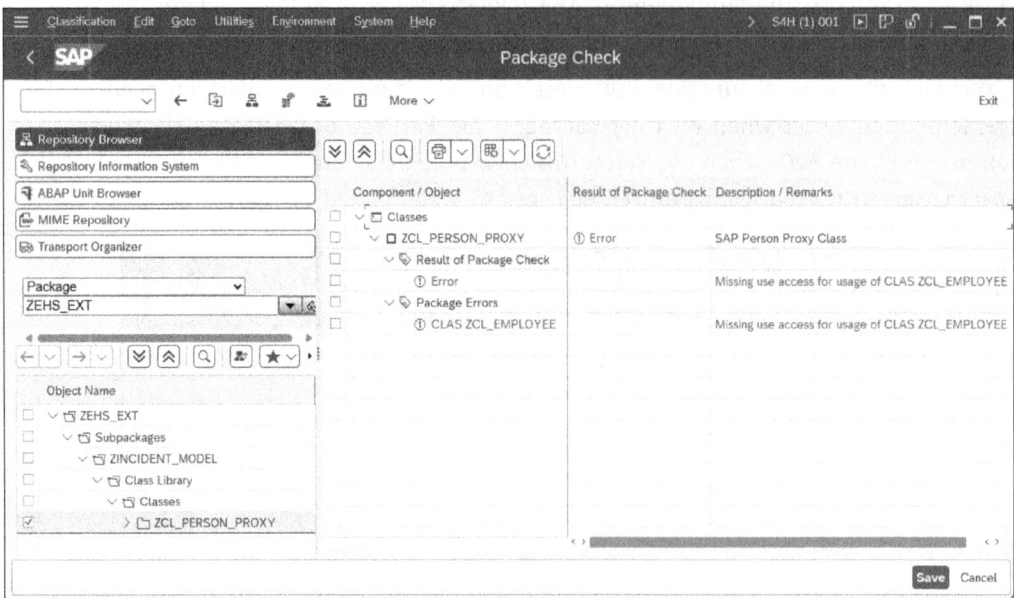

Figure 7.21 Viewing the Results of a Package Check

To correct the package check error, you must go back to the package hierarchy and define the relevant use accesses (see Figure 7.22). Once these use accesses are in place, the package check errors are resolved.

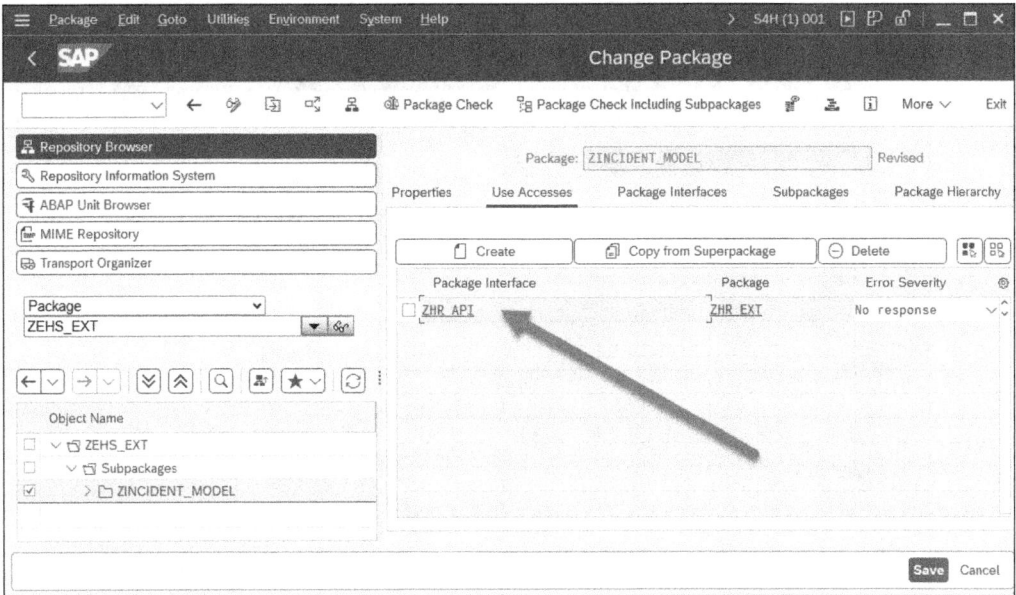

Figure 7.22 Adding Use Accesses to the HR API From the SAP EHS Management Package Hierarchy

In addition to manual package checks performed within the ABAP Workbench, package checks can also be triggered implicitly via the Extended Program Check tool (Transaction SLIN) and also via the CTS system during the transport release process. These checks are put into place to ensure that developers comply with best practices and declare their package dependencies up front. By doing so, development organizations can maintain clear visibility of dependencies and more quickly determine the impacts of migrating code libraries to other SAP Business Suite systems and assessing upgrade issues.

7.2.8   Restriction of Client Packages

The use accesses we introduced in Section 7.2.6 allow client packages to declare which provider packages they intend on using internally. Sometimes, you might want to go in the other direction and formally declare within the provider package which client packages you want to provide access to. While such tight coupling is somewhat rare, there will be times when it makes sense to lock access down to a handful of related packages.

Since use accesses are defined in terms of package interfaces, you must restrict access to client packages at this level. The steps are as follows:

1. Open the target package interface and navigate to the **Properties** tab. In the **General Properties** section, restrict access from client packages by checking the **Enable Restriction of Client Packages** checkbox as shown in Figure 7.23.

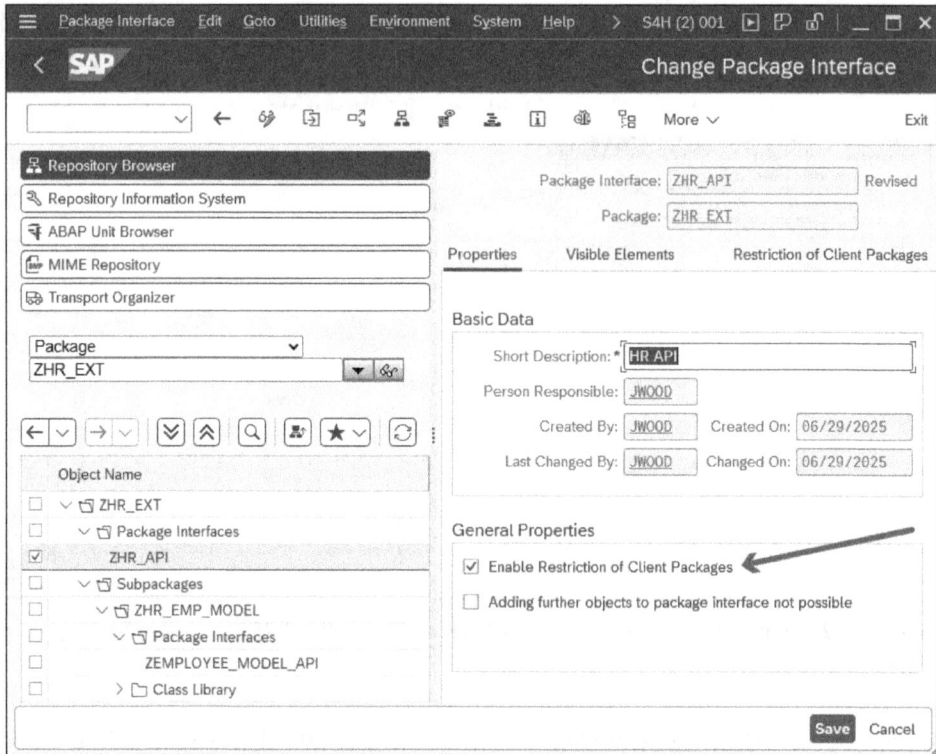

Figure 7.23 Restricting Access to Client Packages (Part 1)

2. As soon as you click on that checkbox, no client packages will be able to access the selected package—even if they have a defined use access. To enable access from client packages, navigate to the **Restriction of Client Packages** tab and plug in the packages you want to provide explicit access to (see Figure 7.24). Here, in addition to specifying the package itself, you also have a couple of checkboxes that you can use to further specify the nature of the relationship:

   – **Point-to-Point Access**
   If you select this checkbox, access is only granted to the target package; none of the package's superpackages receive implicit access.

   – **Include Subpackages**
   If you select this checkbox, access is implicitly granted to the selected package's subpackages. This setting comes in handy when access is provided to a superpackage or main package whose subpackages are generally unknown to the provider package. This way, access is inherited at the subpackage level, and you don't have to constantly adjust the client package accesses within the provider package.

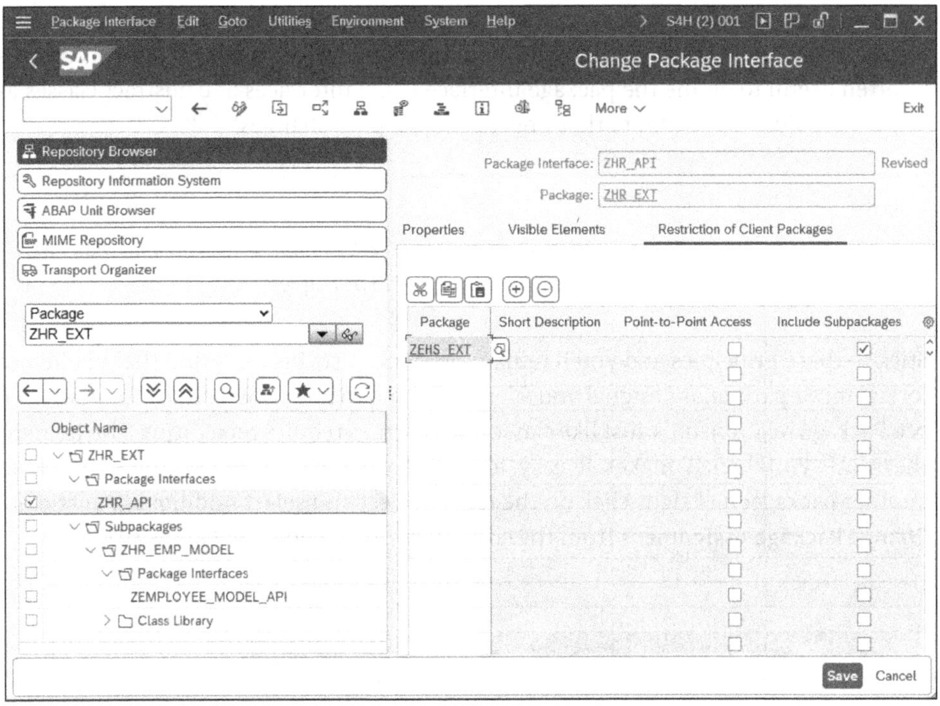

Figure 7.24 Restricting Access to Client Packages (Part 2)

7.3   Package Design Concepts

As you've learned over the course of this chapter, the package concept is quite flexible. While such flexibility is a good thing, it does make it a bit difficult to formulate a standards guide that determines when to use structured or main packages or how deep to define a package hierarchy. The answers to these questions really do depend on the context and, as a result, are subject to change.

Rather than trying to place unnecessary restrictions around the package design process, you should let basic design principles guide your decision making. In his book, *UML Distilled: A Brief Guide to the Standard Object Modeling Language, Third Edition* (Addison-Wesley Professional, 2003), Martin Fowler identifies three basic principles that you can use to help you design your package architectures:

- The *common closure principle* states that development objects within the same package should be changed for the same reasons. This reaffirms the goal of maintaining cohesiveness in software modules (whether they're packages, classes, or something else).

- The *common reuse principle* suggests that development objects within a package should all be reused together.

- The *static dependencies principle* advises you to consider how *stable* your package is if there are many dependencies flowing into it. For example, if ten packages are dependent on a single package, it's important for the interface(s) of that package to remain stable to avoid widespread rippling effects whenever a change occurs. It's often useful to define the package interface using interfaces and abstract classes as they provide the flexibility that's needed to adapt to changes.

Note

Fowler credits Robert C. Martin's book *Agile Software Development, Principles, Patterns, and Practices* (Pearson, 2002) when describing these principles.

Stick to these principles and you'll remain on track. Also, bear in mind that you're not locked into a particular design if you eventually find that it's not working for your project. Package relationships, just like classes, sometimes require refactoring. Fortunately, the ABAP Workbench makes it easy for you to reassign a development object to another package: Just right-click on the target object and select **Additional Functions · Change Package Assignment** from the context menu as shown in Figure 7.25.

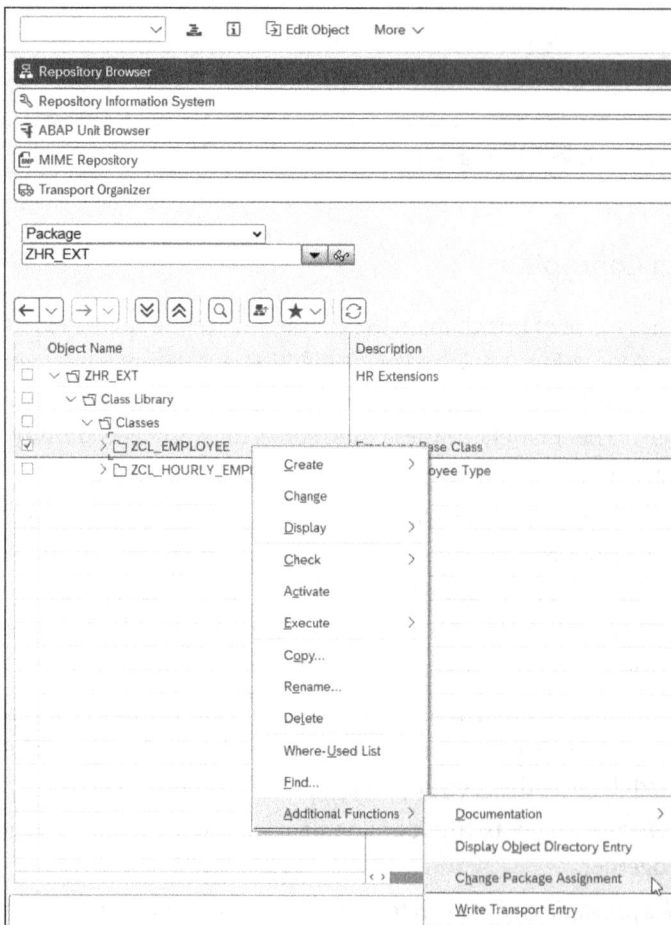

Figure 7.25 Changing the Package Assignment of Development Objects

7.4   UML Tutorial: Package Diagrams

The component design process can become quite involved, being heavily influenced by the subjective whims of developers that often have conflicting design goals. Typically, this process evolves over several iterations that gradually reshape the model to reflect the system that's being implemented. The Unified Modeling Language (UML) supports the documentation of this design process with the *package diagram*. A package diagram allows you to group related classes and interfaces (and indeed, other development objects) into higher-level units called *packages*. Note that the overlap between the term "package" in UML and the ABAP packages is purely coincidental. A UML package is a logical concept that can be implemented in many ways by various programming languages. However, as you'll see, the ABAP package concept does align very closely with the UML package construct.

Figure 7.26 shows an example of a package diagram for a simple online travel reservation application built using the Web Dynpro ABAP web application development framework. Each of the folder-shaped icons in the diagram depict individual packages within the application architecture. The dotted lines between packages depict *dependencies* between the packages. The direction of the line indicates the direction of the dependency. In our example, the Customer UI and Travel Agent UI packages both depend on the WDA Framework and Travel Reservation Model packages. Similarly, objects within the Travel Reservation Model package depend on ABAP Dictionary objects defined within the Travel Reservation Dictionary package.

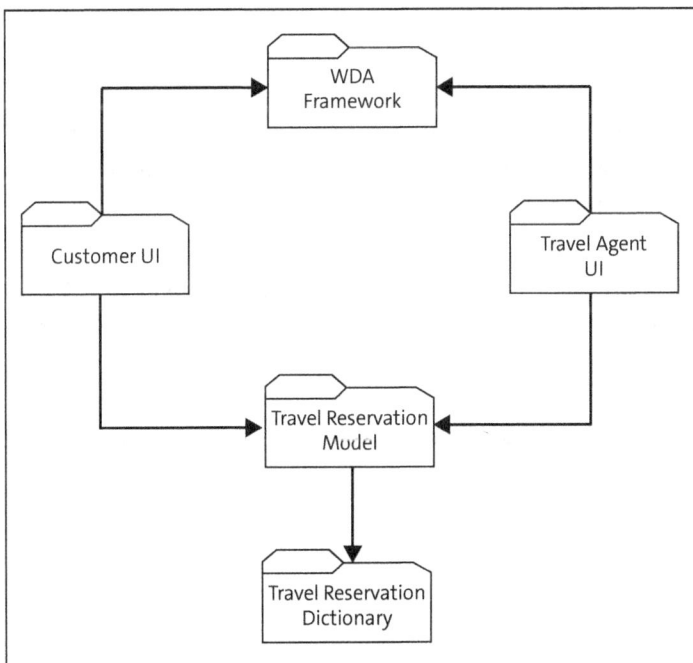

Figure 7.26 Package Diagram for a Web Dynpro ABAP Application

We could have used the formal ABAP package names in this diagram, but as you can see in Figure 7.27, this is optional in UML. Like many UML diagrams, there aren't a lot of notation restrictions for a package diagram. For example, the package diagram in Figure 7.27 expands the basic notation to depict a few of the classes embedded within packages P1 and P2. The + and - visibility tokens indicate whether the classes belong to the public or private interface of the package.

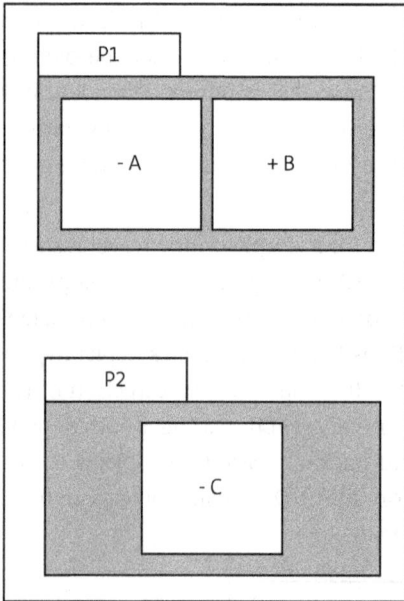

Figure 7.27 Including Classes in Package Diagrams

In the example shown in Figure 7.27, class B has been added to the package interface of package P1. This addition helps you better visualize the nature of the dependency between packages P1 and P2, showing you that class C is dependent upon the publicly exposed class B.

Package diagrams are very useful in illustrating the constituent components of a system design. If you find that the package diagram for your system looks like a plate of spaghetti, then it's likely that your packages are not well encapsulated, and you probably need to refactor your package hierarchy. Consequently, updating your package diagrams periodically is a good way to gauge the effectiveness of your component designs over time.

7.5 Summary

In this chapter, you learned how to perform component-based software development in ABAP. You learned how to apply the package concept to encapsulate related development objects together into logical software units with defined interfaces and dependencies. While this requires some work up front, you'll find it useful in keeping the software catalog clean and easy to manage. Plus, it aligns nicely with the end goal of improving the level of abstraction you want to deal with when building custom software.

In the next chapter, we'll look at the ABAP *class-based exception handling concept* and see how it's used to encapsulate exception scenarios and isolate exception handling logic from the normal program flow.

7

Chapter 8
Error Handling with Exception Classes

Software programs operate in an environment based on rules. However, as the saying goes, there's an exception to every rule. In this chapter, we'll explore ways of dealing with exception cases in ABAP programs using the class-based exception handling concept.

No matter how hard you may try to improve the quality of your code, there's simply no way to avoid every type of error that might occur during the execution of an application. Indeed, some errors are accidentally introduced by programmers trying too hard to make their applications *error-proof*. For example, the error handling logic might obscure the main purpose of the program flow, making the code harder to understand and maintain, and thus more susceptible to errors.

Generally speaking, error-handling logic is a *cross-cutting* concern that becomes tangled within the normal flow of the core application logic. Ideally, you'd like to be able to detangle error-handling logic from the main program flow so that these two orthogonal concerns can be managed separately. In this chapter, you'll learn how to apply the ABAP *class-based exception handling concept* to achieve this separation of concerns.

8.1 Lessons Learned from Prior Approaches

In the early days of ABAP, there wasn't a comprehensive strategy for dealing with exceptions within the ABAP Objects language. Consequently, developers were forced to improvise, weaving custom exception-handling code into their normal program logic. While developers did the best they could with what they had available to them at the time, the resultant solutions were less than optimal. In this section, we'll consider some of the lessons learned from these early approaches to exception handling in ABAP.

8.1.1 Lesson 1: Exception Handling Logic Gets in the Way

Without built-in language support for exception handling, developers were left to build out their exception handling logic using regular procedural code. While this works in principle, the exception logic usually ends up getting in the way of the main program logic. To put this problem into perspective, consider the code excerpt in Listing 8.1. Here, you can see how error-handling logic has been added to a procedural report program that's calling a series of subroutines to perform various tasks. In a contrived example like this, it's not too hard to follow what the program is doing. Still, notice that there are more lines of code devoted to dealing with exceptions than the actual program logic. In larger production programs, this problem becomes even more pronounced.

```
DATA lv_retcode TYPE sy-subrc.
PERFORM sub1 CHANGING lv_retcode.
IF lv_retcode NE 0.
  "Error handling logic...
ENDIF.

PERFORM sub2 CHANGING lv_retcode.
CASE lv_retcode.
  WHEN 0.
    "Operation was successful...
  WHEN 1.
    "Error handling logic...
  WHEN 2.
    "Error handling logic...
  WHEN OTHERS.
    "Error handling logic...
ENDCASE.

PERFORM sub3 CHANGING lv_retcode.
IF lv_retcode NE 0.
  "Yet more error handling logic...
ENDIF.
```

Listing 8.1 Handling Exceptions Using a Manual Approach

Ideally, you'd like to decouple the normal processing logic from the exception logic so that the two separate concerns remain separate. This makes the code easier to read and trace through. We'll explore ways of achieving this throughout this chapter.

8.1.2   Lesson 2: Exception Handling Requires Varying Amounts of Data

In the code excerpt in Listing 8.1, you can see how the various subroutines are passing back a return code value that signifies whether or not an exception occurred. In this contrived example, that's probably all the information you need to deal with the error. However, in many cases, you need much more than just a simple return code; you may also require context about the source of the error, messages that explain what went wrong, and so forth.

Though you could conceivably add this data as exporting parameters to the subroutine/ function/method signature, doing so clutters up the interface quite a bit. This problem is compounded by the fact that different developers may bundle these exception parameters in different ways. For example, most Business Application Programming Interface (BAPI) function modules return an error message table that has the line type

BAPIRET2. Internally, these BAPIs frequently call other standard function modules or subroutines that do not maintain message table parameters of this type. Consequently, additional code must be written in the BAPI function to translate between the various message table types. In Section 8.3, you'll see how to develop exception classes that encapsulate these details much more efficiently.

8.1.3   Lesson 3: The Need for Transparency

Another problem with ad hoc exception handling strategies is that it can be very difficult to identify the types of errors that might occur in a module without digging into the code. For instance, consider the subroutines in Listing 8.1. How would a client know what kind of errors might occur when these are called? Is sub1 dependent on some resource (e.g., a connection to an external SAP HANA database) that might not be available whenever it's called? What happens if sub2 attempts to divide by zero?

From a design perspective, the interface of your modules should be *explicit* about the types of errors that can occur within them. After all, exceptions are part of the application programming interface (API) contract for a module, too. To some degree, certain concepts that we've covered previously provide support for this requirement. For example, you can create named exceptions for methods and function modules using the EXCEPTIONS addition. However, these exceptions are essentially static error codes that have been assigned some semantic meaning inside the method or function module. The meaning of these exceptions becomes obscured outside of the scope of the defining module, especially when new exceptions are added into the mix. Recognizing this, SAP implemented a new class-based concept to deal with exceptions that can be used consistently in all ABAP contexts (e.g., programs and processing blocks). You'll learn more about this concept in Section 8.2.

8.2   The Class-Based Exception Handling Concept

As the name suggests, the class-based exception handling concept uses a special type of ABAP Objects class called an *exception class* to encapsulate exception situations that may occur within a program. These classes are integrated into a framework that makes it easier for you to separate the exception handling aspects of a program from the core functional aspects of the program. This framework is orchestrated by the TRY control structure, which is shown in Listing 8.2.

```
DATA lo_ex TYPE REF TO cx_exception_type.
DATA lo_root TYPE REF TO cx_root.
TRY.
  "Main programming logic goes here...
CATCH cx_exception_type INTO lo_ex.
  "Exception handler block
  lo_ex->...
```

```
CATCH cx_root INTO lo_root.
  "Exception handler block
  lo_root->...
CLEANUP.
  "Optional cleanup block
ENDTRY.
```

Listing 8.2 Basic Form of the TRY...ENDTRY Control Structure

The TRY statement separates the normal application flow from the exception-handling flows by creating separate execution and processing blocks. The TRY block contains the normal application code that may trigger different exceptions along the way. These exceptions are handled by special exception handler blocks called CATCH blocks, which contain code that helps you recover from an exception situation in an application-specific way. After an exception is dealt with, you also have the option of adding a special CLEANUP block to do any necessary cleanup work before the TRY statement returns control to the normal program flow. The basic flow of a TRY statement is depicted in Figure 8.1.

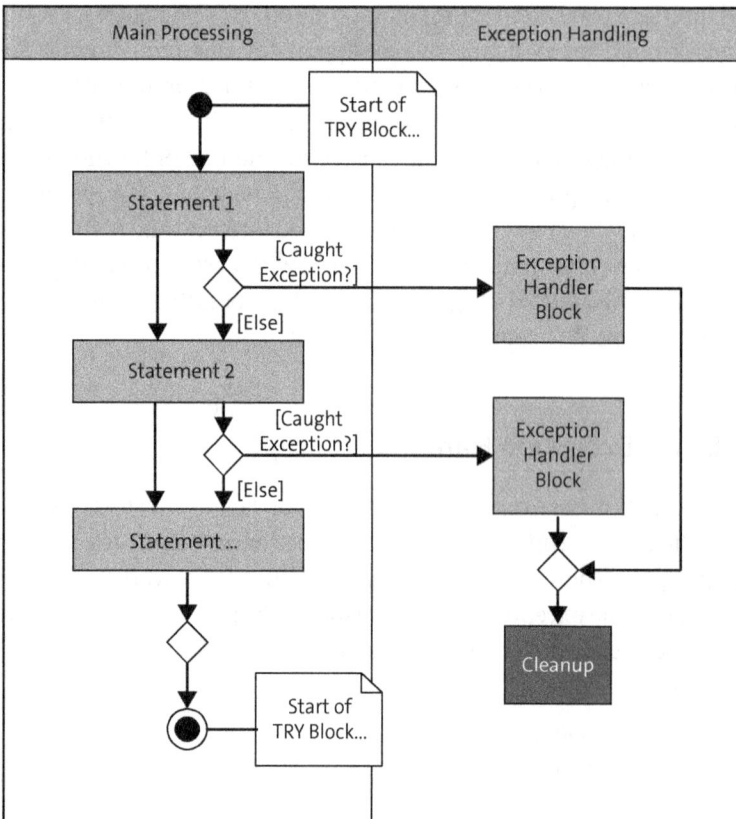

Figure 8.1 Flow Diagram for the TRY Control Statement

In the upcoming sections, you'll see how all this plays out within various ABAP programming contexts.

> **Note**
>
> While the class-based exception handling concept does naturally utilize object-oriented concepts, this does not preclude its use in procedural contexts, too. Class-based exceptions can be incorporated into subroutines, function modules, and even event blocks (e.g., the START-OF-SELECTION event in ABAP report programs).

8.3   Creating Exception Classes

Before we begin looking at how exception classes are used to handle exceptions, it's helpful to first take a moment to understand how these classes are defined. Therefore, in this section, we'll explore the anatomy of exception classes and show you how to build your own custom exception classes.

8.3.1   Understanding Exception Class Types

At the end of the day, exception classes are just like any other ABAP Objects class. They have attributes and methods, and they are maintained in the Class Builder. However, unlike regular ABAP Objects classes that descend from the generic OBJECT type, exception classes descend from one of the three abstract classes defined underneath the abstract CX_ROOT exception class: CX_STATIC_CHECK, CX_DYNAMIC_CHECK, or CX_NO_CHECK (see Figure 8.2). Aside from this constraint, it's object-oriented ABAP as per usual.

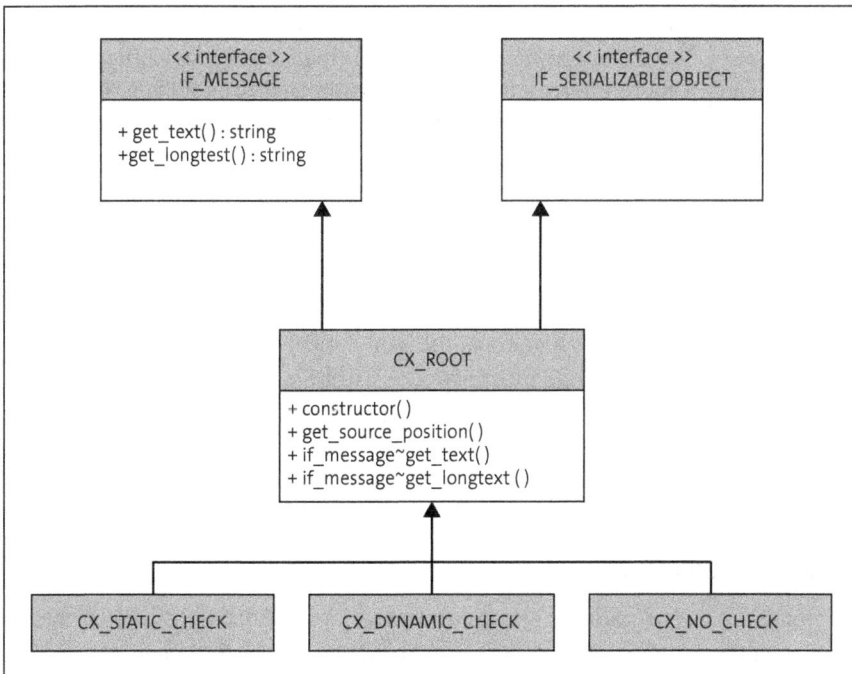

Figure 8.2 Class Diagram for the CX_ROOT Inheritance Tree

Later on, in Section 8.3.2 and Section 8.3.3, you'll find that it's a straightforward exercise to create custom exception types by subclassing one of the three abstract base types. The biggest challenge here is determining which base type to inherit from. Table 8.1 highlights the differences between these types to give you a clearer sense of which type to use in specific circumstances.

Exception Class	Usage Type
CX_STATIC_CHECK	Exceptions of this type represent checked error conditions that may occur within the logic of an application program. Such exceptions must either be explicitly declared in a procedure's interface using the RAISING addition or handled locally within a TRY statement. If an this type of exception is not properly handled, the compiler will issue a warning during the syntax check.
CX_DYNAMIC_CHECK	Exceptions of this type represent unchecked error conditions that likely stem from errors in the program logic. For example, the standard exception class CX_SY_ZERODIVIDE represents a situation where a division operation was attempted with a divisor whose value is 0.
	Realistically speaking, this kind of error should not happen, and if it does, it might not be possible to recover from it gracefully. Moreover, since a mathematics-intensive program could produce this kind of error in almost every statement, it's not practical to handle all the possible exceptions that might occur. Therefore, exceptions of this type do not have to be explicitly handled and are not subject to static syntax checks at compilation time. Of course, failure to properly handle such an exception will ultimately result in a runtime error.
CX_NO_CHECK	Exceptions of this type are similar to those deriving from CX_DYNAMIC_CHECK. The primary difference is that these exceptions will be automatically forwarded if they aren't explicitly handled locally in a TRY statement. In other words, the RAISING clause of a method or subroutine implicitly contains the CX_NO_CHECK addition in its signature, so it's not possible to add additional subordinate classes of this type to the procedure's signature.

Table 8.1 Base-Level Exception Types in ABAP

Though there are many different schools of thought concerning the creation of exception types, we recommend defining most of your custom exception types with the CX_STATIC_CHECK superclass. Using this approach, you improve the readability and documentation of your code by explicitly calling out the types of exceptions that can be raised from within your modules.

8.3.2   Local Exception Classes

Now that you have a sense of how exception classes are defined, let's look at what it takes to create a local exception class. Like any local class types, local exception types are non-reusable types that are unique to a particular application (i.e., a report).

As you can see in Listing 8.3, the syntax used to create a local exception type mirrors that of any local class definition that inherits from some base class type. This minimal syntax is all that's required to create the custom local exception type; the base-level functionality is inherited from CX_STATIC_CHECK in this case. Though it's technically possible to expand the definition of the subtype, SAP recommends that you do not define additional methods and/or redefine inherited methods in local exception classes.

```
CLASS lcx_local_excpetion DEFINITION
  INHERITING FROM cx_static_check.
ENDCLASS.
```

Listing 8.3 Defining a Local Exception Class

The class definition in Listing 8.3 shows that the naming convention for local class types is LCX_{some_meaningful_name}. In this case, the LCX prefix is used to distinguish between local exception classes and regular local classes whose name starts with the LCL prefix.

8.3.3   Global Exception Classes

Most of the time, when you define exception classes, you create them globally in the ABAP Repository so that you can reuse them in other contexts. Global exception classes, like other global class types, are defined using the Class Builder, which adjusts to the *Exception Builder* perspective whenever you're editing an exception class.

As you can see in Figure 8.3, the **Create Class** dialog box looks different when you select the **Exception Class** type. Here, you enter details into the following fields:

- **Class**
  Enter a name for the exception class.

- **Superclass**
  Enter the name of the superclass. This must be defined as one of the three base exception class types (CX_STATIC_CHECK, CX_DYNAMIC_CHECK, or CX_NO_CHECK) or a subclass of those types.

- **Description**
  Enter a meaningful description for the exception class.

You can use the **With messages of message classes as exception texts** checkbox to include support for the integration of messages defined within a message class (i.e., in Transaction SE91). We'll discuss this option in further detail in Section 8.3.5.

Exception classes must be named according to the convention <namespace>CX_{meaningful_name}. For example, when defining an exception class in the default customer namespace, the name would start with the prefix "YCX_" or "ZCX_".

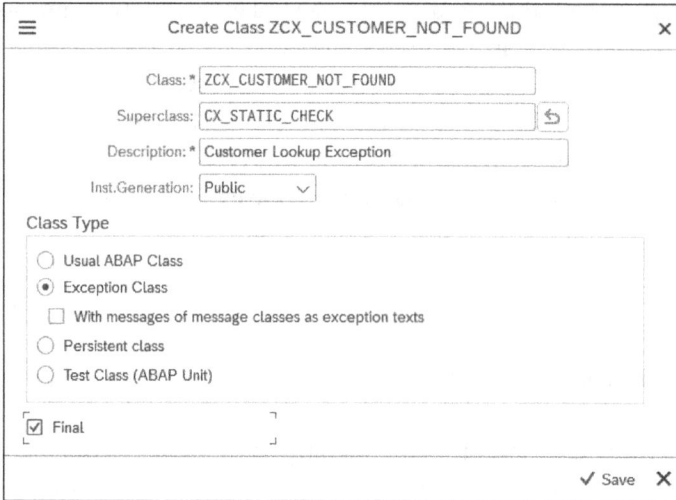

Figure 8.3 Creating Global Exception Classes in the Class Builder

Once an exception type is created, you can edit it using the Exception Builder perspective just as you would any normal global class type. For example, in Figure 8.4, notice that all the same tab pages are provided when you edit exception classes.

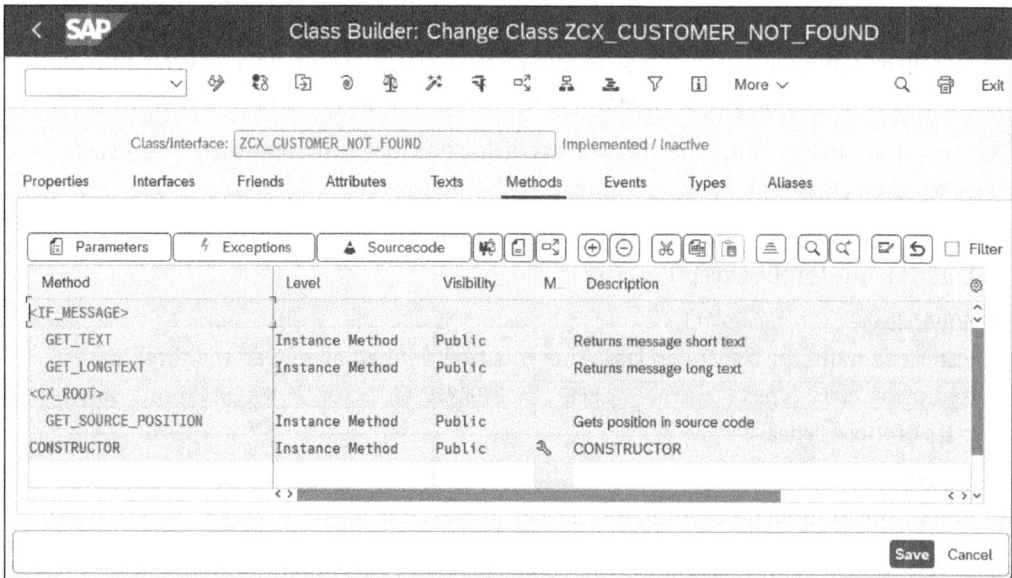

Figure 8.4 Editing an Exception Class Using the Exception Builder

However, there are a couple of nuances to be aware of when editing exception class types:

- The auto-generated `constructor()` method cannot be edited like the `constructor()` methods of regular class types. This is by design—SAP wants to guarantee that the constructor of any exception type contains a consistent interface that makes it possible to create new instances using the `RAISE EXCEPTION` statement.

- Any public attribute defined on the **Attributes** tab will be dynamically added as an importing parameter to the autogenerated `constructor()` method. The implementation of the `constructor()` method will also be adjusted to ensure that the importing parameter value is mapped to the corresponding instance attribute at runtime.

Whereas SAP advises minimal implementation for local exception types, global exception classes can—and often should—be extended to represent distinct exception conditions with greater precision. For example, you can define additional instance attributes to capture more details about the error condition as well as helper methods to lookup these details and generate formatted output messages. While the details will vary from exception class to exception class, the main takeaway is that you shouldn't be afraid to flex your newfound object-oriented programming (OOP) skills to create exception class hierarchies and frameworks that make it easier to deal with exceptions. After all, what's the point of an object-oriented approach if you don't take advantage of object-oriented techniques in the code?

8.3.4   Defining Exception Texts

Ideally, whenever an exception occurs, you want to be able to recover from it gracefully using logic defined within an exception handler block. Unfortunately, this is not always possible. Unexpected exception situations usually require some kind of intervention, whether it's an error message displayed on a screen or a message written in an error log. In either case, you need to produce meaningful error messages that allow someone else to properly investigate the problem. The Exception Builder supports you in this endeavor by allowing you to configure *exception texts* for global classes.

Exception texts are maintained on the **Texts** tab of the Exception Builder (see Figure 8.5). However, behind the scenes, the actual text is stored in the *Online Text Repository* (OTR). The OTR is a central storage repository for texts that are defined within the application server (AS) ABAP. Like most reusable texts, OTR texts are translatable, making them ideally suited for implementing internationalized messages.

Within the Exception Builder, each exception text is defined using a unique exception text ID (i.e., `ZCX_CUSTOMER_NOT_FOUND` in Figure 8.5). The exception text ID correlates to a constant attribute with the same name that has the data type `SOTR_CONC`. These constant attributes belong to the same namespace as normal attributes, so it's a good idea to use the standard naming convention for constants (i.e., the `CO_` prefix) when defining exception text IDs in the Exception Builder. If you look carefully, you'll notice that each constant attribute defined in relation to an exception text ID is initialized with a hexadecimal string value. This value is the globally unique key of the corresponding text object in the OTR.

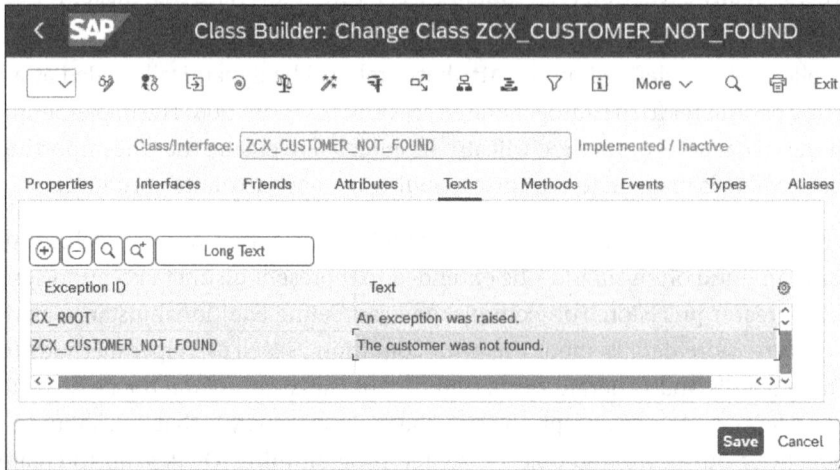

Figure 8.5 Defining Exception Texts in the Exception Builder

You can define text parameters in your exception texts by surrounding elementary attribute names with ampersands (&). For example, the exception text READ_ERROR from exception class CX_SY_FILE_IO shown in Figure 8.6 contains three text parameters: FILENAME, ERRORCODE, and ERRORTEXT. At runtime, when an exception of this type is raised, the correspondingly named instance attributes will be used to generate the READ_ERROR text when the get_text() method is called. This approach helps produce message texts that are meaningful to the end user.

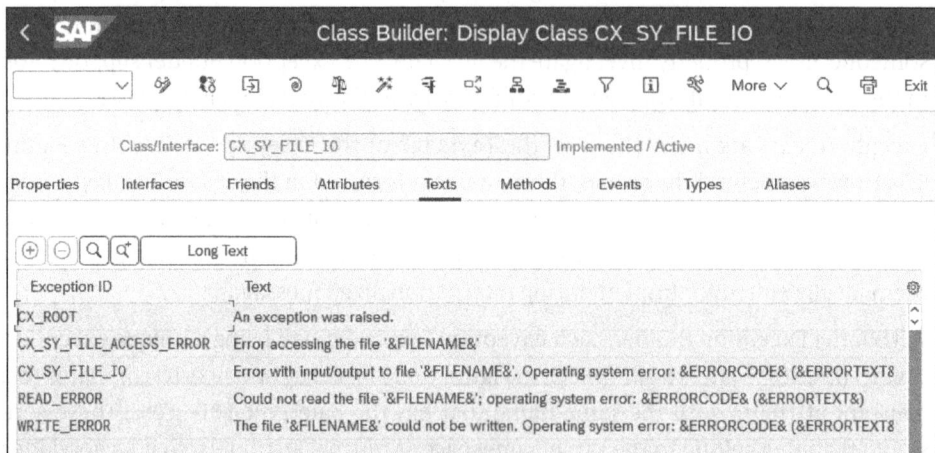

Figure 8.6 Defining Parameterized Texts

8.3.5   Mapping Exception Texts to Message Classes

You have the option of mapping exception texts to messages defined within messages classes. This functionality is enabled up front during the exception class creation process with the selection of the **With messages of message classes as exception texts** checkbox (see Figure 8.3). Alternatively, you can add in this support after the fact for existing exception types by manually implementing the IF_T100_MESSAGE interface, though you may have to forcefully remove any preexisting texts in the class to get it to work.

Once this functionality is enabled, you can maintain texts on the **Texts** tab by *mapping* an exception text to a message number in a message class. Such message classes are maintained outside of the Class Builder using the ABAP Workbench (Transaction SE80) or Message Maintenance (Transaction SE91). As you can see in Figure 8.7, the mapping process is straightforward: You simply enter a **Message Class** and **Message Number**, and then the text is brought into context. If the message in question happens to define attributes, you can also map those attributes to instance attributes from the exception class.

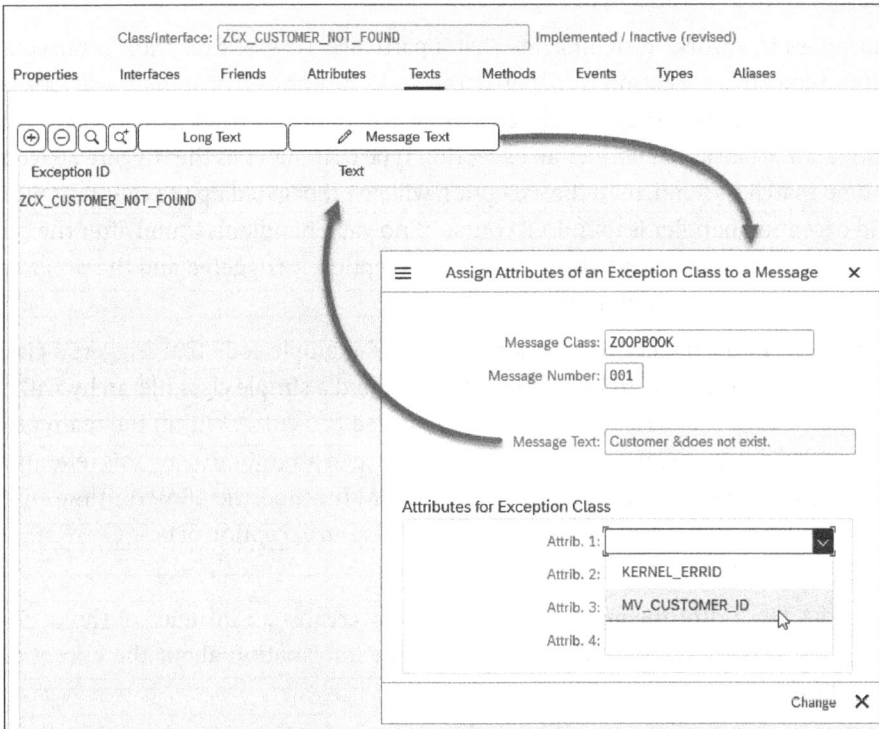

Figure 8.7 Mapping Exception Texts to Messages from Message Classes

The primary benefit of using the message mapping option with your exception classes is that you can leverage a preexisting message base maintained across the development landscape. Such messages can be maintained with long text and translated into other languages using the tools provided in Message Maintenance.

8.4   Dealing with Exceptions

When an ABAP program is executed, there are two different types of exceptions that can occur: exceptions that are raised explicitly using the RAISE EXCEPTION statement and exceptions that are raised implicitly by the ABAP runtime environment. In either case, you want to react to these exceptions and recover as gracefully as possible. In this section, we'll show you how to use the TRY statement to trap exception conditions in ABAP programs.

8.4.1   Handling Exceptions

As you learned in Section 8.2, exceptions within the class-based exception handling concept are handled using CATCH blocks. These CATCH blocks are designed to handle exceptions of a particular type (or, as you'll learn, a family of related types). The exception type is defined in terms of an exception class that's part of an inheritance hierarchy based on the generic CX_ROOT superclass.

So, how does this work? Well, imagine that a particular type of exception is raised at runtime. When this exception is triggered, the ABAP runtime environment will look to see if the statement that triggered the exception is part of a TRY block. If it is, then it will look for a CATCH block that defines an exception type that matches the triggered exception. If no match is found, then the exception will be propagated up the call stack until a valid exception handler is found. Of course, if no valid handler is found after the call stack is completely unwound, then a runtime exception is triggered and the program abends and results in a short dump.

To put all this into perspective, let's consider some example code that triggers a class cast exception at runtime. In Listing 8.4, we've created a simple class hierarchy with a parent class called LCL_PARENT and a child class called LCL_CHILD. Within the main program logic, we create an instance of LCL_PARENT and then attempt to copy this reference into an object reference variable of type LCL_CHILD. At runtime, the ABAP runtime environment will detect the illegal widening cast and raise an exception of type CX_SY_MOVE_CAST_ERROR. From here, several things happen:

1. First, the ABAP runtime environment implicitly creates an instance of the CX_SY_MOVE_CAST_ERROR class and fills it with pertinent information about the exception condition.

2. Next, a check is made to determine if there's an appropriate CATCH block which can handle the error.

3. Finally, since we did have our illegal cast wrapped up in a TRY statement with a corresponding CATCH block, control is immediately transferred to this CATCH block and a reference to the dynamically generated exception object from the first item on this list is passed to the lx_cast_error variable defined within the INTO addition.

```
CLASS lcl_parent DEFINITION.
  PUBLIC SECTION.
    METHODS: a, b.
ENDCLASS.

CLASS lcl_parent IMPLEMENTATION.
  METHOD a.
  ENDMETHOD.

  METHOD b.
  ENDMETHOD.
ENDCLASS.

CLASS lcl_child DEFINITION INHERITING FROM lcl_parent.
  PUBLIC SECTION.
    METHODS: c.
ENDCLASS.

CLASS lcl_child IMPLEMENTATION.
  METHOD c.
  ENDMETHOD.
ENDCLASS.

DATA: lo_parent TYPE REF TO lcl_parent,
      lo_child TYPE REF TO lcl_child,
      lx_cast_error TYPE REF TO cx_sy_move_cast_error,
      lv_program_name TYPE syrepid,
      lv_include_name TYPE syrepid,
      lv_line_number TYPE I,
      lv_text TYPE string,
      lv_long_text TYPE string.

"Attempt a widening case where the dynamic type of the source
"object reference is not compatible with the static type of the
"target object reference:
TRY.
  CREATE OBJECT lo_parent.
  lo_child ?= lo_parent.
CATCH cx_sy_move_cast_error INTO lx_cast_error.
  "Retrieve information about the exception condition:
  lx_cast_error->get_source_position(
    IMPORTING
      program_name = lv_program_name
      include_name = lv_include_name
      source_line  = lv_line_number ).
```

```
    lv_text = lx_cast_error->get_text( ).
    lv_long_text = lx_cast_error->get_longtext( ).
    ...
ENDTRY.
```

Listing 8.4 Handling a Casting Exception Using the TRY Statement

As you can see in Listing 8.4, the CX_SY_MOVE_CAST_ERROR instance contains useful information about the nature of the exception and its origins. Naturally, the type of information within a given exception class will vary depending on the exception type. However, since all exception classes descend from the CX_ROOT superclass, every exception class is guaranteed to provide the get_text() and get_longtext() methods of the implemented IF_MESSAGE interface.

How Do I Know Which Exceptions to Catch?

You might be wondering how we knew to catch the CX_SY_MOVE_CAST_ERROR exception type when performing our cast operation in Listing 8.4. Though some of this knowledge comes from experience working with the class-based exception handling concept, there are some basic rules of thumb to consider as you get started.

For exceptions that are triggered implicitly by the ABAP runtime environment, you can discover which exceptions are triggered for a particular statement by looking at the ABAP Keyword Documentation. For example, if you were typing out the code in Listing 8.4 and wanted to know what would happen if a widening cast failed, you could put your cursor on the ?= operator and press the [F1] key to launch the ABAP Keyword Documentation. Within this context-sensitive documentation, there will be a section called *Exceptions* in which you can discover the types of catchable exceptions triggered by the ABAP runtime environment.

For all other exception class types, you must look within the code to figure out which exception types might be triggered. While this may seem like a daunting task, it's not as bad as it sounds—provided that the code in question was developed using best practices. We'll take a closer look at this in Section 8.5, where you'll learn how exception types are added to method signatures.

Handling Multiple Exceptions Within a Single TRY Block

Technically speaking, you can include as many CATCH blocks as you want inside of a TRY block. For example, if you had a logical unit of work where several different types of exceptions could be triggered, you might build your TRY block in similar fashion to the code excerpt in Listing 8.5. Here, if any of the statements within the TRY block raised an exception of type CX_XSLT_ABAP_CALL_ERROR, CX_XSLT_FORMAT_ERROR, or CX_XSLT_RUNTIME_ERROR, the ABAP runtime environment would kick in as per usual and scan through the CATCH blocks until an appropriate handler was found.

```
TRY.
  oref->method1( ).
  oref->method2( ).
  oref->method3( ).
CATCH cx_xslt_abap_call_error.
  ...
CATCH cx_xslt_format_error.
  ...
CATCH cx_xslt_runtime_error.
  ...
ENDTRY.
```

Listing 8.5 Handling Multiple Exception Types Within a TRY Block

Implementing Generic CATCH Blocks

As you can see in Listing 8.5, you can get pretty granular with your exception handling, defining a CATCH block for any type of exception that might be triggered. Sometimes though, such granularity is overkill. For instance, in the code excerpt in Listing 8.5, each of the exception types are defined as subclasses of a superclass called CX_TRANSFORMATION_ERROR. Depending on the scenario, you may not care *why* a transformation failed—you just want to know an exception occurred so that you can deal with it in a generic way. In these situations, you can refactor the TRY block to look like the code excerpt in Listing 8.6.

```
TRY.
  CALL TRANSFORMATION xsl_test...
CATCH cx_transformation_error.
  ...
ENDTRY.
```

Listing 8.6 Implementing Generic CATCH Blocks

Whenever you set up a CATCH block like the one shown in Listing 8.6, you're informing the ABAP runtime environment that we you to handle exceptions of type CX_TRANSFORMATION_ERROR *and* any exception types that are defined as a descendant of CX_TRANSFORMATION_ERROR. Note that we use the term "descendant" rather than "child" here, because you can traverse far down the exception class hierarchy if you need to.

The declaration of a generic CATCH block like the one shown in Listing 8.6 doesn't preclude you from also defining a selected number of specific exception blocks that have a higher precedence than the generic one. To understand how this works, consider the code excerpt in Listing 8.7. Here, whenever the CALL TRANSFORMATION statement is executed, there could be several exception types triggered. While you want to handle most of these exception types in a generic way, there could be one or two types that you want to handle differently. For instance, if an ABAP call from within the XSLT transformation fails, you might want to handle that differently from everything else. To achieve this separation, you simply define the more specific exception handlers *before* you define the generic exception handlers. That way, whenever an exception is triggered, the ABAP runtime environment will work its way through the CATCH blocks until it finds the most accurate match.

```
TRY.
  CALL TRANSFORMATION xsl_test...
CATCH cx_xslt_abap_call_error.
  ...
CATCH cx_transformation_error.
  ...
ENDTRY.
```

Listing 8.7 Picking and Choosing Specific Exception Types to Listen For

Note

When you define generic CATCH blocks like the ones demonstrated in Listing 8.7, you must declare generic exception types *after* any CATCH blocks that define exception handlers for subordinate classes. If you think about it, this makes sense, as the system wouldn't reach the more specific exception handlers if it finds a matching exception handler for the superordinate class first. Still, if all of this seems confusing, don't worry; the compiler will tell you where you've gone wrong.

Before we move on from this topic, we should warn you that while generic CATCH blocks are highly useful, it's important not to get too carried away and start ignoring exceptions or throwing away useful information. It's important to strike a balance and consider from a defensive programming perspective how you can reasonably handle the types of exceptions that might crop up.

8.4.2   Cleaning Up the Mess

After you've recovered from an exception situation in a CATCH block, you may need to perform some additional cleanup tasks before you hand control back over to the normal program flow. For example, consider a program where you're writing some data to an output file. Inside a TRY block, you open a file and start writing records to it. However, at some point an I/O exception occurs and processing halts in the TRY block before you get a chance to close the file. In this case, you can use the optional CLEANUP block to close the file since this block is guaranteed to be called by the ABAP runtime environment before the TRY statement is exited. A simplified example of this scenario is shown in Listing 8.8.

```
TRY.
   OPEN DATASET lv_file FOR OUTPUT IN TEXT MODE ENCODING DEFAULT.

   LOOP AT lt_extract INTO ls_record.
     TRANSFER ls_record TO lv_file.
   ENDLOOP.

   CLOSE DATASET lv_file.
CATCH cx_sy_file_io.
   "Process I/O errors here...
CLEANUP.
   "Make sure that the file gets closed:
   CLOSE DATASET lv_file.
ENDTRY.
```

Listing 8.8 Recovering from an Exception Using the CLEANUP Block

Note that the CLEANUP block in a TRY statement is guaranteed to be called whenever an exception occurs regardless of whether the system can actually locate a suitable exception handler in that TRY statement. Prior to exiting, the CLEANUP block is executed to clean up any local resources used within the context of the current TRY statement. As such, you should only use the CLEANUP block for its intended purpose. You're also not allowed to execute statements that alter the control flow of a program, such as RETURN and STOP.

8.5   Raising and Forwarding Exceptions

As we noted in Section 8.4, exceptions can be raised either implicitly by the ABAP runtime environment or explicitly using the RAISE EXCEPTION statement. In this section, we'll take a closer look at these two different exception types and see how they're propagated to exception handlers.

8.5.1   System-Driven Exceptions

As you observed in the code examples in Section 8.3, there are many ABAP statements that may trigger an exception at runtime. For example, if an attempt to write to a file fails via the TRANSFER statement, the ABAP runtime environment will automatically raise an exception of type CX_SY_FILE_IO. Similarly, a division operation which attempts to divide by zero will raise an exception of type CX_SY_ZERODIVIDE.

Conceptually speaking, system-driving exceptions such as CX_SY_FILE_IO are classified as *unchecked exceptions* because they're not checked by the ABAP compiler at compile time. The ABAP compiler has no way of knowing if an I/O error might occur at runtime. From a development perspective, the main takeaway is that you can't rely on the compiler to warn you that you might have a logic error. Instead, you should follow best practices for defensive programming and make sure you handle these exceptions on your own. If you're new to the class-based exception handling concept, we recommend reading through the ABAP Keyword Documentation to make sure that you're familiar with the types of exceptions that specific statements might trigger.

8.5.2   Raising Exceptions Programmatically

While system-driven exceptions may occur from time to time, most exceptions that you deal with are explicitly triggered from ABAP code using the RAISE EXCEPTION statement. Here, the focus is on dealing with application-level logic errors that might occur.

To illustrate how this works, imagine that you're tasked with building an API to look up customer details, such as a customer's credit rating. The input to this method is the customer's ID number; the output is the customer's credit score. Listing 8.9 shows a preliminary version of this method. Here, the first step is to look up the customer master record using the ID number provided via the importing IV_CUSTOMER_ID parameter. However, look at what happens whenever the provided customer ID is invalid. Since it's not possible for the method logic to continue without a valid customer number, a dummy credit rating (i.e., -1) is returned to the caller. Here, it's the responsibility of the caller to inspect the value and determine whether the credit rating was valid.

```
CLASS lcl_customer DEFINITION.
  PUBLIC SECTION.
    CLASS-METHODS:
      get_credit_rating IMPORTING iv_customer_id TYPE kunnr
                        RETURNING VALUE(rv_rating) TYPE i.
ENDCLASS.

CLASS lcl_customer IMPLEMENTATION.
  METHOD get_credit_rating.
    "Read the customer master record from the database:
    SELECT ...
      FROM but000
     WHERE partner EQ iv_customer_id.

    IF sy-subrc NE 0.
      rv_rating = -1.
      RETURN.
    ENDIF.
    ...
  ENDMETHOD.
ENDCLASS.
```

Listing 8.9 Handling Errors in Application Logic (Part 1)

As you can see in Listing 8.9, the exception handling approach taken is conceptually similar to the anti-pattern related to return code passing we observed in Section 8.1.2. In Listing 8.9, we're effectively re-purposing the RV_RATING parameter as a return code of sorts to inform the user of an error. Of course, you could also define discrete exporting parameters to carry exception details separately, but the main problem persists—you need a better mechanism of raising a red flag and making sure that the caller reacts appropriately. This is where the RAISE EXCEPTION statement comes into play. With the RAISE EXCEPTION statement, you can communicate these situations explicitly by raising specific types of exceptions.

Listing 8.10 shows the basic syntax for the RAISE EXCEPTION statement. One on hand, the RAISE EXCEPTION statement behaves similarly to the CREATE OBJECT statement in that it creates an instance of whatever exception type you specify using the TYPE addition (e.g., CX_EXCEPTION_TYPE). Much like the CREATE OBJECT statement, you have the option of specifying exporting parameters that are passed into the constructor of the exception object. On the other hand, the RAISE EXCEPTION statement also behaves like a control statement in the sense that it interrupts the normal program flow by causing the ABAP runtime environment to unwind the call stack in search of an appropriate CATCH block to handle the exception.

```
RAISE EXCEPTION TYPE cx_exception_type
    [EXPORTING
        f1 = a1
        f2 = a2
        ...].
```

Listing 8.10 Basic Syntax of the RAISE EXCEPTION Statement

Now that you know how to raise exceptions, let's look at how you could rework the get_credit_rating() method from Listing 8.9 to include support for class-based exceptions. In Listing 8.11, we start by defining a simple exception class called LCX_CUSTOMER_NOT_FOUND. Then, after the introduction of the new exception class, the next major change you'll notice in the refactored code is the amendment to the signature of the get_credit_rating() method. Here, we're using the RAISING addition to identify the exception type(s) that might be raised during this processing of this method. Though this is an optional addition to the code, it improves the readability of the get_credit_rating() method.

```
CLASS lcx_customer_not_found DEFINITION
    INHERITING FROM cx_static_check.
ENDCLASS.

CLASS lcl_customer DEFINITION.
    PUBLIC SECTION.
        CLASS-METHODS:
```

```
      get_credit_rating
        IMPORTING iv_customer_id TYPE kunnr
        RETURNING VALUE(rv_rating) TYPE i
          RAISING lcx_customer_not_found.
ENDCLASS.

CLASS lcl_customer IMPLEMENTATION.
  METHOD get_credit_rating.
    "Read the customer master record from the database:
    SELECT ...
      FROM knb1
     WHERE kunnr EQ iv_customer_id.

    IF sy-subrc NE 0.
      RAISE EXCEPTION TYPE lcx_customer_not_found.
    ENDIF.
    ...
  ENDMETHOD.
ENDCLASS.
```

Listing 8.11 Handling Errors in Application Logic (Part 2)

Within the method implementation itself, you can see how we're using the RAISE EXCEPTION statement to raise the exception whenever the lookup on the customer master data fails. As soon as we raise this exception, the method processing will halt, and the exception will begin bubbling up the call stack until an appropriate exception handler is found. This forces clients to deal with the exception head on, as demonstrated in the sample code in Listing 8.12.

```
DATA: lv_customer TYPE kunnr VALUE '1234567890',
      lv_credit_rating TYPE i.
      lo_customer_error TYPE REF TO lcx_customer_not_found.
TRY.
  lv_credit_rating =
    lcl_customer=>get_credit_rating( lv_customer ).
CATCH lcx_customer_not_found INTO lo_customer_error.
  "Handle the error in an application-specific way:
  MESSAGE lo_customer_error TYPE 'E'.
ENDTRY.
```

Listing 8.12 Handling Application-Specific Exception Types

If you look closely at the CATCH block in Listing 8.12, you can see how we're using a special variant of the MESSAGE statement to output the exception message. When this statement is evaluated at runtime, the ABAP runtime environment will silently invoke the `if_message~get_text()` method to fetch the exception text and display it on the screen. This makes it very easy for certain types of applications to relay error messages to end users.

Raising Exceptions with the COND and SWITCH Statements

You can use the `COND` and `SWITCH` constructor operators to evaluate logical conditions within the context of an initialization operation. Though semantically similar to the conditional `IF` and `CASE` statements, the usage context for the `COND` and `SWITCH` statements is limited to introducing conditional logic in variable assignment expressions.

Depending on the type of assignment you're performing, the input data being evaluated might not match any particular pattern. In this case, you may be unable to reasonably initialize the target variable and must raise an exception. When this occurs, you can raise an exception using the overloaded `THROW` statement.

The code excerpt in Listing 8.13 demonstrates how this syntax works for the `COND` statement. Here, we're evaluating a date value which provided in plain text format and trying to convert it to an internal ABAP date type. This contrived example only supports two input formats for the date: YYYYMMDD or MM/DD/YYYY. If the date value doesn't match up with this format, then an exception of type `LCX_INVALID_DATE_FORMAT` is raised using the `THROW` statement. Although we're not passing any parameters into the exception class's constructor method, we could have done so by entering the parameters within the parentheses after the exception class name in the expression. This is similar to how the `NEW` operator works, as discussed in Chapter 2.

```
CLASS lcx_invalid_date_format DEFINITION
        INHERITING FROM cx_no_check.
ENDCLASS.

DATA(lv_raw_date) = `06/02/2015`.
DATA(lv_date) =
  COND d( WHEN cl_abap_matcher=>matches(
              pattern = `\d{8}`
                text = lv_raw_date ) EQ abap_true
          THEN lv_raw_date
        WHEN cl_abap_matcher=>matches(
              pattern = `\d{2}[/.-]\d{2}[/.-]\d{4}`
                text = lv_raw_date ) EQ abap_true
          THEN
            lv_raw_date+6(4) && lv_raw_date(2) &&
              lv_raw_date+3(2)
          ELSE
            THROW lcx_invalid_date_format( ) ).
```

Listing 8.13 Raising an Exception in a COND Statement Using the THROW Command

The code excerpt in Listing 8.14 shows how you can achieve similar results using the SWITCH statement. Since the syntax is largely the same in both cases, we won't re-hash the syntactical particulars of this statement.

```
CLASS lcx_language_unknown DEFINITION
      INHERITING FROM cx_no_check.
ENDCLASS.

DATA(lv_message) =
  SWITCH string( sy-langu
    WHEN 'E'
      THEN `Welcome to ABAP Objects`
    WHEN 'S'
      THEN `Bienvenido a Objetos ABAP`
    WHEN 'D'
      THEN `Willkommen in ABAP Objects`
    ELSE
      THROW lcx_language_unknown( ) ).
```

Listing 8.14 Raising an Exception in a SWITCH Statement Using the THROW Command

8.5.3   Propagating Exceptions

In the customer credit rating check scenario introduced in Section 8.5.2 and Listing 8.11, it's clear that the get_credit_rating() method can't do much if the provided customer ID is invalid. Therefore, rather than swallowing up the exception or terminating silently in the background, the method explicitly raises an exception using the RAISE EXCEPTION statement. This exception propagates the call stack so that the callers can deal with it. Here, notice that we refer to *callers* in plural since the immediate callers may also find that they too are unable to deal with the exception.

To put this into perspective, consider the LCL_CUSTOMER_REPORT class in Listing 8.15. This contrived class can be used to generate a customer output report leveraging the features defined in the LCL_CUSTOMER utilities class we created in Section 8.5.2. If you drill through the code, you'll find that the method that performs most of the heavy lifting — process_customer()—may encounter the LCX_CUSTOMER_NOT_FOUND exception defined in Listing 8.11. When this occurs, there's really no point in continuing with the processing of the customer, so the method forwards it up the call stack using a hybrid form of the RAISE EXCEPTION statement. For all other exception types that might occur, it's assumed that the method is able to handle these internally, so we're using a generic catch block for CX_ROOT to prevent them from propagating back to the caller.

```
CLASS lcl_customer_report DEFINITION.
  PUBLIC SECTION.
    CLASS-METHODS:
      execute.
```

```
    PRIVATE SECTION.
      TYPES: BEGIN OF ty_customer,
               kunnr TYPE kunnr,
               ...
             END OF ty_customer.
      DATA mt_customers TYPE STANDARD TABLE OF ty_customer.

      METHODS:
        read_customer_data,
        process_customers,
        process_customer IMPORTING is_customer TYPE ty_customer
                         RAISING lcx_customer_not_found.
ENDCLASS.

CLASS lcl_customer_report IMPLEMENTATION.
  METHOD execute.
    DATA lo_report TYPE REF TO lcl_customer_report.
    CREATE OBJECT lo_report.

    lo_report->read_customer_data( ).

    lo_report->process_customers( ).
  ENDMETHOD.

  METHOD read_customer_data.
    ...
  ENDMETHOD.

  METHOD process_customers.
    FIELD-SYMBOLS <ls_customer> LIKE LINE OF me->mt_customers.
    DATA lx_root TYPE REF TO cx_root.

    LOOP AT me->mt_customers ASSIGNING <ls_customer>.
      TRY.
        process_customer( <ls_customer> ).
      CATCH cx_root INTO lx_root.
        MESSAGE lx_root TYPE 'E'.
      ENDTRY.
    ENDLOOP.
  ENDMETHOD.

  METHOD process_customer.
    DATA lx_customer_error TYPE REF TO lcx_customer_not_found.
```

```
    TRY.
      "Fetch the customer's credit rating:
      IF lcl_customer=>get_credit_rating( is_customer-kunnr )
          GT 550.
        ...
      ENDIF.

      "Perform other options which might raise
      "different exceptions that we want to handle internally...
    CATCH lcx_customer_not_found INTO lx_customer_error.
      "Forward the exception on to the caller:
      RAISE EXCEPTION lx_customer_error.
    CATCH cx_root.
      "Handle the error locally...
    ENDTRY.
  ENDMETHOD.
ENDCLASS.
```

Listing 8.15 Propagating Exceptions Using the RAISE EXCEPTION Statement

Any time you propagate exceptions from a method, it's a good idea to declare your intentions up front by including the target exception types in the RAISING clause of the method definition. Though you've already seen some examples of this, Listing 8.16 illustrates the syntax more clearly. Here, you can see that a given method can define many exception types as part of its signature.

```
METHOD some_method RAISING cx_ex1 cx_ex2 ...
```

Listing 8.16 Basic Syntax of the RAISING Addition

In practice, you should keep the total number of exceptions within the RAISING clause to a handful of types. If you find yourself defining more than a few exception types within a method's signature, your method is probably doing too much and lacks cohesion.

For global classes maintained in the form-based view of the Class Builder tool, you can add exceptions to a method signature by selecting the method on the **Methods** tab and clicking on the **Exceptions** button, as shown in Figure 8.8.

This brings up the editor page shown in Figure 8.9. From here, you can add the necessary global exception types to the signature by filling in the exception class names in the **Exception** column.

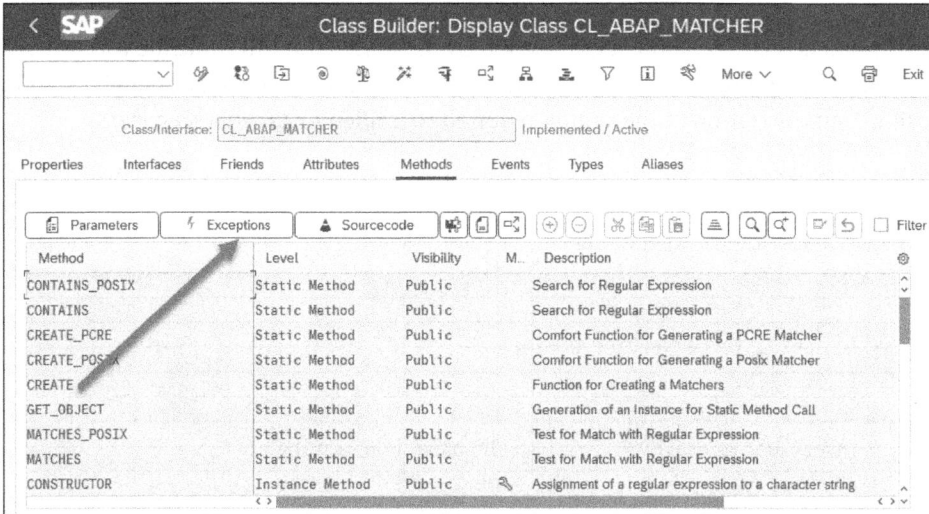

Figure 8.8 Adding Exceptions to the Signature of Methods in a Global Class (Part 1)

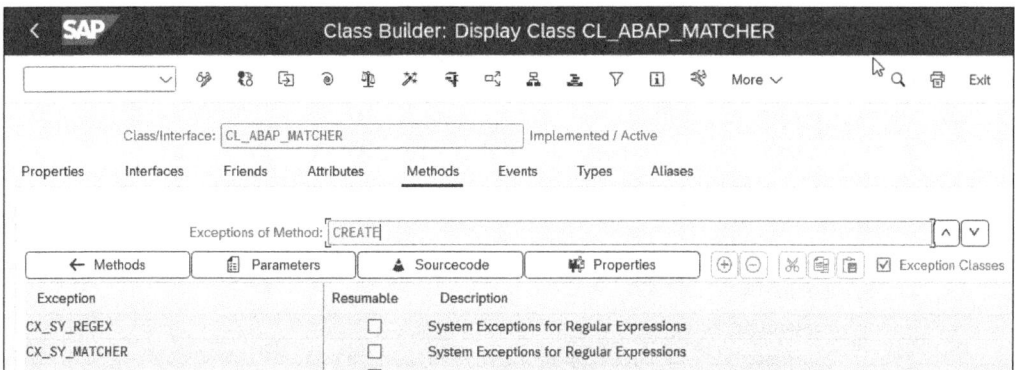

Figure 8.9 Adding Exceptions to the Signature of Methods in a Global Class (Part 2)

8.5.4   Resumable Exceptions

The class-based exception handling concept includes support for *resumable exceptions*. As the name suggests, resumable exceptions are exceptions that are dealt with so cleanly that you can allow the program flow to *resume* after you clean up the mess in a CATCH block.

To demonstrate how this works, let's look at an example. In Listing 8.17, we've developed a simple report program that contains a couple of local test classes that are used to upload documents to a specialized document store. Though database agnostic, our document upload service is optimized for the SAP HANA database, making use of native features such as stored procedures. Within the code, checks are made to determine if the SAP HANA database is available (e.g., by reading the value of the built-in `cl_db_sys=>is_in_memory_db` attribute). If SAP HANA isn't available, then a resumable exception is raised to allow clients to determine if they want to continue without SAP HANA or cease processing. Though admittedly contrived, this example gives you a useful demonstration of the syntax required to implement such a scenario.

```
REPORT zresumable_test.
CLASS lcx_doc_service_error DEFINITION
  INHERITING FROM cx_static_check.
ENDCLASS.

CLASS lcl_persistence_service DEFINITION.
  PUBLIC SECTION.
    METHODS:
      insert_document IMPORTING iv_file_name TYPE string
                                iv_mime_type TYPE string
                                iv_payload TYPE xstring
                      RAISING RESUMABLE(cx_sy_sql_error).
    ...
ENDCLASS.

CLASS lcl_persistence_service IMPLEMENTATION.
  METHOD insert_document.
    IF cl_db_sys=>is_in_memory_db EQ abap_true.
      "Call stored procedure to insert the document using AMDP...
    ELSE.
      RAISE RESUMABLE EXCEPTION TYPE cx_sy_sql_error
        EXPORTING
          sqlcode = 900
          sqlmsg = `SAP HANA is not available. ` &&
                   `Will process through OpenSQL instead.`.
    ENDIF.

    "Insert the document using OpenSQL instead...
    ...
  ENDMETHOD.
ENDCLASS.

CLASS lcl_document_service DEFINITION.
  PUBLIC SECTION.
    METHODS:
      constructor,
      upload_document IMPORTING iv_file_name TYPE string
                                iv_mime_type TYPE string
                                iv_payload TYPE xstring
                      RAISING lcx_doc_service_error.
```

```
    PRIVATE SECTION.
      DATA mv_session_id TYPE guid_16.
      DATA mo_persistence TYPE REF TO lcl_persistence_service.

      METHODS:
        log IMPORTING iv_message TYPE csequence.
ENDCLASS.

CLASS lcl_document_service IMPLEMENTATION.
  METHOD constructor.
    TRY.
      me->mv_session_id =
       cl_system_uuid=>create_uuid_x16_static( ).
      me->mo_persistence = NEW lcl_persistence_service( ).
    CATCH cx_uuid_error.
    ENDTRY.
  ENDMETHOD.

  METHOD upload_document.
    DATA lx_sql_error TYPE REF TO cx_sy_sql_error.
    TRY.
      mo_persistence->insert_document(
        iv_file_name = iv_file_name
        iv_mime_type = iv_mime_type
        iv_payload = iv_payload ).

      log( |File "{ iv_file_name }" was uploaded.| ).
    CATCH BEFORE UNWIND cx_sy_sql_error INTO lx_sql_error.
      "Test the nature of the exception to determine
      "if we should resume or abort:
      IF lx_sql_error->sqlcode EQ 900.
        log( lx_sql_error->sqlmsg ).
        RESUME.
      ELSE.
        RAISE EXCEPTION TYPE lcx_doc_service_error
          EXPORTING
            previous = lx_sql_error.
      ENDIF.
    ENDTRY.
  ENDMETHOD.

  METHOD log.
    WRITE: / iv_message.
  ENDMETHOD.
ENDCLASS.
```

```
START-OF-SELECTION.
DATA(lo_doc_service) = NEW lcl_document_service( ).
lo_doc_service->upload_document(
  EXPORTING
    iv_file_name = 'Test.txt'
    iv_mime_type = 'text/plain'
    iv_payload = CONV xstring( `This is a test.` ) ).
```

Listing 8.17 Working with Resumable Exceptions

If you look over the example code in Listing 8.17, you can see several syntax elements on display that make this flow work:

- Within the definition section of the LCL_PERSISTENCE_SERVICE class, notice how the signature of the insert_document() method includes the RAISING RESUMABLE(cx_sy_sql_error) addition. This addition declares that the insert_document() method may raise a resumable exception of type CX_SY_SQL_ERROR.

- Next, in the insert_document() method itself, we have a condition check on the cl_db_sys=>is_in_memory_db attribute to determine if the AS ABAP is running on top of the SAP HANA database. If not, an exception of type CX_SY_SQL_ERROR is raised using the RAISE RESUMABLE EXCEPTION statement. The RESUMABLE addition tells the ABAP runtime environment that we *may* want to resume processing after this exception.

- In the calling upload_document() method of class LCL_DOCUMENT_SERVICE, we've wrapped the call to insert_document() inside of a TRY statement. Notice how we're using the BEFORE UNWIND addition of the CATCH statement to inform the ABAP runtime environment that we want to keep the processing context from which the exception was raised so that we can pick up where things left off after we deal with the exception.

- Finally, within the CATCH block, you can see how we're assessing the nature of the error and determining if we should resume. In the case of an SAP HANA unavailable error, you would use the RESUME statement to let the processing continue to the failover section of the persistence service, where you use regular OpenSQL to store the uploaded document.

Figure 8.10 contains a Unified Modeling Language (UML) sequence diagram that illustrates the exception flow. As you can see, most of the magic happens between the point that you raise the resumable exception and the point that the corresponding CATCH block resumes processing with the RESUME statement. After flow is resumed, the insert_document() method picks up right where it left off—with its local variables intact—and proceeds with the failover logic.

In summary, resumable exceptions provide a clean mechanism for you to trap error conditions, recover from them, and move on as if the exception hadn't happened in the first place. This functionality comes in handy in situations where you need to build failover logic into your programs. With resumable exceptions, you can capture the exception conditions in an exception object, pass the information on to a CATCH block, and let the CATCH block determine whether you wish to proceed. This is preferable to having lots of IF statements scattered throughout the code to assess these conditions and redundantly react to them.

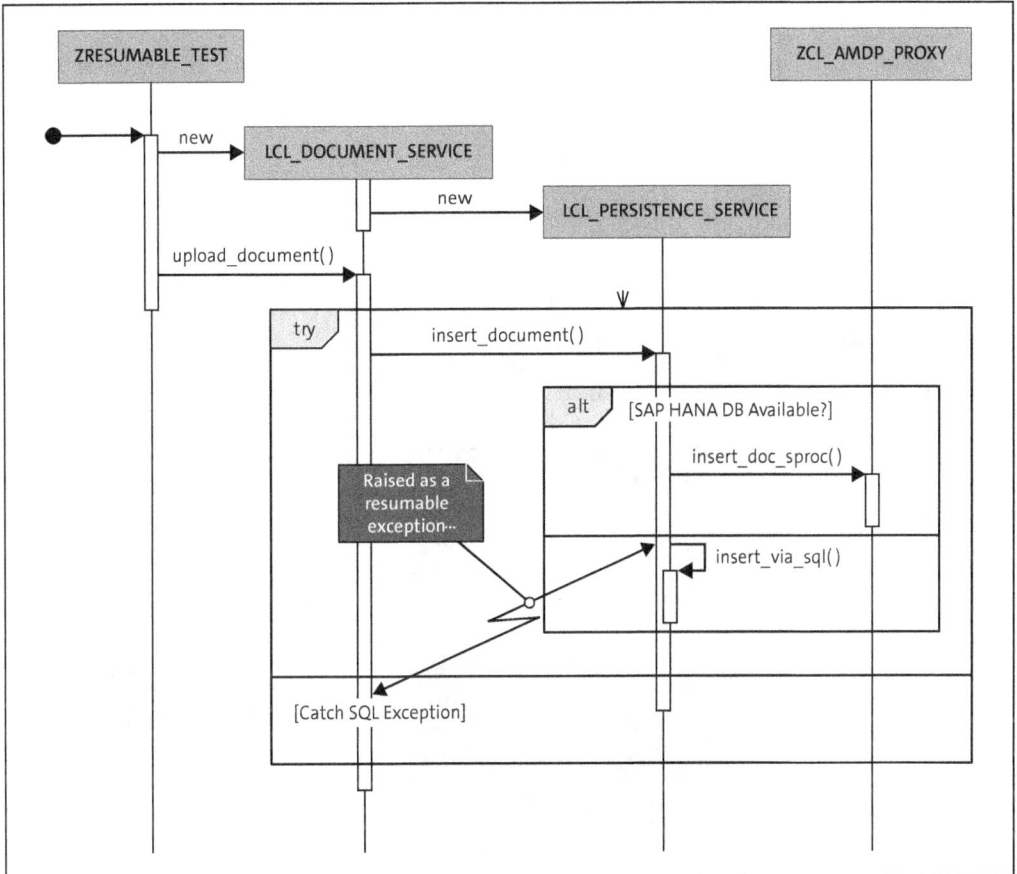

Figure 8.10 UML Sequence Diagram Showing Resumable Exception Flow

8.6   UML Tutorial: Activity Diagrams

The UML activity diagram is a behavioral diagram that depicts the high-level flow within a block of code. Activity diagrams share certain similarities with flowcharts used in the procedural world. However, as you'll soon learn, there are certain things you can do with activity diagrams that you cannot do with flowcharts.

Figure 8.11 shows an example of an activity diagram depicting the flow of a simple ABAP extract program. The flow begins at the *initial node* action and proceeds to the first action, called Receive Query Parameters. Notice that the action names we have used here are generic; in the ABAP extract program, the Receive Query Parameters action would encompass the generation of a selection screen and the entry of selection parameters by a user. You can trace the control flow of an activity diagram by following the directed edges between actions. Eventually, the program flow proceeds all the way down to the *activity final* action (see Figure 8.11).

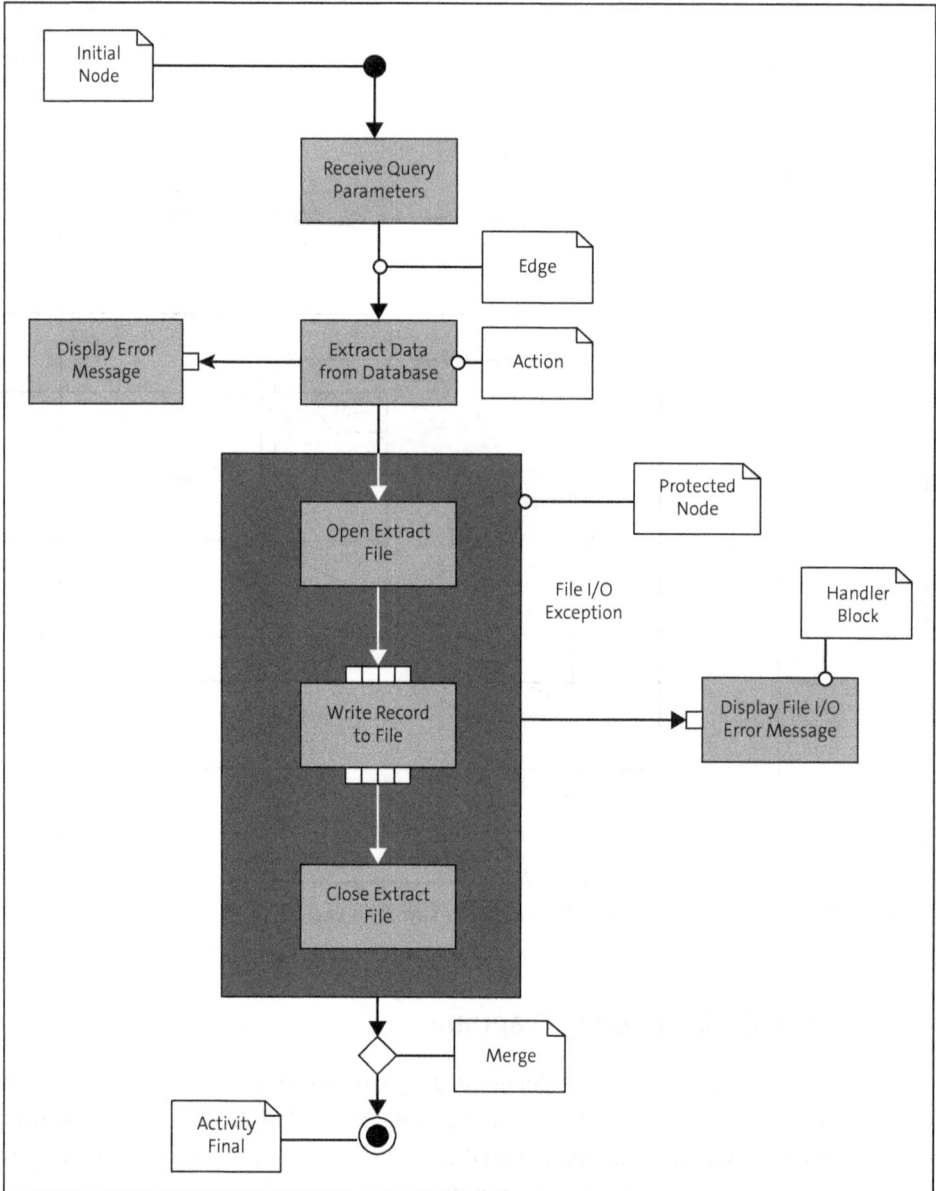

Figure 8.11 Example of a UML Activity Diagram

One significant addition to activity diagrams in the UML 2.0 standard was the specification of *protected nodes* and *handler blocks*. As you can see in Figure 8.11, the action Extract Data from Database is a protected node that might trigger an exception (i.e., Selection Failed). If no data is found in the database for the given selection criteria, control is transferred to the Display Error Message handler block.

It's also possible to group together multiple actions inside of a protected node. For example, all of the file I/O actions in Figure 8.11 were grouped together in a protected node that reacts to file I/O exceptions. Notice that the flow from each handler block leads into a diamond-shaped node called a *merge*. Merges provide a convenient way of channeling multiple input flows into a common output flow.

Notice that the Write Record to File action's boundary is depicted with a dotted line rather than the solid line used for the other normal actions. This dotted line marks an *expansion region* in the activity diagram. In the flow, the Write Record to File expansion region (along with the inputting tokens shown as small pins along the top of the action) represents a loop that takes the extract records from the database lookup and iteratively writes them to the extract file. This kind of notation is more elegant than the typical use of conditionals in flowcharts to determine whether there are more records to process.

One of the beauties of activity diagrams is that they're extremely easy to read, oftentimes requiring little to no translation for non-technical members of the team. They're an excellent communication tool for describing and refining a program flow with functional team members. Typically, once the process flow within an activity diagram is agreed upon, you can put the design in more technical terms with an interaction diagram, such as a sequence diagram.

8.7 Summary

In this chapter, you observed how the class-based exception handling concept greatly simplifies the process of dealing with errors that may occur within an application. The definition of a common framework for dealing with exceptions is essential for the development of reusable components since it provides a consistent model for relaying exceptions to users and clients.

In the next chapter, we'll take a look at ABAP Unit and see how a test-driven approach to development can significantly improve the quality of your ABAP code.

Chapter 9
Unit Tests with ABAP Unit

ABAP Unit is a framework that helps developers verify outcomes and validate the results of paths throughout their code. In this chapter, you'll learn more about this framework and how you can leverage it to develop and release defect-free applications.

A professional developer's goal for every project is to write elegant, agile, and defect-free code. In practice, however, this goal can be difficult to achieve—especially if you only rely on informal testing processes. Fortunately, there are tools that can help improve your code quality and boost your confidence when making future changes. One of the most effective tools in the modern developer's toolbox is a robust *automated unit testing framework*.

Unit testing frameworks like ABAP Unit enable developers to write individual *unit tests* that target and verify the intended behavior of a specific *unit* of code. A code unit can be defined as the smallest testable part of an application; more commonly, it's a method or function of a larger code container, such as a class. If that definition seems narrow, you can organize unit tests into groups in ABAP Unit to adjust the extent of what you're testing. You can then incorporate each group into a test run and set it to run manually or automatically at specific points throughout the software development lifecycle to ensure that changes in code did not introduce any defects or software regressions.

One of the main goals of every unit test is to confirm that your individual modules fulfill the terms of their application programming interface (API) contracts. Verifying this behavior early on helps to eliminate the tedious module-level bugs that prohibit integration and functional tests from running smoothly. In this chapter, you'll learn how the ABAP Unit test tool supports you in the process of developing and executing unit tests.

9.1   ABAP Unit Overview

In 1998, Kent Beck designed the first unit testing framework that could be used to provide common elements necessary for building and running automated unit tests. The framework, created for the Smalltalk language, was called SUnit. Since that time, this same testing model (colloquially known as xUnit these days) has been adapted to create testing frameworks for other languages such as Java (JUnit) and .NET (NUnit). With the SAP NetWeaver 2004 release, SAP introduced ABAP Unit as a testing framework for ABAP Objects. In this section, we'll introduce the basic concepts of ABAP Unit and review how it works.

9.1.1   Unit Testing Terminology

To understand how to use ABAP Unit, it's important to know the basic terms used throughout the framework (see Table 9.1). These terms—and concepts they represent—are largely based on concepts outlined in the core xUnit framework.

Term	Description
Test class	A test class defines an environment for running multiple related unit tests (implemented as test methods).
Test method	Test methods are special-instance methods of a test class that can be invoked to produce test results. In the xUnit framework, a test method represents a single unit test.
Fixture	A fixture defines an environment for running unit tests in the proper context. Fixtures are configured in special callback methods defined within a test class. You can insert code in these methods to obtain and clean up resources (e.g., object instances and file handles) used in the unit test methods.
Test task	A test task groups test classes together, allowing their methods to be executed together in a single test run.
Test run	A test run controls the execution of a test task. Test runs produce test results that you can view in the **ABAP Unit: Result Display** screen.
Assertion	Inside a test method, individual logical tests are made to assess the correctness of a particular piece of functionality. These logical tests that verify conditions with a true/false proposition are known as *assertions*.

Table 9.1 Basic ABAP Unit Terminology

9.1.2   Understanding How ABAP Unit Works

The ABAP Unit test framework is tightly integrated into both the ABAP Workbench and ABAP development tools, making it very easy to set up and execute tests for a given ABAP program. The tests themselves are written in ABAP and are nothing more than local classes. This means that no additional language or interface skills are required to begin working with the ABAP Unit framework.

In Section 9.2, you'll learn how to create unit test classes in the ABAP Workbench. Although there is some nuance involved in setting up test classes, developing the tests themselves is straightforward. All you have to do to create a test is define a parameterless instance method. Within those test methods, you'll perform individual tests and then validate the results using assertions from class CL_AUNIT_ASSERT.

Each unit test class contains *fixtures* that help set up and tear down the objects and resources needed to execute each test. Prior to the execution of each test method, the runtime environment checks to see if the test class contains a parameterless method called setup(). If the method is found, the runtime environment will call it *before* it calls the test method to ensure that the test is set up properly. Similarly, after the test is completed, the teardown() callback method is called on the test class instance to clean up and/or release any resources used to run the test. The full lifecycle is illustrated in the Unified Modeling Language (UML) sequence diagram in Figure 9.1.

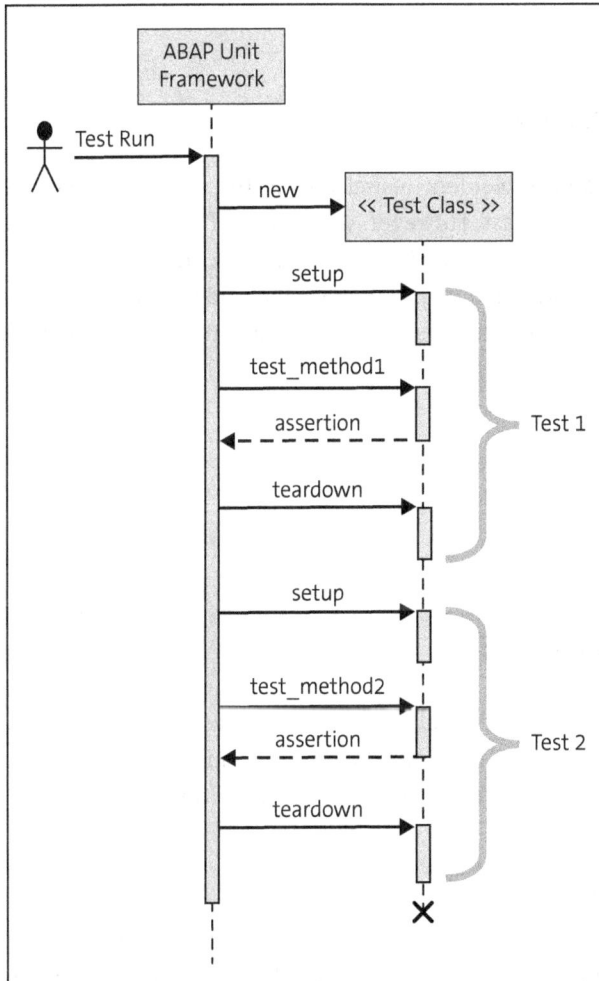

Figure 9.1 Understanding the Execution Flow for ABAP Unit Test Classes

The outcomes of the various tests are shown in the **ABAP Unit: Result Display** screen. The details shown provide information about what went wrong and where. These details are also context-sensitive, allowing you to navigate to the source of the problem within the ABAP Workbench.

9.1.3   ABAP Unit and Production Code

The goal of the ABAP Unit framework is to allow tests to be crafted and executed in a way that ensures code performs as expected. These tests should run throughout the full development lifecycle to ensure that defects or software regressions are not introduced when code is promoted or transported to a production environment. In a live environment, tests could potentially cause problems if allowed to execute and, for this reason, ABAP does not generate byte code for ABAP Unit tests in production systems. This ensures that the tests cannot be executed in a business-critical environment and that your test code does not become a strain or cause adverse side effects.

9.2   Creating Unit Test Classes

For the most part, you define and implement test classes in the same way that you would build a regular ABAP Objects class. However, you must define test classes and test methods using the FOR TESTING addition as shown in Listing 9.1. This addition effectively divides an application into two separate parts—test code and production code—so that ABAP understands which code should not be generated in production systems, as mentioned in Section 9.1.3.

```
CLASS ltc_my_test_class DEFINITION
      FOR TESTING.
[Risk Level {Critical|Dangerous|Harmless}]
[Duration {Short|Medium|Long}]
[...]
ENDCLASS.
```

Listing 9.1 Basic Form of ABAP Unit Test Classes

The following sections describe the pieces that you will need to assemble in order to create a unit test class. We also discuss naming conventions and introduce the test class generation wizard, which helps bootstrap your unit test development.

9.2.1   Unit Test Naming Conventions

As you've learned, it's important to carefully consider the naming structure of your class, method, and variable names in ABAP. The same goes for unit test construction. One important reason for this outside the scope of readability and maintenance is the fact that if a unit test fails, the class and method names will appear in the failure message. If the class and method name provide adequate information on what is being tested, developers will be able to better identify and resolve problems with the code.

Although there are no binding naming conventions, SAP does provide suggestions for ABAP Unit test class prefixes (see Table 9.2).

Suggested Prefix	Type of Class
LTC_	Local ABAP Unit test class
LTD_	Local test double
LTH_	Local test utilities class

Table 9.2 SAP-Suggested ABAP Unit Test Class Prefixes

It's ultimately up to the developer to write clean, understandable, and organized unit tests. Many unit test naming strategies have been adopted throughout the development community across various technologies. These strategies were created to allow unit test names to express specific requirements, identify the targeted unit of work (methods, functions, classes, or combinations thereof) and state the expected results. If you don't already have naming conventions in place for your unit test methods, you can explore online resources to find a convention that suits your needs.

9.2.2   Generating Test Classes for Global Classes

You can generate unit test classes for global classes using the test class generation tool in the Class Builder. To access it, place your cursor on the name of the object you wish to create a unit test class for and select the menu path **Utilities • Test Classes • Generate**. Alternatively, you can display the context menu and select **Create • Generate Test Class**. This launches the **Test Class Generation** wizard shown in Figure 9.2.

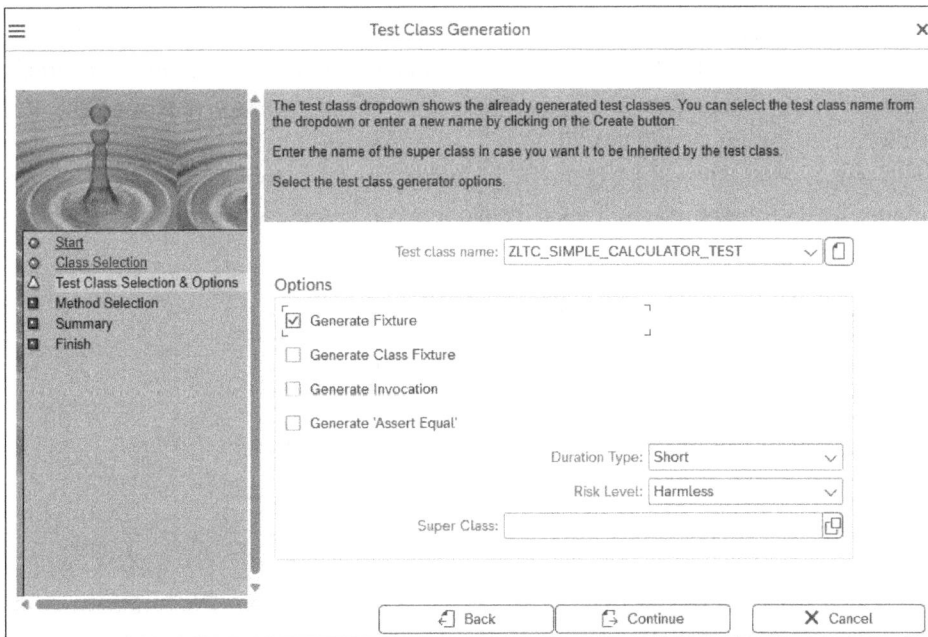

Figure 9.2 Working with the Test Class Generation Wizard

Here, you're presented with options for specifying the unit test class attributes, generating fixtures, test methods, and so on. While you can quickly bootstrap a unit test class this way, you're still responsible for building setup and teardown operations, along with implementing all your unit test methods and assertions within them.

Once your test class is generated, you can access it from the Class Builder by selecting **Goto · Local Test Classes** from the Class Editor menu. The test class will also be available in the Object Navigator (Transaction SE80) in the **Local Classes** folder under the object with which the class was associated.

9.2.3   Defining Test Attributes

When defining a test class, you *must* specify two test attributes used by the ABAP runtime environment during the execution of a test: the *risk level* and the *execution duration*. These attributes are defined as statements in systems with release 7.02 or higher (see Listing 9.1) and as special *pseudo comments* in older systems (see Listing 9.2). You include them after the CLASS ... FOR TESTING statement.

```
CLASS ... FOR TESTING.
"#AU Risk_Level Critical|Dangerous|Harmless
"#AU Duration Short|Medium|Long
```

Listing 9.2 Test Attributes in Systems with Releases Earlier than 7.02

The RISK LEVEL attribute describes the effects that a test could have on the system. It's possible that test methods may invoke functionality that could make changes to system settings and/or the database. The three risk level values you can assign are CRITICAL, DANGEROUS, and HARMLESS, with HARMLESS meaning that the test would *not* make changes to persistent data or impact the system in any negative way (see Table 9.3). However, every test should be written with the objective of being harmless to the environment within which it's executed. By leveraging the dependency injection techniques we'll discuss in Section 9.4.1, developers can prevent unwanted consequences from running a unit test.

Risk Level	Potential Side Effects
CRITICAL	Could alter system settings, Customizing, and so on.
DANGEROUS	Could change records in the database.
HARMLESS	No side effects; the test is innocuous.

Table 9.3 Risk Level Attribute Values

You can restrict test classes that introduce any risk over that of HARMLESS for execution based on client Customizing settings (defined in the Implementation Guide [IMG] or in

Transaction SAUNIT_CLIENT_SETUP). This way, for example, you can protect a *golden client* from test side effects that could impact other project efforts.

The DURATION attribute specifies the expected execution duration of a test class. This attribute helps the ABAP runtime environment know when a test has run too long (perhaps due to an error in the test code, such as an infinite loop). The possible values of this attribute are SHORT, MEDIUM and LONG, and they have default values of 1 minute, 5 minutes, and 60 minutes, respectively. These default values can also be adjusted in Transaction SAUNIT_CLIENT_SETUP.

Your goal when writing unit tests is to keep them quick and lean. Any test that takes longer than a few seconds should be flagged for examination, as both the test and the target class under test could need refactoring. As you build up your unit test arsenal, you should be confident that all of them could be executed at any time. If these tests are HARMLESS and have a SHORT duration, you can confidently make changes in the code and execute unit tests to ensure that no regressions or defects were introduced.

9.2.4 Building Test Methods

Test methods are defined as parameterless instance methods in a test class. The signature of these methods also requires the FOR TESTING addition (see Listing 9.3).

```
CLASS ltc_my_test_class DEFINITION FOR TESTING
      RISK LEVEL HARMLESS
      DURATION SHORT
      FINAL.
  PRIVATE SECTION.
METHODS:
        test_method1 FOR TESTING.
        test_method2 FOR TESTING.
```

Listing 9.3 Defining Test Methods

Each test method in a test class corresponds to a single unit test. As described previously, a test method should concentrate on testing a single software unit (e.g., a method or a function module) rather than testing an entire application. It's important to keep unit tests granular so that you can focus on potential bugs that might creep into different parts of the program. Most of the time, implementing a test method will simply consist of a single call to a module of the program under test followed by a status check using utility methods defined in class CL_ABAP_UNIT_ASSERT.

If you look closely at the code in Listing 9.3, you'll notice that the test methods have been defined in the private section of the test class. This is by design because test classes implicitly share a friendship relationship with the test driver of the ABAP runtime environment. Consequently, you should prefer to define your test methods in the private or protected (if it's inherited) sections of your test class.

9.2.5   Working with Fixtures

Test classes group related test methods (i.e., unit tests) together into a logical unit. In these classes, you can define special fixture methods that help set up and tear down unit tests. These methods have predefined names that are automatically recognized by the ABAP runtime environment. Each method is defined in the private section of the class and has no parameters. Table 9.4 describes the fixture methods supported by the ABAP Unit framework.

Method Name	Usage Type
setup()	This instance method is called prior to the invocation of every test method in the test class.
teardown()	This instance method is called after every invocation of a test method in the test class.
class_setup()	This class method is called once before any test methods are called in the test class.
class_teardown()	This class method is invoked after all of the test methods in the test class have been called.

Table 9.4 Fixture Methods and Their Usages

Fixture methods are an excellent place to define common initialization code that is relevant for all the test methods in a test class. In particular, the instance methods setup() and teardown() provide a useful hook for implementing code that ensures that each test is executed independently using the proper runtime configuration.

9.2.6   Working with Test Seams

In release 7.50 of the ABAP programming language, SAP introduced a pair of statements: the TEST-SEAM and TEST-INJECTION statements. These statements make it easy to mock external dependencies such as databases and interfaces.

To illustrate how these new statements work, consider the sample code in Listing 9.4. In this case, we have a simple class called LCL_CLIENT that defines a method called get_partners(). Within the get_partners() method, let's imagine that we're using a remote function call (RFC) or Business Application Programming Interface (BAPI) function to compile the list of business partners—perhaps from an external customer relationship management (CRM) system. Since this is a heavy-duty lookup that could potentially fail or time out, we want to mock this lookup in our code. Put differently, we want to fake the lookup and just produce some reasonable test data that we can then use to validate the rest of our logic.

Within our sample client class, all we need to do is use the TEST-SEAM statement to mark off the section of the code that we want to mock later on. In Listing 9.4, we called that test seam partner_lookup.

```
CLASS lcl_client DEFINITION.
  PUBLIC SECTION.
    METHODS:
      get_partners RETURNING VALUE(rt_partners) TYPE /bobf/t_frw_key.
ENDCLASS.

CLASS lcl_client IMPLEMENTATION.
  METHOD get_partners.
    TEST-SEAM partner_lookup.
      CALL FUNCTION...DESTINATION
        TABLES
          et_partners = rt_partners.
    END-TEST-SEAM.
  ENDMETHOD.
ENDCLASS.

CLASS ltc_test_client DEFINITION
  FOR TESTING RISK LEVEL HARMLESS DURATION SHORT FINAL.
  PRIVATE SECTION.
    METHODS:
      test_partner_lookup FOR TESTING.
ENDCLASS.

CLASS ltc_test_client IMPLEMENTATION.
  METHOD test_partner_lookup.
    TEST-INJECTION partner_lookup.
      TRY.
        "Instead of calling the RFC/BAPI function, just generate some
        "arbitrary test data:
        DATA(lv_guid1) = cl_system_uuid=>create_uuid_x16( ).
        APPEND lv_guid1 TO rt_partners.

        DATA(lv_guid2) = cl_system_uuid=>create_uuid_x16( ).
        APPEND lv_guid2 TO rt_partners.
      CATCH cx_uuid_error INTO DATA(lo_uuid_error).
      ENDTRY.
    END-TEST-INJECTION.

    DATA(lo_client) = NEWlcl_client( ).
    DATA(lt_partners) = lo_client->get_partners( ).
```

```
    cl_abap_unit_assert=>assert_equals(...)
  ENDMETHOD.
ENDCLASS.
```

Listing 9.4 Working with Test Seams

With our test seam in place, we can then mock the call to the external function in our test class using the TEST-INJECTION statement shown in Listing 9.4. Now, when the test is executed, this inclusion will cause the runtime environment to bypass the RFC call and swap in the arbitrary code we included to fill the RT_PARTNERS table. From there, it's business as usual as we test the LCL_CLIENT class and validate the logic in our get_part-ners() method.

9.2.7   Defining Reusable Test Classes

There is often a need to design reusable pieces of logic that may be used by many local unit test classes. These *global test classes* can take the form of a service or helper class to facilitate unit test creation or act as a parent to a unit test class to reuse logic by inheritance.

The only rules when creating a *service class* are that the service class must still contain the FOR TESTING option and that it will not implement any fixtures or unit test methods. The service class is essentially a standard ABAP class that needs to be associated with ABAP Unit.

A *parent unit test class* is created as an ABSTRACT class. It also contains the FOR TESTING option and implements at least one fixture or test method. Local test classes may now inherit from this parent unit test class and reuse its logic while still maintaining their local relationship to the code they were intended to test. If a parent unit test class is not declared as ABSTRACT, then ABAP Unit will report a warning. For more information, consult the documentation "ABAP Test and Analysis Tools: Correct Use of Global Test Classes," available at *http://help.sap.com*.

9.3   Assertions in ABAP Unit

An assertion is as a verification of a condition using a true/false proposition. This condition is evaluated by code encapsulated within what ABAP Unit refers to as a *constraint*. The CL_ABAP_UNIT_ASSERT class contains many methods to test constraints, and these methods should cover most of your needs when verifying outcomes within unit tests (see Table 9.5 for some examples). All these methods expect a piece of data to be verified that is *actually* returned from the code being tested (the act parameter). The assert_equals() also requires an expected piece of data (the exp parameter), which is used in a comparison against that data passed into the act parameter (see Listing 9.5).

```
cl_abap_unit_assert=>assert_equals(
  EXPORTING
    act = lv_actual
    exp = lv_expected
    msg = 'Unit Test Test Failed' ).
```

Listing 9.5 Assertion Example

When all assertions within a test method evaluate true, the unit test has successfully passed, and based on the tested condition, the code path is functioning as expected. If any assertion fails in a unit test, that unit test will not pass, and an error will occur in the **ABAP Unit: Result Display** screen when executing the test. The severity level and any text passed into the msg optional import parameters of the assertion will show under the **Failures and Messages** panel in the upper-right portion of the screen (Section 9.7).

Method Name	Description
assert_equals()	The two objects being evaluated must have a comparable type. This assertion will return true if the objects are equal and may be enhanced to perform *complex comparisons* by creating a class that implements IF_AUNIT_OBJECT (logic for comparison resides in the equal_to() method) and passing that class as the argument of the exp parameter of assert_equals().
assert_bound()	Returns true if the reference of a reference variable has the ABAP property IS BOUND.
assert_initial()	Returns true if an object has the ABAP property IS INITIAL.
assert_subrc()	Returns true if a specific value is returned from SY-SUBRC.
assert_that()	Evaluates a specific condition and is used in testing custom constraints (Section 9.3.1). The exp parameter expects an instance of a class that implements IF_CONSTRAINT.
fail()	Forces the test to fail.

Table 9.5 Examples Methods of CL_ABAP_UNIT_ASSERT

9.3.1   Creating and Evaluating Custom Constraints

If you ever need to evaluate a condition outside the scope of the methods present within CL_ABAP_UNIT_ASSERT, you can apply *custom constraints* by creating a class that implements the IF_CONSTRAINT interface (see Listing 9.6). This custom constraint class must, in turn, implement two methods defined by IF_CONSTRAINT. These are the is_valid() method, which contains the evaluation logic of the constraint and returns a Boolean result on whether the condition was met, and the get_description() method, which returns a message describing the test results. To utilize the user-defined constraint within unit tests, you can invoke the assert_that() method of CL_ABAP_UNIT_

ASSERT and pass the custom constraint's methods as the exp parameter. Any necessary parameters of your custom constraint must be passed via the constructor of the class and set as attributes.

```
CLASS ltc_my_constraint DEFINITION.
    PUBLIC SECTION.
            INTERFACES if_constraint.
ENDCLASS.
```

Listing 9.6 Defining a Custom Constraint

9.3.2   Applying Multiple Constraints

If you require more complex assertion, you can also combine multiple custom constraints via logical operators (AND, OR, NOT, XOR, etc.) in the CL_AUNIT_CONSTRAINTS class of ABAP Unit. These logical operators are represented as static methods of the CL_AUNIT_CONSTRAINTS class and return instances of objects that implement the IF_CONSTRAINT interface. The resulting combined constraint joined by this logical operator may, in turn, be passed to the exp parameter of the assert_that() method of CL_ABAP_UNIT_ASSERT (see Listing 9.7).

```
DATA:
  " lo_first_constraint and lo_second_constraint implement
  " IF_CONSTRAINT interface
  lo_first_constraint     TYPE REF TO zltc_first_constraint,
  lo_second_constraint    TYPE REF TO zltc_second_constraint,
  lo_compound_constraint  TYPE REF TO if_constraint,
  lo_actual_results       TYPE REF TO data.

* Create instances of both custom constraints
CREATE OBJECT lo_first_constraint.
CREATE OBJECT lo_second_constraint.

* Make call to code under test to retrieve actual results
  ...

* Combine constraints using static method of cl_aunit_constraints
lo_compound_constraint = cl_aunit_constraints=>and(
  c1 = lo_first_constraint
  c2 = lo_second_constraint ).

* Utilize compound constraint in assertion
cl_abap_unit_assert=>assert_that(
  act = lo_actual_results   " Data to check from code under test
  exp = lo_compound_constraint ).
```

Listing 9.7 Combining Custom Constraints with CL_AUNIT_CONSTRAINTS

To leverage CL_AUNIT_CONSTRAINT, create an instance of each custom constraint, passing any required parameters via the class constructors (see Listing 9.7). A call can then be made to the required logical operator static method of CL_AUNIT_CONSTRAINT using your custom constraint objects as parameters. The static method will return a new instance of an object that implements IF_CONSTRAINT. This new instance will allow for a check against both custom constraints and evaluate those constraints based on the logical operation used to join them. The result is a class that can be used as the exp parameter of the assert_that() method of CL_ABAP_UNIT_ASSERT.

Multiple Assertions in a Unit Test

There are differing opinions on the number of assertions that should be present within each unit test or unit test method. One idea is that each test method should contain only one assertion to keep the code clean and easy to understand. You should minimize assertions whenever possible, and one assert per unit test is a good guideline to follow.

However, an alternative view is to restrict your unit tests to evaluate a single concept. If you're testing two independent outcomes, then two tests are necessary. However, if you're testing a single concept that requires multiple assertions to cover specific dimensions of the outcome, a single test method or unit test will suffice.

9.4   Managing Dependencies

One of the positive side effects of adopting a unit testing framework and writing tests against ABAP objects is that these objects will require a break from outside dependencies to be properly tested. Database operations, function module calls, authorization checks, and other services will need to be abstracted, encapsulated into specialized ABAP objects, and set up to be *injectable* into ABAP classes.

This exercise is known as dependency injection. In the following sections, we'll review some options for resolving dependencies in code and creating modularized, testable ABAP objects.

9.4.1   Dependency Injection

Dependency injection comes in many forms (see the upcoming text box) and allows you to set up encapsulated ABAP objects and services for any dependent class. The goal is to incorporate better software design and, in the context of ABAP Unit, to better isolate a unit of code by letting you inject test doubles in place of normal production objects during unit testing.

These test doubles can take the form of a test stub or mocked object (among others), are usually simplified versions of their production counterparts, and are set to return predictable data so that only the code under test may be targeted and evaluated properly.

This helps to ensure that each test is independent and repeatable in any environment, which is crucial for any unit test to be valid.

For more information on resolving dependencies and using test doubles or mocked objects in ABAP, check out *ABAP to the Future* (SAP PRESS, 2022).

Methods of Dependency Injection

The following is a list of standard techniques to decouple your object-oriented application code from encapsulated services and other dependencies:

- **Constructor injection**
  Needed ABAP objects or dependencies are passed as parameters to the class constructor.
- **Setter injection**
  This requires that the ABAP object provides setter methods for each dependency.
- **Interface injection**
  An interface defines the injection method or the setter methods of the class dependencies.

Comparison of these methods is beyond the scope of this book, but you can find more information in the documentation "Managing Dependencies in ABAP Unit Testing," available at *http://help.sap.com*, or refer to other online resources.

9.4.2   Private Dependency Injection

Another dependency injection method available in ABAP is *private dependency injection*, a technique that is better suited to substituting dependencies while writing unit tests in ABAP Unit. By associating a test class as a FRIEND to a class under test, the test class will have access to private and protected attributes of the class under test so that dependencies may be injected. This allows dependencies to be broken while protecting the class being tested from any outside manipulation. For this method to work properly, dependencies in the class under test must be resolved or set in methods outside of those that leverage them. In other words, the class constructor() would establish these dependencies and set them to private or protected attributes so that other methods of the class may use these objects in their logic.

You can find more information about private dependency injection in the documentation "ABAP Test and Analysis Tools," available at *http://help.sap.com*. Sample code is provided to give an example of how to implement this form of dependency injection.

9.4.3   Partially Implemented Interfaces

Release 7.40 added an improvement that removed the requirement for test classes to completely implement an inherited interface. Using the PARTIALLY IMPLEMENTED addition, you can now create classes that act as test doubles, and you only need to implement the pieces of the interface that the code being tested requires (see Listing 9.8). This removes that hassle of having to implement the entire contract that an interface presents or add the pragma ##needed to each empty method implementation. Again, this method of implementing an interface in a class can *only* be utilized in test classes.

```
CLASS ltd_test_double DEFINITION
     FOR TESTING.
     INTERFACES if_interface PARTIALLY IMPLEMENTED.
ENDCLASS.
```

Listing 9.8 Basic Form of a Partially Implemented Interface

9.4.4 Working with Test Doubles

In Section 9.2.6, we showed you how the new TEST-SEAM statement could be used to mock data in a static way. When working with dependency injection, there will be times when you want to swap out implementations in a more dynamic fashion. Enter *test doubles*.

In basic terms, a test double is an object that stands in for a real object during a test. It's used to isolate the unit under test by controlling its dependencies. In that sense, there's nothing really fancy about test doubles from an object-oriented perspective—although there are plenty of resources online that show how developers build elaborate test frameworks using test doubles.

For developers working in SAP S/4HANA systems, one very interesting application for test doubles is testing database lookups using views from *core data services* (CDS). Imagine that you're working on a particularly complex set of CDS views and want to verify that the join conditions work and that the resultant data is formatted properly. However, if the CDS views use tables that don't have much data available, they might be difficult to test. Alternatively, perhaps the volume of data available makes it prohibitive to test as part of a larger test suite. In these situations, you can use the CDS test double framework to mock data in the underlying tables of your CDS views, which allows you to run different kinds of tests.

To put these concepts into perspective, consider the LTC_PARTNER_TEST class in Listing 9.9. This class is being used to test the SAP standard CDS view I_BusinessPartner. To facilitate the test, we're utilizing framework class CL_CDS_TEST_ENVIRONMENT to create a CDS test environment—see methods class_setup() and setup() for details. Then, in the test_partner_search() method, we're injecting data from table BUT000 as a test double to run our test. Once the test double is in place, the query against the I_Business-Partner view will be hijacked by the CDS test environment to pull the partner data from the test double rather than going directly to the BUT000 table. From there, it's business as usual in terms of testing the results of the query using assertions.

```
CLASS ltc_partner_test DEFINITION FINAL FOR TESTING
  DURATION SHORT
  RISK LEVEL HARMESS.
  PRIVATE SECTION.
    ...
ENDCLASS.

CLASS ltc_partner_test IMPLEMENTATION.
  METHOD class_setup.
    so_cds_environment =
      cl_cds_test_environment=>create( i_for_entity = 'I_BusinessPartner' ).
  ENDMETHOD.

  METHOD class_teardown.
    so_cds_environment->destroy( ).
  ENDMETHOD.

  METHOD setup.
    so_cds_environment->clear_doubles( ).
  ENDMETHOD.

  METHOD test_partner_search.
    "Create some static test partner data:
    DATA lt_partners TYPE STANDARD TABLE OF but000.
    APPEND INITIAL LINE TO lt_partners ASSIGNING FIELD-SYMBOL(<ls_partner>).
    <ls_partner>-partner = '1234567890'.
    <ls_partner>-type = '2'.
    <ls_partner>-name_org1 = 'Angus Animation'.
    <ls_partner>-crusr = <ls_partner>-chusr = sy-uname.
    <ls_partner>-crdat = <ls_partner>-chdat = sy-datum.
    <ls_partner>-crtim = <ls_partner>-chtim = sy-uzeit.

    APPEND INITIAL LINE TO lt_partners ASSIGNING <ls_partner>.
    <ls_partner>-partner = '2345678901'.
    <ls_partner>-type = '2'.
    <ls_partner>-name_org1 = 'Acme Anvil Corporation'.
    <ls_partner>-crusr = <ls_partner>-chusr = sy-uname.
    <ls_partner>-crdat = <ls_partner>-chdat = sy-datum.
    <ls_partner>-crtim = <ls_partner>-chtim = sy-uzeit.

    APPEND INITIAL LINE TO lt_partners ASSIGNING <ls_partner>.
    <ls_partner>-partner = '3456789012'.
    <ls_partner>-type = '2'.
    <ls_partner>-name_org1 = 'Coyote Enterprises'.
```

```
    <ls_partner>-crusr = <ls_partner>-chusr = sy-uname.
    <ls_partner>-crdat = <ls_partner>-chdat = sy-datum.
    <ls_partner>-crtim = <ls_partner>-chtim = sy-uzeit.

    "Inject the partner test data into the CDS environment as a test double:
    DATA(lo_partner_test_data) = cl_cds_test_data=>create( i_data = lt_partners
).
    DATA(lo_partners_double) = so_cds_environment->get_double( i_name = 'BUT000'
).
    lo_partners_double->insert( lo_partner_test_data ).

    "Perform the partner search using the CDS view:
    SELECT * FROM I_BusinessPartner INTO TABLE @DATA(lt_results).

    "Check the results:
    DATA(lv_count) = lines( lt_results ).
    cl_abap_unit_assert=>assert(...
  ENDMETHOD.
ENDCLASS.
```

Listing 9.9 Working with CDS Test Doubles

Although the example in Listing 9.9 is somewhat contrived, it sets the stage for much more complex testing scenarios where you're mocking data from a larger variety of tables and run through lots of different experiments. Working with test doubles is much more effective than trying to restage different data permutations when you're testing large-scale data solutions.

9.4.5   Other Sources of Information

The purpose of this section was to review some important software patterns that will aid in unit testing, but all this information may be overwhelming for the uninitiated. We also understand that it would require a substantial refactoring effort to reshape legacy code that was not crafted with dependency encapsulation and injection in mind. It's not required that you completely understand or implement all of the techniques right away. These are complicated topics that go beyond the scope of this book. The good news is that there are many wonderful sources of information on topics such as dependency injection, software design patterns and *SOLID design principles* available online and in print. Here's a selection of our recommendations:

- "ABAP Unit (Release 7.4)" documentation, available at *http://help.sap.com*
- *Clean Code: A Handbook of Agile Software Craftsmanship* by Robert C. Martin (Prentice Hall, 2009)
- *ABAP to the Future* by Paul Hardy (SAP PRESS, 2022)

- "The Principles of OOD," available at *http://www.butunclebob.com/ArticleS.Uncle-Bob.PrinciplesOfOod*

- *Head First Design Patterns: A Brain-Friendly Guide* by Freeman et al. (O'Reilly Media, 2004)

SOLID Design Principles

SOLID is a mnemonic acronym for the first five object-oriented design (OOD) principles created by Robert C. Martin (also known as Uncle Bob):

- **S** (The Single-Responsibility Principle): A class should have only a single responsibility.
- **O** (The Open-Closed Principle): Objects or entities should be open for extension but closed for modification.
- **L** (The Liskov Substitution Principle): Objects in a program should be replaceable with instances of their subtypes without altering the correctness of that program.
- **I** (The Interface Segregation Principle): Many client-specific interfaces are better than one general-purpose interface.
- **D** (The Dependency Inversion Principle): You should depend on abstractions and not upon concretions. Dependency injection, covered in Section 9.4, is an example of this principle.

Refer to *http://www.butunclebob.com/ArticleS.UncleBob.PrinciplesOfOod* for more information.

9.5   Case Study: Creating a Unit Test in ABAP Unit

Now that you have a basic understanding of how to create unit test classes, consider this example. The simple calculator class ZCL_SIMPLE_CALCULATOR in Listing 9.10 contains basic arithmetic operations in the form of methods. Accompanying this class is an ABAP Unit test class named ZLTC_SIMPLE_CALCULATOR_TEST with unit tests to cover one addition and subtraction condition along with testing whether the CX_SY_ZERODI-VIDE exception is properly thrown by the divide() method when the divisor parameter is set to zero. The setup() fixture method within the unit test class instantiates and initializes the ZCL_SIMPLE_CALCULATOR object, which is then used by each test method. The teardown performs cleanup after every unit test has executed.

We adopted the MethodName_StateUnderTest_ExpectedBehavior naming convention for each of the unit tests outlined in Listing 9.10 but, as described in Section 9.2.1, the development community has introduced other viable strategies.

```
* Class Under Test
CLASS zcl_simple_calculator DEFINITION
  PUBLIC
  FINAL
```

```
  CREATE PUBLIC.
  PUBLIC SECTION.
    TYPES:
      ty_quotient type p LENGTH 7 DECIMALS 4 .
    METHODS add
      IMPORTING
        iv_first_addend  TYPE int1
        iv_second_addend TYPE int1
      RETURNING
        VALUE(rv_sum) TYPE int2.
    METHODS subtract
      IMPORTING
        iv_minuend     TYPE int1
        iv_subtrahend  TYPE int1
      RETURNING
        VALUE(rv_difference) TYPE int1.
    METHODS multiply
      IMPORTING
        iv_first_factor  TYPE int1
        iv_second_factor TYPE int1
      RETURNING
        VALUE(rv_product) TYPE int4.
    METHODS divide
      IMPORTING
        iv_dividend  TYPE int1
        iv_divisor   TYPE int1
      RETURNING
        VALUE(rv_quotient) TYPE ty_quotient
      RAISING
        cx_sy_zerodivide.
ENDCLASS.

CLASS zcl_simple_calculator IMPLEMENTATION.
  METHOD add.
    rv_sum = iv_first_addend + iv_second_addend.
  ENDMETHOD.
  METHOD divide.
    rv_quotient = iv_dividend / iv_divisor.
  ENDMETHOD.
  METHOD multiply.
    rv_product = iv_first_factor * iv_second_factor.
  ENDMETHOD.
  METHOD subtract.
    rv_difference = iv_minuend - iv_subtrahend.
```

9

```
    ENDMETHOD.
ENDCLASS.

* Unit Test Class
CLASS zltc_simple_calculator_test definition FOR TESTING
  DURATION SHORT
  RISK LEVEL HARMLESS.
  PRIVATE SECTION.
    DATA mo_calculator TYPE REF TO zcl_simple_calculator.
    METHODS setup.
    METHODS add_2plus2_sum4 FOR TESTING.
    METHODS subtract_6minus2_dif4 FOR TESTING.
    METHODS divide_zerodivisor_exception FOR TESTING.
    METHODS teardown.
ENDCLASS.

CLASS ZLTC_SIMPLE_CALCULATOR_TEST IMPLEMENTATION.
  METHOD setup.
    CREATE OBJECT mo_calculator.
  ENDMETHOD.

  METHOD add_2plus2_sum4.
    DATA lv_expected      TYPE int2 VALUE 4.
    DATA lv_sum           TYPE int2.
    lv_sum = mo_calculator->add(
      EXPORTING
        iv_first_addend   = 2
        iv_second_addend  = 2 ).
    cl_abap_unit_assert=>assert_equals(
      EXPORTING
        act = lv_sum
        exp = lv_expected
        msg = 'Calculator Addition Test Failed' ).
  ENDMETHOD.

  METHOD subtract_6minus2_dif4.
    DATA lv_expected    TYPE int2 VALUE 4.
    DATA lv_difference  TYPE int2.
    lv_difference = mo_calculator->subtract(
      EXPORTING
        iv_minuend    = 6
        iv_subtrahend = 2 ).
    cl_abap_unit_assert=>assert_equals(
      EXPORTING
```

```
      act = lv_difference
      exp = lv_expected
      msg = 'Calculator Subtraction Test Failed' ).
  ENDMETHOD.

  METHOD divide_zerodivisor_exception.
    TRY.
        mo_calculator->divide(
          EXPORTING
            iv_dividend = 5
            iv_divisor  = 0 ).
        cl_abap_unit_assert=>fail(
          msg   = 'CX_SY_ZERODIVIDE was not raised' ).
      CATCH cx_sy_zerodivide.
    ENDTRY.
  ENDMETHOD.

  METHOD teardown.
    CLEAR mo_calculator.
  ENDMETHOD.
ENDCLASS.
```

Listing 9.10 A Simple Unit Test Example

Since CL_ABAP_UNIT_ASSERT has not defined an assertion method to cover exceptions, the divide_zerodivisor_exception() method in the unit test class in Listing 9.10 had to be implemented in an atypical manner. The method wraps a forced exception in a TRY block and sets the unit test to failed (via the fail() method of the CL_ABAP_UNIT_ASSERT class) if the exception is not thrown and caught, which would skip the fail method call entirely and move right to the CATCH statement.

9.6   Executing Unit Tests

After you've created your unit tests in ABAP Unit, you can run them in several different ways. In the following subsections, we'll look at options for performing unit tests individually using the ABAP Workbench and in batch via the ABAP Unit Test Browser or the Code Inspector tool.

9.6.1   Integration with the ABAP Workbench

As we stated previously, the ABAP Unit test tool is tightly integrated into the ABAP Workbench. Therefore, it's easy to start test runs using standard menu options. For example, to initiate a test run for the ZLTC_SIMPLE_CALCULATOR_TEST test class defined in

Listing 9.10, follow the context menu path of the ZLTC_SIMPLE_CALCULATOR by right-clicking the host object and selecting **Execute • Unit Tests**. Upon execution, you'll be taken to the result display view shown in Figure 9.3 to review the results of the test run.

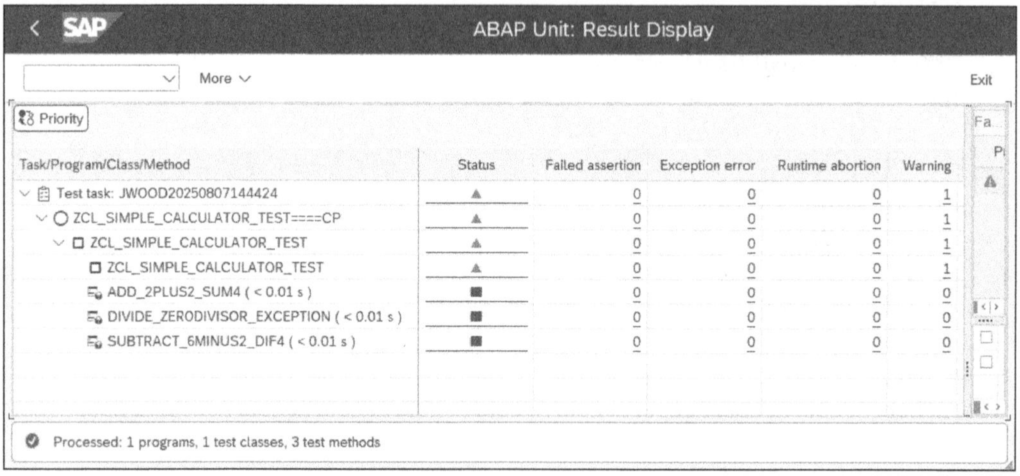

Task/Program/Class/Method	Status	Failed assertion	Exception error	Runtime abortion	Warning
⌄ 🗐 Test task: JWOOD20250807144424	▲	0	0	0	1
⌄ ○ ZCL_SIMPLE_CALCULATOR_TEST====CP	▲	0	0	0	1
⌄ ☐ ZCL_SIMPLE_CALCULATOR_TEST	▲	0	0	0	1
☐ ZCL_SIMPLE_CALCULATOR_TEST	▲	0	0	0	1
🖳 ADD_2PLUS2_SUM4 (< 0.01 s)	■	0	0	0	0
🖳 DIVIDE_ZERODIVISOR_EXCEPTION (< 0.01 s)	■	0	0	0	0
🖳 SUBTRACT_6MINUS2_DIF4 (< 0.01 s)	■	0	0	0	0

Processed: 1 programs, 1 test classes, 3 test methods

Figure 9.3 Executing the Simple Calculator Unit Tests

There's no predefined sequence in which the test methods are executed; after all, they are meant to be run independently. If the test succeeds, then a success message will appear in the status bar at the bottom of the screen. However, if there are errors in the unit test, then the ABAP Unit interface will be displayed. We'll look at the results of a test run with errors in Section 9.7.

9.6.2   Creating Favorites in the ABAP Unit Test Browser

Although the ABAP Unit Test Browser is integrated into the ABAP Workbench, before you can make use of it, you'll need to select **Utilities • Settings...** from the main menu path and select the **Workbench (General)** tab. On that tab, check the **ABAP Unit Test Browser** checkbox under **Browser Selection**.

With this browser enabled, you can group unit tests together into *favorites* and run them as a whole or individually. First, select **ABAP Unit Browser** from the navigation menu on the left and make sure that the **Favorite** is the value selected within the **Choose Selection Criteria** field. You can now create a new favorite by clicking on the **Create Favorite** button and assigning it a name (**Favorite**) and title (**Title**; see Figure 9.4 and Figure 9.5).

Once you create the favorite group, you can add unit tests by switching to edit mode and clicking on the **Add Elements** button from the top of the right-hand navigation panel. You will be presented with a search screen to locate any objects and their associated unit tests to add to your favorite group.

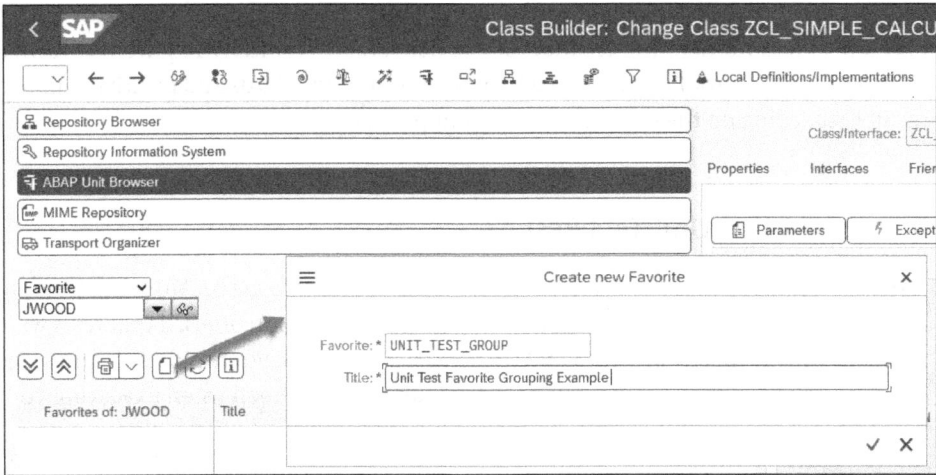

Figure 9.4 Creating a New ABAP Unit Browser Favorite Group

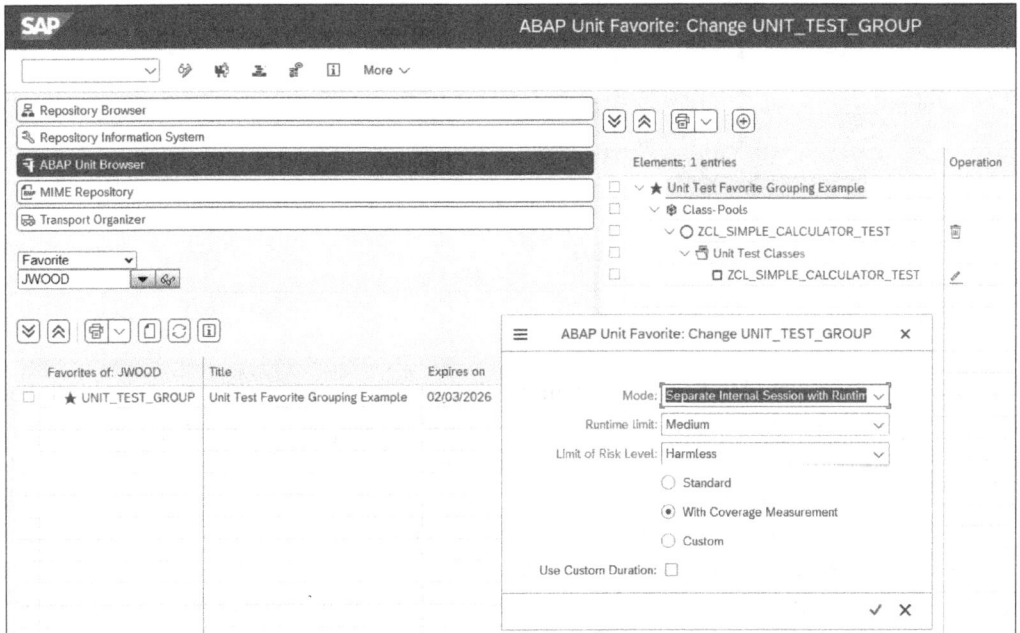

Figure 9.5 Show/Edit Options for an ABAP Unit Browser Favorite

9.6.3   Integration with the Code Inspector

You can also integrate ABAP Unit tests inside the Code Inspector tool (Transaction SCI). This tool performs additional static checks for ABAP Repository objects, including the verification of naming conventions for variables, the proper use of ABAP statements, and so on. Although the configuration and use of this tool is outside of the scope of this book, it's very useful for implementing additional quality assurance steps in your development cycle. The integration of ABAP Unit inside the Code Inspector allows developers to automate the creation of the deliverables typically required by formal code reviews (i.e., proof of adherence to project coding standards and positive unit test results), speeding up the overall development process.

9.7   Evaluating Unit Test Results

Although unlikely, changes could be made to the ZCL_SIMPLE_CALCULATOR class (refer to Listing 9.10) that would cause a method to return incorrect outcomes. If unit tests were present to cover these outcomes and their expected results, you would see a message like the one in Figure 9.6 in the **ABAP Unit: Result Display** screen when executing your unit tests against the simple calculator class.

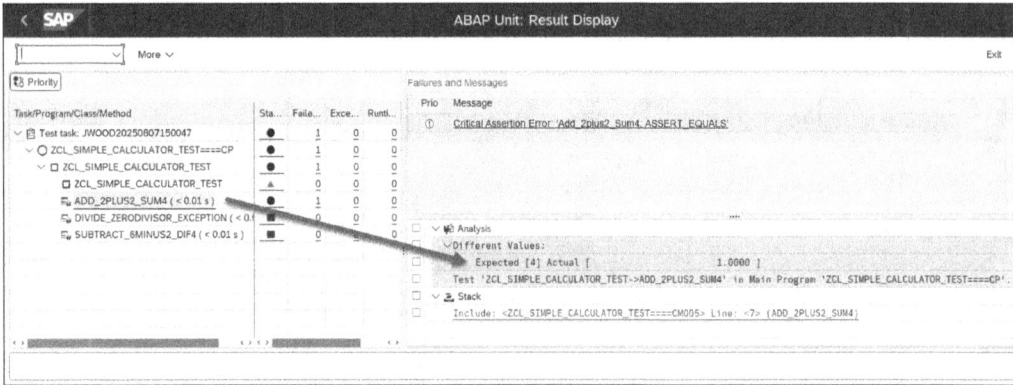

Figure 9.6 When a Unit Test Fails

ABAP Unit would indicate a problem by showing an error status for the failed test. The top-right **Failures and Messages** panel would also display the severity level and any message (both optional import parameters of methods of the CL_ABAP_UNIT_ASSERT class) associated with the assertion(s) that failed. None of the unit tests created in this chapter passed in a severity level as a parameter, so the default of CRITICAL was set. The bottom-right panel displays an analysis of the test with the expected and actual results (in this case) of the assertion along with a stack trace. If the test task had been larger, there could have been more errors generated from other test methods. Therefore, the stack information provided in the bottom-right pane can be very useful for determining where a particular assertion failed. Clicking on any line within the stack trace allows you to dig into the test code to begin diagnosing why the unit test failed.

9.8   Measuring Code Coverage

As more unit tests are written to cover specific code paths and outcomes, the degree of the application's code that is being tested by coded unit tests, or *code coverage*, increases. There are tools available that interpret and visualize this coverage, which helps identify areas of an application that are lacking in unit tests or may not have any unit test coverage at all. Using this information, developers are better able to target code for future unit testing development efforts.

One thing to understand about code coverage is that it's *not* an indication of the *quality* of the unit tests covering the application code. There's always the potential for a poorly written unit test to give the illusion that specific lines of code are less prone to software regressions. A unit test could even be created without an assertion and, thus, not test any condition of a unit of code but still contribute to the overall code coverage of that application.

While it would be optimal to achieve 100% code coverage across an application, it's often more prudent to focus on writing quality unit tests that address the many outcomes of a unit of code. Since a program can have many more potential outcomes than lines of code, some code may require full coverage many times over. In other words, the goal in writing unit tests should not just be to attain a specific percentage of coverage but to carefully consider the outcomes that should be tested when designing software units.

Your tests are just as important as any other code and should give you confidence when refactoring. If your code coverage is high but bugs still make their way to production or you are hesitant to refactor, then your unit test quality is not being held in high enough regard.

You can visualize code coverage by navigating to the ABAP Workbench (Transaction SE80), right-clicking the object, and following the context menu path **Execute · Execute Tests With · Coverage**. Under the **Coverage Metrics** tab (see Figure 9.7), you can view the degree of code coverage by branch, procedure, and statement along with the option to drill down into code coverage by statement by double clicking on any of the methods of ZCL_SIMPLE_CALCULATOR class.

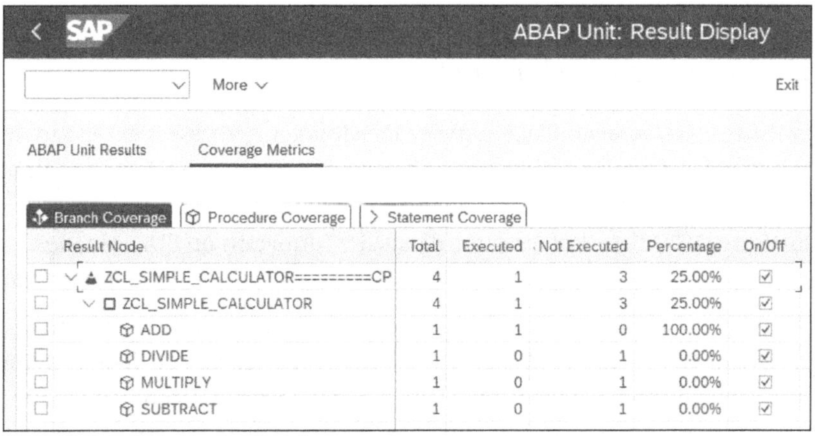

Result Node	Total	Executed	Not Executed	Percentage	On/Off
ZCL_SIMPLE_CALCULATOR=========CP	4	1	3	25.00%	☑
ZCL_SIMPLE_CALCULATOR	4	1	3	25.00%	☑
ADD	1	1	0	100.00%	☑
DIVIDE	1	0	1	0.00%	☑
MULTIPLY	1	0	1	0.00%	☑
SUBTRACT	1	0	1	0.00%	☑

Figure 9.7 Simple Calculator Code Coverage Example

9.9   Moving Toward Test-Driven Development

An advanced technique for developing unit tests involves writing a test *before* any new functionality or improvement is crafted. This software design technique is known as *test-driven development* (TDD) and allows for a unit test to drive development in small increments. TDD requires a developer to first write a failing unit test targeting a specific outcome in an automated testing environment such as ABAP Unit. When the test is complete, the developer writes the minimal amount of code needed to make that unit test pass. This process of red (writing a small failing test) to green (writing the minimal functional code needed to make the test pass) to refactor (improve the design of the code while making sure that all tests continue to pass) is repeated until a new piece of functionality is complete.

Although the full-blown use of TDD (and indeed, the extreme programming methodology from which it came) may be too controversial for your development team, the value of quality automated unit tests cannot be underestimated. It's imperative that you define your ABAP Unit tests as quickly as possible so that you can incorporate them into your normal development process. These tests will help keep you on target by providing immediate feedback whenever your individual modules begin to deviate from the terms of their API contracts. Unit tests will also shed light on areas of your design that need some work. For example, if you find that a given module is difficult to test, there's likely something wrong with it.

Finally, unit tests should inspire you with the confidence to take risks in your development. For instance, in Chapter 5, we discussed the concept of refactoring to improve the design of existing code. Without unit testing, you might be hesitant to perform certain refactorings for fear of breaking some unforeseen dependent code. Similarly, you might also be cautious about implementing enhancements for the same reasons. However, with unit tests, you can apply the changes and know immediately whether you broke something in the system without having to conduct a full-scale regression test.

9.10   Behavior-Driven Development

The *behavior-driven development* (BDD) software development process is similar to the TDD process but with a focus on incorporating the domain (or business) experts along with the software developers when developing and writing unit tests. We'll provide a brief introduction to BDD here, but for more detailed information on BDD, see the blog post "Introducing BDD" by Dan North, available at *http://dannorth.net/introducing-bdd/*.

The goal of BDD is to adopt natural language in both unit test method naming and acceptance test definitions so that all parties involved in development can understand the intent and the expected *behavior*. Unit test method names are generally written based on the outcomes that *should* occur. For example, take `return_new_object_instance()`, where the word "should" is either implied or commented in ABAP due to the 30-character length restriction for method names. This naming convention helps everyone understand what is being tested and the expected results (or what you *should* see as returned data in some cases).

In BDD, the following template is used to define the functional acceptance test scenarios and act as a guide in the creation of unit tests:

- Given some initial context (unit test setup),
- When an event occurs (calling the code under test),
- Then ensure some outcome(s) (a unit test assertion).

The idea is to allow for the mapping of these scenario fragments to actual code. This natural way of describing a test scenario helps *all* parties verify that the unit tests are meeting specifications and facilitates collaborative efforts in software development.

9.11   UML Tutorial: Use Case Diagrams

Even if you haven't spent much time working with UML before, it's likely that you may have heard the term *use case* used in various contexts at one time or another. Use cases are an important part of the UML standard, though ironically, the UML specification has very little to say about how to go about actually defining one. Instead, it focuses on the use case diagram, which only tells you a very small part of the story.

In *Writing Effective Use Cases* (Addison-Wesley Professional, 2000), Alistair Cockburn defines a use case as something that "...captures a contract between the stakeholders of a system about its behavior." In other words, you can think of a use case as a method for capturing the functional requirements of a system or module. A use case is fairly succinct, describing a single interaction scenario between a requesting user or system (referred to as an *actor*) and the system under discussion. Each use case defines a *main success scenario* that defines how an actor can achieve their goal. At each step within the main success scenario, it's likely that something might occur to cause the flow of the use case to deviate. These deviation scenarios are referred to as *extensions*. Separating these extension scenarios from the main success scenario makes the use case much easier to read.

Use case development is a collaborative process that requires a lot of communication within a project team. Most of the time, this process is driven heavily by business analysts who may not be familiar with UML. Therefore, use case scenarios are often best represented in text form. You'll see an example of this form in Section 9.11.2.

9.11.1   Use Case Terminology

Before you see an example use case, it's important that you understand some basic terminology. Table 9.6 provides a description of some of the most common terms used in use case parlance.

Term	Description
Actor	A user or system that interacts with the system under discussion. From the perspective of the system under discussion, an actor is defined in terms of the role(s) it plays in the system.
Primary actor	The primary actor is the actor that initiates the use case scenario.
Scope	The scope describes the system under discussion.
Preconditions	Preconditions describe what must be true before the use case can begin. For example, a precondition of a web application might be that the user has been properly authenticated. In this case, the precondition simplifies the prose in the use case scenario since you don't have to include steps to verify that a user is authenticated before executing a given step.
Guarantees	A guarantee describes the invariants maintained by the system throughout a use case scenario. For example, a use case scenario describing a transfer of funds between two accounts in a banking system would have guarantees that ensure that both the source and target account are debited and credited correctly.
Main success scenario	The primary scenario of the use case that describes how an actor will reach their goal. You can think of this as the "sunny day" scenario for the use case.
Extension scenarios	Extension scenarios are scenarios that describe alternative behavior within the main success scenario.

Table 9.6 Some Basic Use Case Terms

9.11.2   An Example Use Case

As we stated previously, there are no hard-and-fast rules for defining use cases. The use case example shown in Table 9.7 highlights some of the more common elements used when defining use case documents.

Use Case: Student Registering for a Training Class Online	
Primary Actor	Student
Scope	Online course registration website
Preconditions	Student has logged onto course website
Main Success Scenario	
1. Student browses the course catalog and selects the course they want to attend.	
2. Student clicks a button to register for the class.	

Table 9.7 An Example Use Case Document

Use Case: Student Registering for a Training Class Online
3. Student fills in basic contact information (e.g., name and email).
4. Student fills in payment information (e.g., credit card details).
5. Student submits the registration request.
6. System verifies that seats are available.
7. System verifies payment information, authorizing the purchase.
8. System displays success confirmation on the screen.
9. System sends a follow-up email confirming the registration.

Extensions	
6a. No seats are available	• System displays message indicating class is full. • Returns to main success scenario at step 1.
7a. Payment information is valid	• Student can select another form of payment or cancel the process.

Table 9.7 An Example Use Case Document (Cont.)

If we've done our job right, the use case in Table 9.7 should be very easy to follow. Here, we've documented a use case for registering for a training class online. Initially, we define the primary actor, the system under discussion, and some basic preconditions for executing the use case. Next, we proceed into the main success scenario, which is defined as a sequence of numbered steps. As you can see, each step is described using action words that are direct and to the point. To keep things succinct, you can reference other use cases by underlining a particular bit of action text. This is demonstrated in the first step, where we defer a detailed discussion to the course catalog search to a separate use case. The use case in Table 9.7 also contains a couple of extension scenarios. These exception scenarios describe what happens if the class is full or if the provided payment details are invalid.

Keep in mind that the example shown in Table 9.7 is just one way of documenting a use case. Generally speaking, a use case is good so long as it accurately describes an interaction with the system. When you read a use case document, you should be able to quickly ascertain the *who*, the *what*, the *when*, the *where*, and the *why* of a particular interaction within the system. When it comes to use case documentation, less is more.

9.11.3   The Use Case Diagram

Figure 9.8 shows an example of a use case diagram for the use case outlined in Table 9.7. As you can see, the graphical notation for use cases in the UML is fairly simple, showing the relationships between actors and use cases. The use cases are drawn within a rectangular box that represents the boundaries of the system. Internally, use cases can define *include* relationships to depict their dependencies on other use cases.

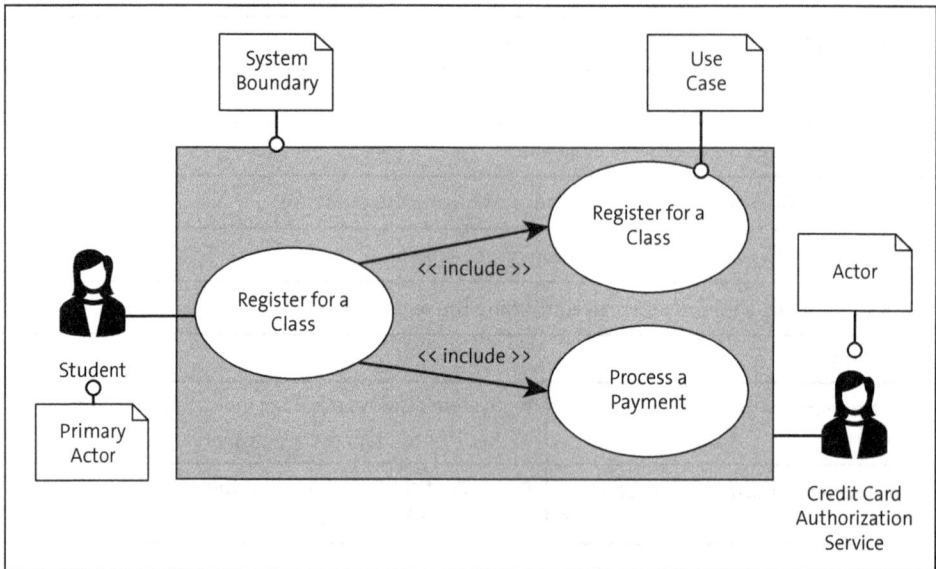

Figure 9.8 A Use Case Diagram Example

In his book *UML Distilled: A Brief Guide to the Standard Object Modeling Language, Third Edition* (Addison-Wesley Professional, 2003), Martin Fowler suggests that one way to look at use case diagrams is as a type of graphical table of contents for a set of use case documents. For example, the use case diagram in Figure 9.8 shows a high-level overview of the course registration system, its use cases, and the relevant actors interacting with those use cases. For more information about any particular use case, you would consult detailed use case documentation such as that shown in Table 9.7.

9.11.4   Use Cases for Requirements Verification

Use cases are an excellent method for capturing functional requirements. Unfortunately, they are not widely used in SAP projects. Consequently, as an ABAP developer, you might be asking yourself why you should care about use cases. After all, most of the time the documentation of functional requirements falls under the purview of the business analysts working on a project.

Typically, in most waterfall methodologies employed on SAP projects, developers do not enter the software development process until a functional specification is written. Here, developers are often expected to simply read through the functional specification and start the design process. However, before proceeding too far down this path, a smart developer will check back with the business analysts to make sure that their interpretation of the requirements is consistent with the vision of the business analysts so that nothing is lost in the translation of the functional requirements.

Use cases can be a very effective tool for documenting such interpretations. Moreover, by spending just a little bit of extra time documenting use cases, developers can make life easier for themselves and others by distilling the requirements into a form that's straightforward and easy to interpret. This documentation becomes a vital part of a technical design document, saving future developers from having to try to interpret a complex functional design from scratch.

9.11.5   Use Cases and Testing

Use cases can also come in handy when you're ready to start developing unit or functional tests. Each action step in a main success or extension scenario usually represents a unit of work that should be tested independently. At the very least, they should give you an excellent start for narrowing down your test scenarios. When you compare it to the alternative of combing through a large functional specification document in search of test scenarios, you can see where the effort of documenting use cases is justified.

9.12   Summary

Unit tests are the last quality assurance checkpoint a development object must pass through before it's turned over to the wider project community. Consequently, it's important that you get them right so you can deliver quality development objects. The design of automated unit tests using the ABAP Unit testing framework simplifies this endeavor by facilitating the creation of robust test cases that produce repeatable results.

In the next chapter, we'll take a look at an ABAP-based framework that makes heavy use of object-oriented concepts: the *Business Object Processing Framework* (BOPF). Here, we'll learn how to build large-scale business objects that can be used to achieve component-based application designs.

Chapter 10

Business Object Development with BOPF

Up until now, we've looked at techniques for encapsulating business logic on a micro scale, focusing our attention on a handful of classes at a time. In this chapter, we'll broaden our scope and look at ways of encapsulating business logic within coarse-grained reusable business objects.

Enterprise software development resembles one big game of chess in that its pieces are constantly in motion. Objectives change, processes evolve, and the only way for IT departments to keep pace is to have a good set of flexible business objects that can be molded, adapted, and mixed and matched to develop business solutions. Business objects are rather like building blocks that can be stacked on top of one another to construct new solutions.

Knowing what you now know about object-oriented design, you probably already have lots of ideas for creating new business objects and implementing component-based architectures. However, while it's certainly possible to build such solutions from scratch, it turns out that SAP provides a very powerful framework that makes this process a whole lot easier: the *Business Object Processing Framework* (BOPF). In this chapter, you'll get to know the BOPF and see how it can be used in tandem with your new-found object-oriented skills to build powerful business solutions.

10.1 What Is BOPF?

If you haven't heard of the BOPF before, then a brief introduction is in order. As the name suggests, the BOPF is a framework for working with business objects (BOs). This framework was designed from the ground up to manage the entire lifecycle of BO development, saving you from having the reinvent the wheel each time you need to develop a new BO.

At a high level, we can organize the services and functionality of the BOPF into two basic categories:

- **Design time**
  At design time, the BOPF provides a series of workbench tools that are used to model and construct BOs. Graphical editor screens and wizards guide developers through the various stages of the BO development process.

- **Runtime**

  At runtime, the BOPF provides a runtime framework which manages business object instances. There's built-in functionality for automatic persistence, transaction and lifecycle management, caching, and much, much more.

The block diagram in Figure 10.1 illustrates how these pieces fit together within an app.

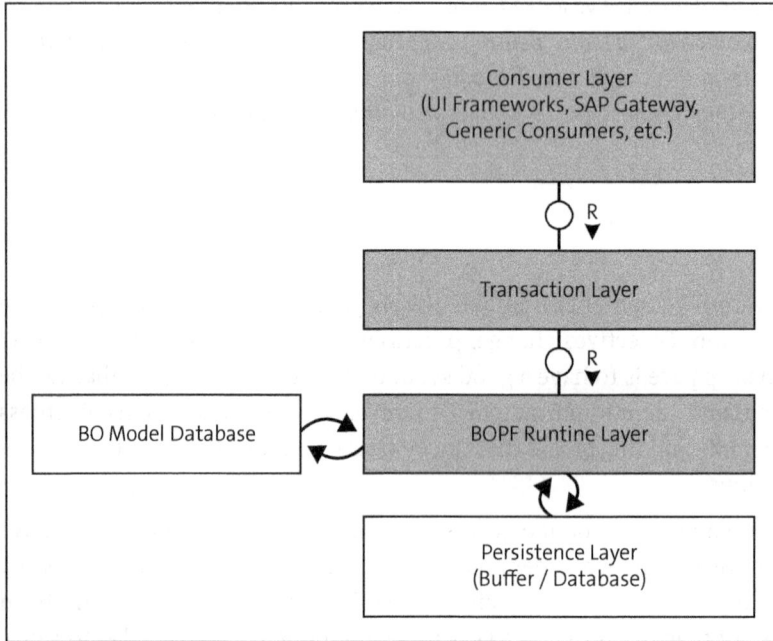

Figure 10.1 High-Level BOPF Architecture

As you can see in Figure 10.1, BOPF app design calls for a layered approach to development:

- **Consumer layer**

  At the consumer layer, clients can utilize the BOPF's object-oriented application programming interface (API) to interface with BOs. Because this API is generic, it's very easy to develop universal consumer frameworks that sit on top of the BOPF. A couple of the more notable consumer frameworks at the time of writing include the *FPM-BOPF integration* (FBI) and *Gateway-BOPF integration* (GBI) frameworks, which integrate the BOPF with the *floorplan manager* (FPM) and SAP Gateway toolsets, respectively. These integration frameworks make it very easy to develop user interface (UI)-centric apps using the FPM and SAPUI5 frameworks.

- **Transaction layer**

  Client sessions are managed through a centralized transaction layer that handles low-level transaction handling details, such as object locking and commit handling.

- **BOPF runtime layer**

  The core of the BOPF functionality exists within the BOPF runtime, which is implemented via a series of standard-delivered classes provided by SAP. This layer contains all the functionality needed to instantiate BOs, handle lifecycle events, and respond to client-level interactions.

- **Persistence layer**

  As you might expect, the persistence layer provides the low-level functionality needed to persist BO instances to the database. It provides services to implement performance optimizations such as buffering and caching. SAP continues to deliver innovations to this layer to improve performance and expand its feature set.

Though it can be hard to wrap your head around these concepts without seeing how they play out from a technical point of view, the major takeaway for now is that the BOPF introduces consistency and predictability into the BO development process. With the BOPF, everything has its place, and the framework makes it difficult for developers to stray too far off the beaten path.

Besides improving the quality of the components being developed, such standardization also leads to interesting innovations as patterns emerge. You'll see some examples of this throughout the course of this chapter. For now, though, let's turn our attention to the inner workings of BOPF BOs.

10.2   Anatomy of a Business Object

According to SAP, BOs within the BOPF are "a representation of a type of uniquely identifiable business entity described by a structural model and an internal process model." This is to say that BOPF BOs:

- Have a structured component model.
- Have a well-defined process model that governs the BO lifecycle, behaviors, and so on.

BOPF BOs are similar to other popular component models in the enterprise software space such as Enterprise JavaBeans (EJBs) in the Java world and COM+ in the Microsoft .NET world. Closer to home, the BOPF shares some similarities with previous ABAP-based business object frameworks like the *business object layer* (BOL) and *generic interaction layer* (GenIL) frameworks.

In the following sections, we'll delve into the various types of elements you can create when modeling BOPF BOs. However, before we go there, let's first take a moment to understand how business entities are modeled within the BOPF. It's helpful to have a visual frame of reference to work with. Consider the /BOBF/DEMO_SALES_ORDER demo BO shown in Figure 10.2. You can review the setup of this BO in your own local system using Transaction /BOBF/CONF_UI.

305

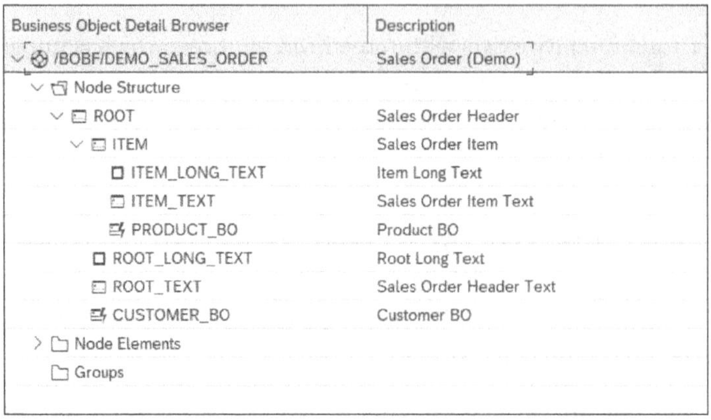

Business Object Detail Browser	Description
⌄ ⚙ /BOBF/DEMO_SALES_ORDER	Sales Order (Demo)
⌄ 🗂 Node Structure	
⌄ 🗅 ROOT	Sales Order Header
⌄ 🗅 ITEM	Sales Order Item
🗅 ITEM_LONG_TEXT	Item Long Text
🗅 ITEM_TEXT	Sales Order Item Text
🗄 PRODUCT_BO	Product BO
🗅 ROOT_LONG_TEXT	Root Long Text
🗅 ROOT_TEXT	Sales Order Header Text
🗄 CUSTOMER_BO	Customer BO
⟩ 🗀 Node Elements	
🗀 Groups	

Figure 10.2 Understanding the Node Hierarchy Concept of BOPF BOs

As you can see in Figure 10.2, the basic modeling unit for BOPF BOs is the *node*. Within a BO, nodes are organized into a hierarchical node structure that mirrors the structure of elements in an XML document. This is to say that there are defined parent-child relationships that give the BO its structure. For instance, in the /BOBF/DEMO_SALES_ORDER BO, data is organized under the top-level ROOT node. Underneath the ROOT node, a node hierarchy is built out with child nodes such as ITEM and ROOT_TEXT. This process of nesting child nodes continues until all aspects of the business entity in question are properly modeled.

In the sections that follow, you'll learn how to use node hierarchies like the one shown in Figure 10.2 to model the various aspects of real-world business entities.

10.2.1   Nodes

At this point, you know that BOPF BOs are organized hierarchically into nodes. However, since the term "node" is rather generic, you're probably wondering what a node is exactly. The answer's complicated by the fact that the term takes on different meanings depending on the context:

- **Design time**
  At design time, nodes are used to model the data and behavior of an individual aspect of a business object. For instance, in the sample /BOBF/DEMO_SALES_ORDER BO, the ITEM node defines all aspects of a sales order line item: data, behaviors, linkages to other sales order-related entities, and so on.

- **Runtime**
  At runtime, nodes are containers (think internal tables) that group together object-like instances called *node rows*. Depending on the design-time definition of a node, you can perform operations on individual node rows or the node collection as a whole. The diagram in Figure 10.3 illustrates what an instance of the /BOBF/DEMO_SALES_ORDER BO might look like from a runtime perspective.

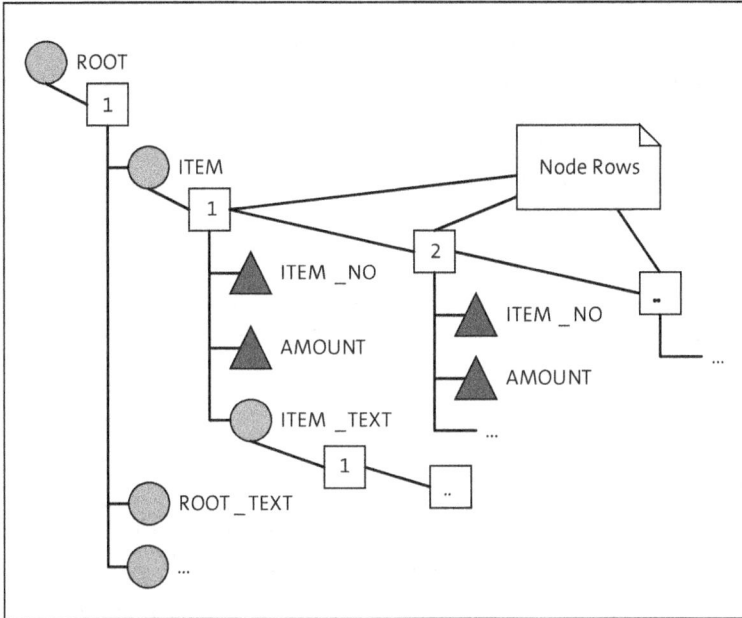

Figure 10.3 Snapshot of a /BOBF/DEMO_SALES_ORDER Instance at Runtime

As you can see in Figure 10.3, each node and node row within a BO instance defines one or more *attributes*. For example, the ITEM node defines attributes such as ITEM_NO and AMOUNT. Collectively, these attributes describe the various aspects of the sales order line item that we want to capture in our model.

Node and Attribute Types

From a data modeling perspective, nodes and node attributes are organized into two distinct categories: *persistent* nodes and attributes and *transient* nodes and attributes. As the name suggests, persistent nodes and attributes are *persisted* to the system database. From a developer perspective, you needn't worry about how this persistence takes place; this will be handled automatically on your behalf by the BOPF runtime environment. The important point to remember from a modeling perspective is that persistent nodes and their corresponding attributes are mapped to database tables and should therefore adhere to good normalization practices.

Transient nodes and attributes represent elements of the data model that are looked up and/or calculated on the fly. For example, say that we wanted to include within our data model details about the sales staff member who recorded an order. In an enterprise resource planning (ERP) system like SAP, basic employee details for this staff member are likely already recorded in the human resources (HR) database. Rather than redundantly store such data within the context of the sales order BO, it makes sense to read the data from the HR system on the fly at runtime. Not only does this eliminate data redundancy, but it also ensures that the data remains fresh over time.

10

Depending on the usage scenario, you can establish such lookups at the node level or the attribute level. In our HR lookup scenario, it would make sense to pull the data into a separate transient node. On the other hand, if we're only looking up a handful of fields, we might just make those fields transient attributes. For example, in the ITEM node of the there's a persistent attribute called PRODUCT_ID that contains a foreign key to a product object maintained elsewhere. If we wanted to pull in the product description into our model, we wouldn't want to redundantly store that description in a persistent attribute. Instead, we'd perform an ad hoc lookup for the product description on demand and copy the value into a transient attribute. A similar approach would be used for calculated values such as the overall value of the sales order, which is calculated by summing up the AMOUNT attribute for each of the node rows within the ITEM node.

Understanding the Node Data Model

Figure 10.4 illustrates the data model of a BOPF node in further detail. Here, there are several attributes that you must configure to build out the data model:

- Whenever a node is initially created, you must specify at the outset whether or not the node will be classified as persistent or transient. If the latter is selected, the **Transient Node** checkbox will be checked.
- Within the **Data Model** section, node attributes are defined as follows:
  - The **Data Structure** field points to an ABAP Dictionary structure that contains all the persistent attributes defined within the node.
  - The (optional) **Transient Structure** field points to an ABAP Dictionary structure that contains all the transient attributes defined within the node.
  - In Section 10.3, you'll learn that the BOPF client API doesn't distinguish between persistent and transient attributes. As a result, the API works with the combined structure type specified in the **Combined Structure** field. This combined structure collects persistent and transient attributes in a singular structure.
  - The **Combined Table Type** field defines a table type whose line type is defined in terms of the combined structure type. This table type is used in API calls that pull back multiple node rows at a time.
- Lastly, in the **Data Access** section (not shown), you'll see the transparent dictionary table where persistent node data is stored in the **Database Table** field. These auto-generated tables will be defined in terms of the persistent attribute structure contained in the **Data Structure** field. The key for any such table is an auto-generated universally unique identifier (UUID) field called DB_KEY. For subnodes, there will also be a foreign key field called PARENT_KEY which can be used to traverse from the child node to its parent. Though such relationships are normally navigated using a node element called an *association*, you can use this field to implement SQL-based lookups of node data as needed.

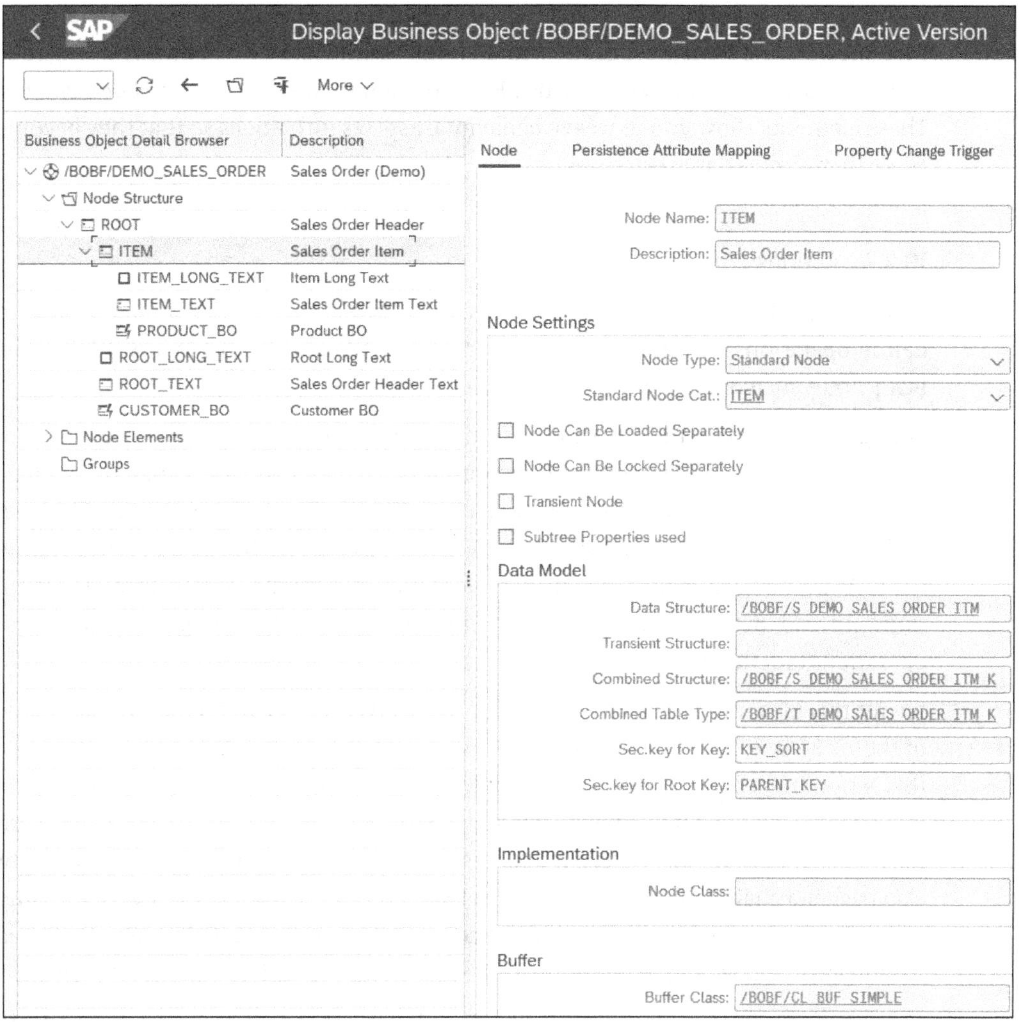

Figure 10.4 Understanding the Node Data Model

Putting It All Together

Before we move on, let's briefly summarize what we've learned in this section. So far, you know that BOs within the BOPF are organized into a node structure or hierarchy. Each node within this hierarchy is meant to model a particular aspect of a business entity. At the node level, you can fill out this model by defining persistent and/or transient attributes as you see fit.

When you think about it, this node-based modeling process is not unlike the one you use to build object models in a pure object-oriented context. Indeed, in many respects, it's appropriate to think of a node definition as being rather like a class definition. Taking this a step further, BOs are similar to composite classes in that they group together nodes and subnodes via associations.

If you get the node and data model right up front, the remaining elements required to accurately model a business entity should follow quite naturally. With that being said, let's peel back another layer of the BOPF component model and look at *node elements*. These elements allow you to weave behavioral aspects into nodes so that they become more than simple data structures.

10.2.2   Actions

If you think of nodes as being like classes, then it follows that you should be able to define operations within a node to encapsulate node-specific behaviors. Within the BOPF node, such operations are called *actions*.

Though conceptually similar to methods in a pure object-oriented context, the scope of the operations carried out with BOPF actions tends to be much larger than the average instance method in an object-oriented class. This is because most attribute level updates are brokered through standard API methods defined by the BOPF core framework. As a result, you won't see lots of getter and setter methods in BOPF node definitions. Instead, BOPF actions are utilized more for encapsulating larger-scale business operations together in a callable package.

To put this idea into perspective, consider the DELIVER action defined in the ROOT node of the /BOBF/DEMO_SALES_ORDER sample BO shown in Figure 10.5. As the name suggests, this action triggers the delivery of one or more sales orders. Internally, this processing will result in the update of pertinent node attributes (e.g., the DELIVERY_STATUS attribute) as well as other downstream updates. In a real-world scenario, this action might also trigger a chain of BOPF action calls between the sales order BO, delivery BOs, and so on. To the outside world, though, the complexities of this task flow are abstracted behind an action with an obvious purpose: to process the delivery of a sales order.

Though there's nothing stopping you from defining lots of fine-grained actions within a node, it's important to remember that while BOs bear many similarities to classes, the interface cut is different. With BOs, the goal is to develop a model that closely resembles a business entity. Therefore, the individual nodes and the actions they define should be defined in higher-level terms. That doesn't mean that the technical layers don't exist; it's just that you want to insulate clients from those details as much as you can.

Shifting away from the conceptual side of things, let's take a closer look at an action definition from a technical perspective. If you look at the DELIVER action shown in Figure 10.5, you can see that there are several attributes that contribute to the design of an action's signature.

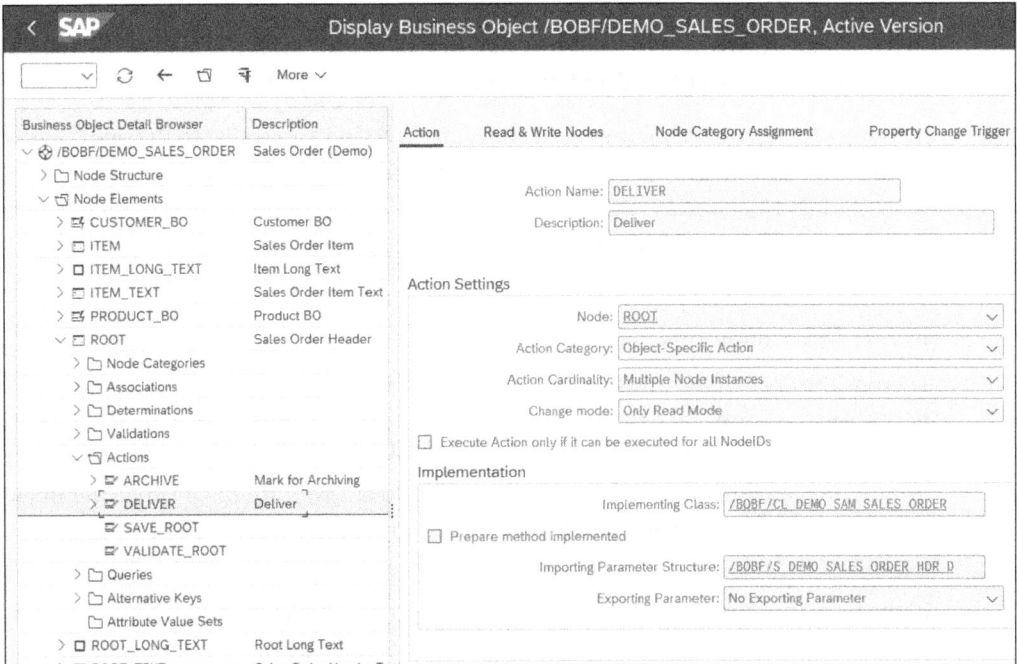

Figure 10.5 Defining an Action in the BOPF

Some of the more notable attributes here include:

- **Node**
  This attribute binds the action with its corresponding node. While such a designa-
  tion might seem obvious within the graphical design time tools, this specification is
  needed by the BOPF runtime to facilitate action processing at runtime.

- **Action Cardinality**
  This attribute identifies the number of node row instances expected to be processed
  by the action. Most of the time, you'll choose between the **Single Node Instance** and
  Multiple Node Instances options depending on whether or not you plan to perform
  bulk operations. The other option is the **Static Action** option, which effectively models the action as a static action. This option would make sense if you wanted to
  define utility actions and the like.

- **Implementing Class**
  In this attribute, you plug in an ABAP Objects class that defines the action logic. This
  class must implement the /BOBF/IF_FRW_ACTION interface. The actual ABAP implementation code goes into the execute() method, which comes pre-configured with
  contextual information that can be used to determine the node row(s) to be processed, among other things.

- **Importing Parameter Structure**
  Use this optional attribute to expand the signature of a BOPF action to include
  input/output parameters as needed. All the fields are encapsulated inside of an

ABAP Dictionary structure that's defined outside of the BOPF. At runtime, this parameter structure is passed by reference to the action implementation class so that attributes can be read and updated where appropriate.

In many respects, the attributes used to define an action are similar to the ones used to define a method of an ABAP Objects class in the form-based view of the Class Builder. Once the basic signature is defined, the focus is on the underlying action implementation class, which is implemented using regular ABAP Objects code.

10.2.3   Determinations

There will be key milestone events you need to react to during the lifecycle of a business object instance. For example, if data changes in one node, the update could have a cascading effect on other related nodes. Or, you might need to take care of some last-minute housekeeping right before a BO instance is saved to ensure that the object instance is persisted correctly. For these situations and others, the BOPF object model provides you with *determinations*.

In the help documentation for the BOPF Enhancement Workbench, SAP defines a determination as "an element assigned to a business object node that describes the internal changing logic on the business object. Like a database trigger, a determination is automatically executed by the BOPF as soon as the BOPF triggering condition is fulfilled."

The internal changing logic referenced in this definition is realized in the form of an ABAP Objects class that implements the /BOBF/IF_FRW_DETERMINATION interface. When determinations are defined, these implementation classes are registered to run whenever specific triggering conditions are met. Table 10.1 shows the types of triggering conditions or patterns supported by the BOPF object model at the time of writing.

Triggering Condition/Pattern	Description
Derive dependent data immediately after modification	This pattern performs cascading updates whenever a node instance is created, updated, or deleted. Some possible scenarios where this pattern would apply include: ■ Setting default values for attributes of newly created node rows. ■ Updating related attributes and/or subnodes whenever node attributes change. ■ Applying state transition rules to ensure that the object remains in a consistent state.
Derive dependent data before saving	This pattern interjects custom logic right before a node instance is saved. You might apply last-minute updates to a node row (e.g., updating date/time audit fields) or fire events that trigger downstream processing outside of the BOPF.

Table 10.1 Determination Patterns Within the BOPF Object Model

Triggering Condition/Pattern	Description
Fill transient attributes of persistent nodes	Determinations following this pattern are used to perform lookups and calculations of transient attributes defined within a node. This pattern can also set runtime properties on BOPF nodes as needed.
Derive instances of transient nodes	Use this pattern in cases where you need to allocate transient nodes on demand. Determinations following this pattern will be launched during the retrieve operation to build the requested transient node instances on the fly.

Table 10.1 Determination Patterns Within the BOPF Object Model (Cont.)

Aside from selecting the appropriate determination pattern, the process for defining a determination is straightforward. As you can see in Figure 10.6, a determination definition mainly consists of the **Determination Name**, the appropriate pattern (**Determination Cat.**), and an implementation class (**Class/Interface**). As noted earlier, the implementation class is a regular ABAP Objects class that implements the /BOBF/IF_FRW_ DETERMINATION interface.

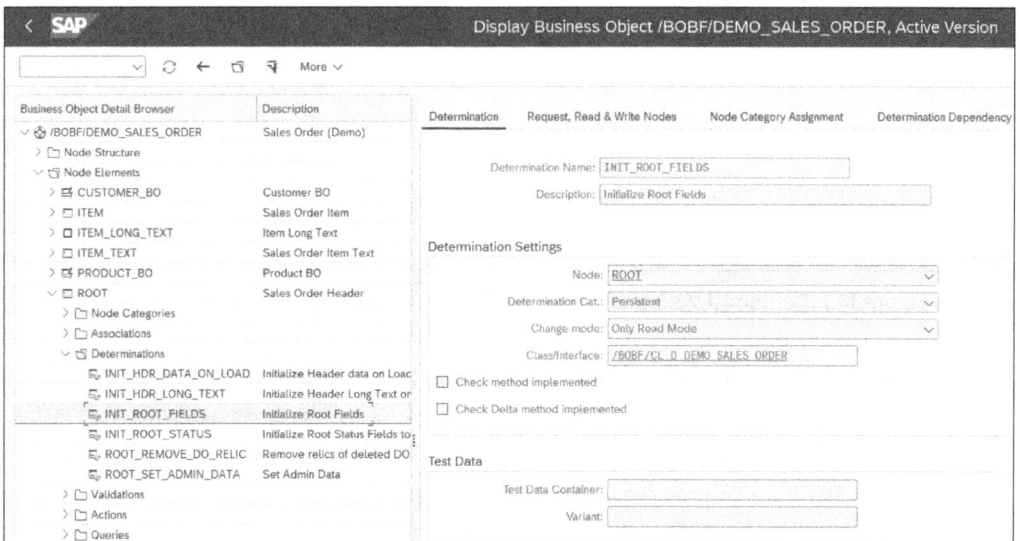

Figure 10.6 Defining a Determination in the BOPF

10.2.4 Validations

In Section 10.3, you'll learn that updates to nodes and attributes are carried out via a generic object-oriented API. Here, a client is generally free to pass in whatever data they like—even if the data is incorrect. If you think back on our encapsulation discussions from Chapter 3, you might naturally assume that this is a violation of basic encapsulation rules. However, as it turns out, there's a method to the BOPF's madness here.

Rather than encoding validation rules into individual setter methods, the BOPF allows you to group business validation rules together in separate node elements called *validations*. Within these validations, you can enforce your consistency checks and ensure that the integrity of BO nodes and the BO as a whole remains intact. At specific milestone events (e.g., right before a BO instance is saved), the BOPF runtime automatically triggers these validations and prevents any invalid data updates from slipping through the cracks. As such, validations represent the last line of defense for BO nodes.

In Figure 10.7, you can see an example of a validation called CHECK_ITEM within the /BOBF/DEMO_SALES_ORDER BO. This validation performs a consistency check to ensure that ITEM node instances remain consistent.

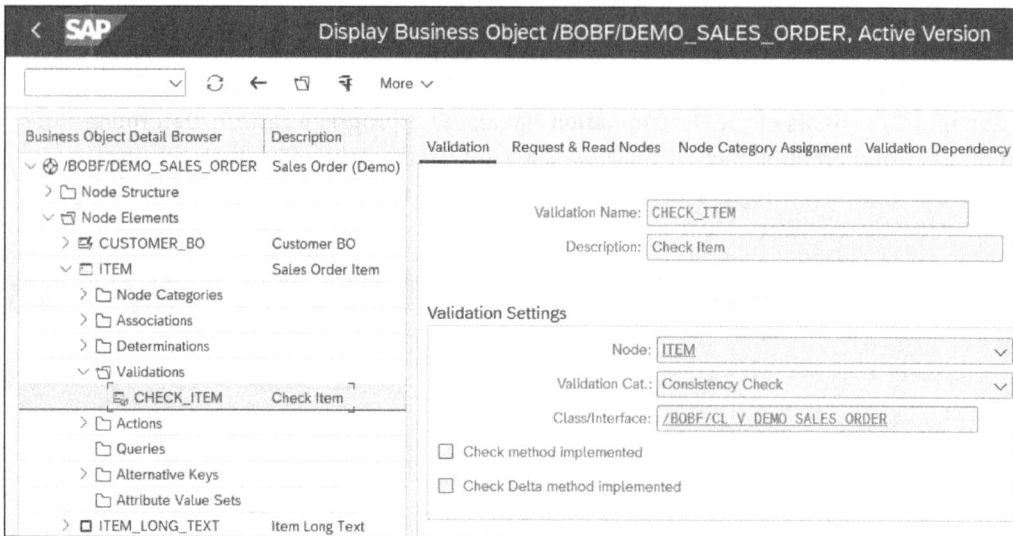

Figure 10.7 Defining a Validation in the BOPF

As you can see in the screenshot, a validation definition consists of three main attributes:

- **Validation Name**
  This attribute defines the validation name, which is an identifier maintained internally within the BOPF. Since validations are not callable objects, the name has little meaning outside of the BOPF framework from a runtime perspective.

- **Validation Cat. (validation category)**
  Here, you can choose between one of two options. The default **Consistency Check** option classifies the validation as an entity used to perform consistency checks before a node instance is saved. The **Action Check** option ensures that a node instance is ready to have an action performed against it.

- **Class/Interface**
  This attribute defines the validation implementation class, which is a plain ABAP

Objects class that implements the /BOBF/IF_FRW_VALIDATION interface. Within this validation class, you can perform the relevant checks and report any errors that might crop up using the EO_MESSAGE and ET_FAILED_KEY exporting parameters.

10.2.5   Associations

When processing BOs, you must be able to navigate through the nodes that make up the BO's node hierarchy. While this may seem like a given, the implementation details are more complex than you might think. Consider the fact that the BOPF runtime must navigate through these relationships by evaluating node rows cached in shared memory. For this task, the BOPF runtime needs to know how to relate a pair of nodes with one another. Within the object model, these relationships are specified in the form of *associations*.

Figure 10.8 illustrates what an association definition looks like in Transaction /BOBF/ CONF_UI. This association defines the relationship between the sales order's ROOT node and the ITEM child node. As you can see in the **Association Settings** section, the details specified here include the source and target nodes in the relationship (**Source Node** and **Associated Node**), the **Cardinality** of the relationship, and the **Resolving Node**. Naturally, the details will vary depending on the types of nodes that are being associated with one another.

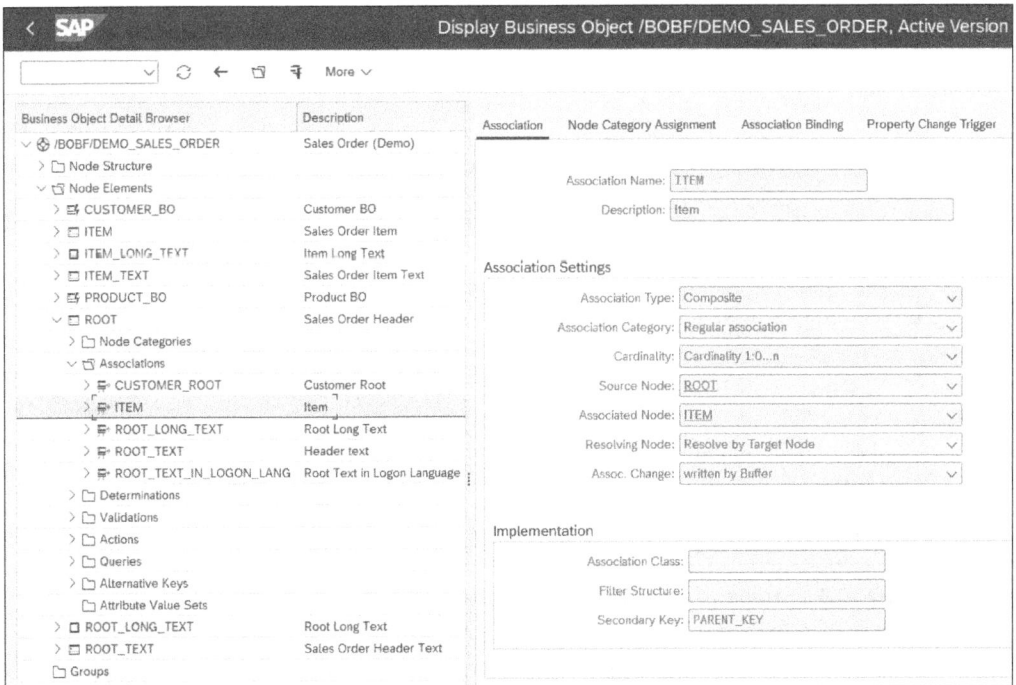

Figure 10.8 Defining an Association in the BOPF

Associations like the ITEM association shown in Figure 10.8 are defined automatically by the BOPF design time tools as you create new BOs. In general, there are three types of associations that will be created automatically by the BOPF design time tools:

- Simple associations from a given node to each of its child nodes. These associations will have a name that corresponds with the name of the child node.

- Associations from a child node to its parent node. These associations will go by the name TO_PARENT.

- Associations from a child node to the root node. These associations will go by the name TO_ROOT.

In addition to the standard associations provided, you also have the option of creating custom associations that simplify node traversal. For example, consider the ROOT_TEXT_IN_LOGON_LANG association shown in Figure 10.9. This association essentially serves the same purpose as the default ROOT_TEXT association, linking the ROOT node with its child ROOT_TEXT node. The difference in this case is that the list of node rows returned by the association is filtered by the user's logon language. In essence, this is a convenience association that saves clients from having to parse through ROOT_TEXT rows to find the instance matching the user's logon language.

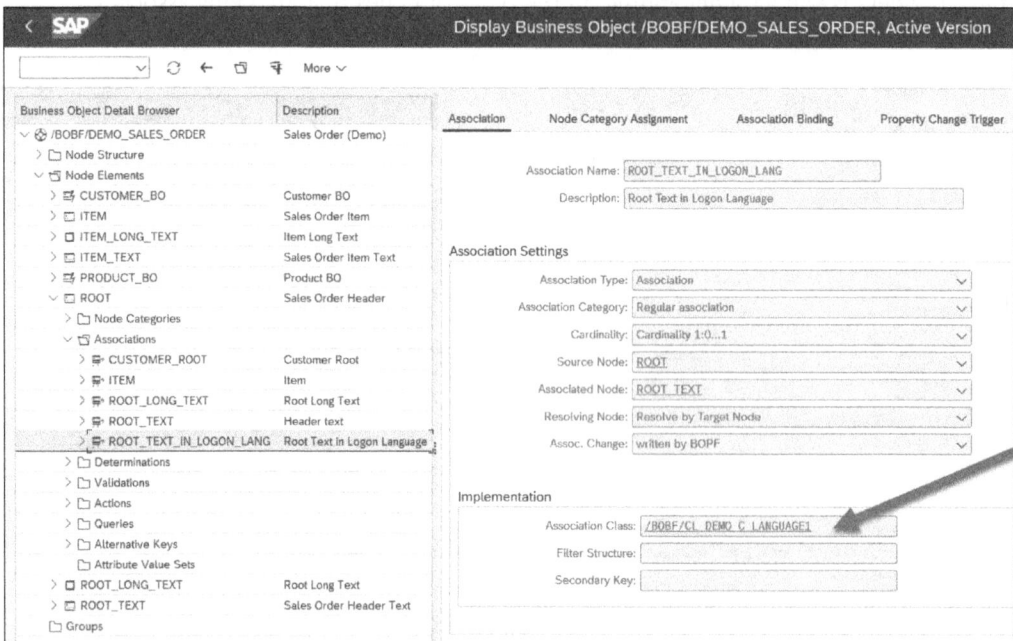

Figure 10.9 Example of a Custom Association

In order to implement custom associations like the ROOT_TEXT_IN_LOGON_LANG association shown in Figure 10.9, you must specify the filter criteria in one of two ways:

- If the join conditions are relatively static in nature, then you can define them graphically on the **Association Binding** tab shown in Figure 10.9.

- Otherwise, you must specify the join/filter conditions in an association class that implements the /BOBF/IF_FRW_ASSOCIATION interface. This class is then plugged into the **Association Class** field in the **Implementation** section shown in Figure 10.9 (e.g., the /BOBF/CL_DEMO_C_LANGUAGE1 class).

Before we wrap this section on associations, we should point out that the BOPF also allows you to define associations between nodes of different BOs as needed. An example of this is the CUSTOMER_ROOT association shown in Figure 10.10. This association, called a cross-business object (XBO), allows you to traverse from the sales order ROOT node to the ROOT node of the /BOBF/DEMO_CUSTOMER BO.

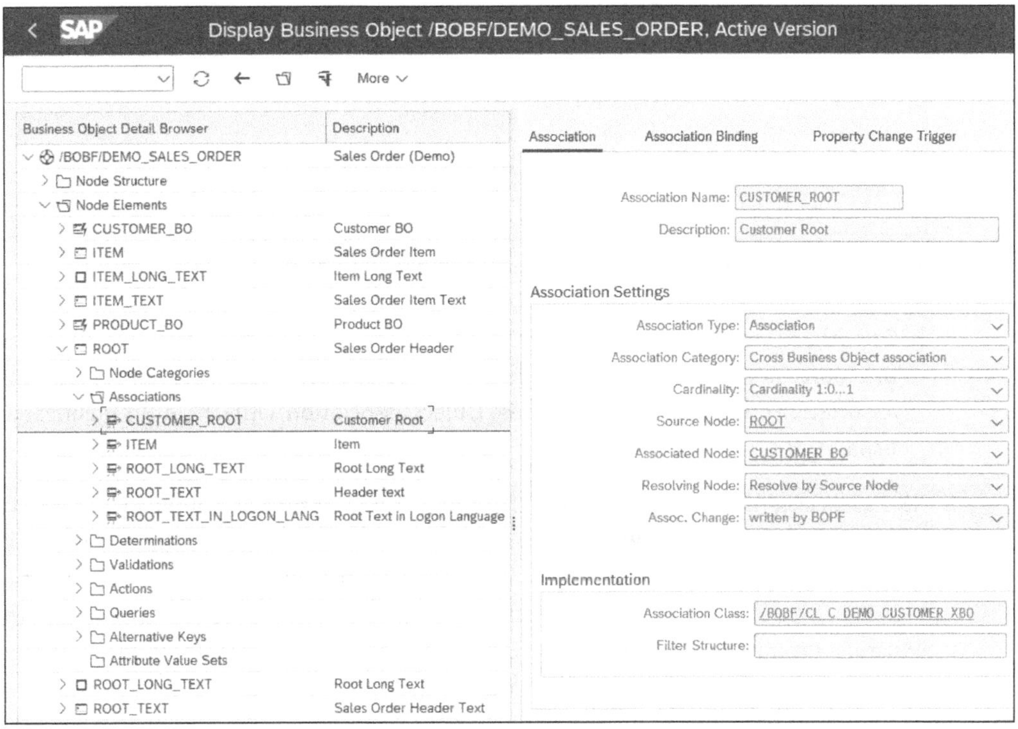

Figure 10.10 Example of an XBO Association

From a configuration perspective, you can define XBO associations as follows:

1. First, map the ROOT node of the linked BO as a child node of the source node (e.g., the ROOT node of the /BOBF/DEMO_SALES_ORDER BO in this case). As you can see in Figure 10.11, the node type is **Business Object Representation Node**. This type allows you to model a transient node that points to the ROOT node of the associated BO.

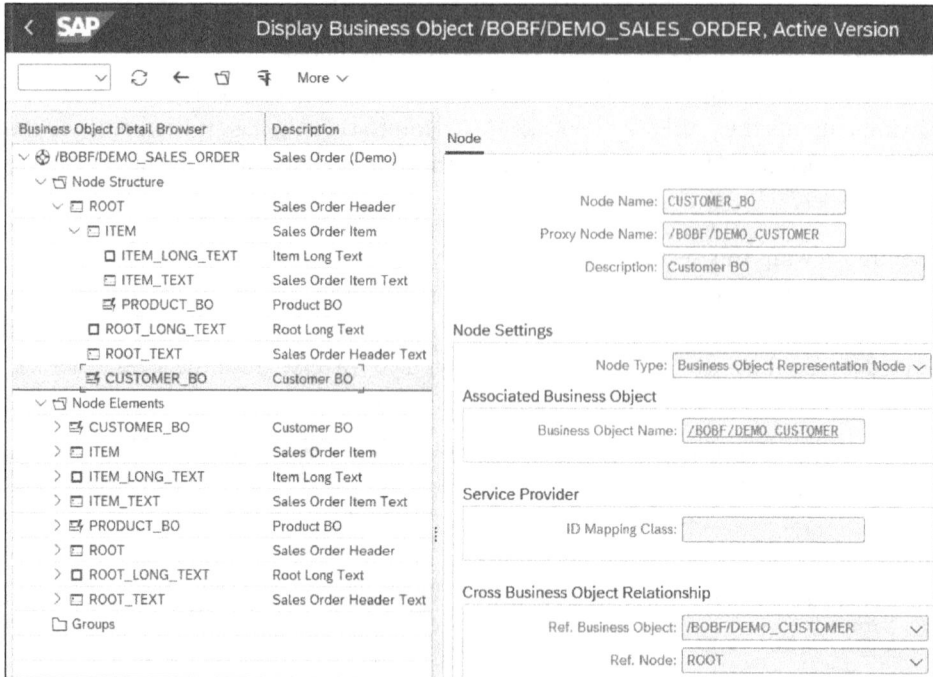

Figure 10.11 Example of an XBO Representation Node Definition

2. Once the BO representation node is in place, define the XBO association by mapping the source node to the delegate node as shown earlier in Figure 10.10. In this case, the **Association Category** is **Cross Business Object association**. Otherwise, it's business as usual.

XBO associations like the CUSTOMER_ROOT association represent one of the primary ways that we can achieve reuse within the BOPF. Here, rather than redundantly define customer details within the /BOBF/DEMO_SALES_ORDER BO, it makes sense to simply link the two BO instances together and create a composite. Over time, as the BO library expands, productivity will increase since developers won't constantly be rebuilding and reapplying business logic in new contexts.

10.2.6   Queries

The last BO node element that we'll look at is *queries*. As you would expect, queries allow you to encapsulate node and BO lookup logic in callable modules that clients can use to search for BO instances. Within the BOPF object model, there are two different types of queries that you can define:

- **Node attribute queries**
  These are modeled queries whose logic is defined within the BOPF runtime. No custom coding is required to implement this type of query.

- **Custom queries**

  Custom queries handle complex query requirements that exceed the capabilities of the canned node attribute queries. The query logic is encapsulated in a custom ABAP Objects class that implements the /BOBF/IF_FRW_QUERY interface.

Figure 10.12 demonstrates the definition of a query within Transaction /BOBF/CONF_UI. As you can see, the attributes required to define a query are straightforward. For modeled queries, you don't have to specify anything, as the BOPF framework will implement all the necessary details for you. For implemented queries, the primary attribute you must specify is the **Query Class**.

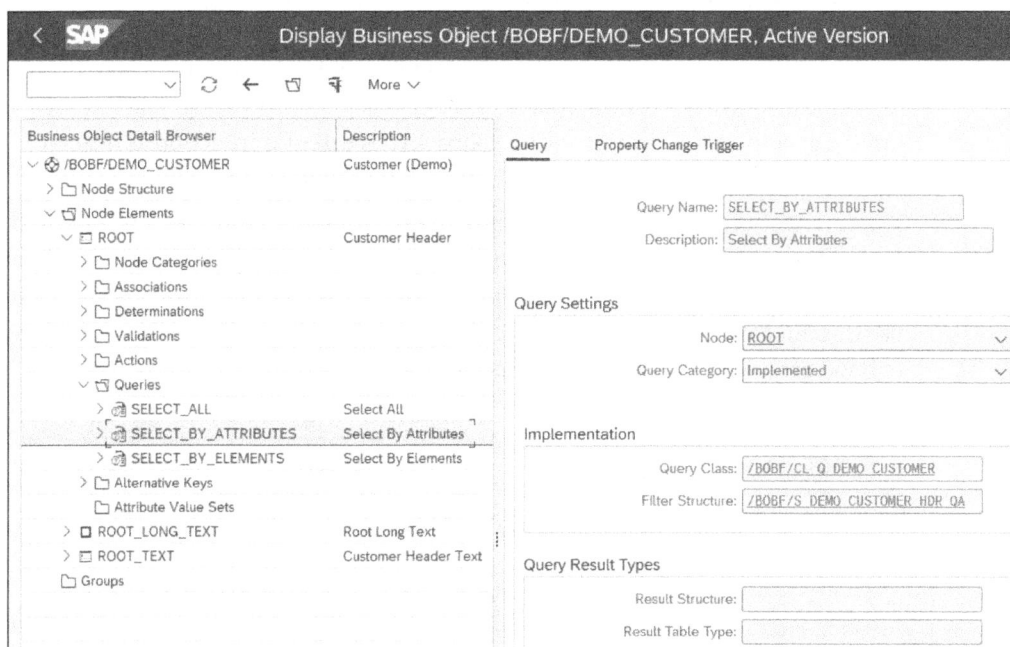

Figure 10.12 Defining a Query in the BOPF

Within a query implementation class, you can invoke other (modeled) queries or dynamically generate SQL code to lookup the relevant entries. The results of the query are passed back via an exporting parameter called ET_KEY which, as you'd expect, contains the keys of the node rows that you find during the course of the query execution. In Section 10.3, you'll learn how to use these keys in BOPF client API calls to fetch and update BOs as needed.

10.3   Working with the BOPF Client API

Now that you're familiar with the makeup of BOPF BOs, you'll see how you can create instances of these BOs and manipulate them from within ABAP programs. SAP has provided you with a rich and flexible object-oriented API that's easy to work with once you get the hang of it.

10.3.1 API Overview

For the most part, the BOPF API can be distilled down to three main object types (see the Unified Modeling Language [UML] class diagram in Figure 10.13):

- **/BOBF/IF_TRA_SERVICE_MANAGER**
  This interface defines the core API methods needed to interface with BO node elements. Here, methods are provided to execute queries, traverse through node hierarchies, execute actions, perform node updates, and more.

- **/BOBF/IF_TRA_TRANSACTION_MGR**
  This interface provides a façade around BOPF transaction managers, which are used to manage transactions. Such transactions could contain a single step (e.g., updates to a single node row) or multiple steps (e.g., adding and updating multiple node rows or calling actions). During the course of the transaction processing, you can use the interface methods to commit and/or rollback the current transaction as needed.

- **/BOBF/IF_FRW_CONFIGURATION**
  This interface provides you with access to the BOPF configuration store where all the design-time BO configuration metadata is stored. You can use the API methods defined by this interface to introspect BOs and determine node data types. You can use this information in conjunction with the *ABAP Runtime Type Services* (RTTS) API to build generic BO processing logic as needed.

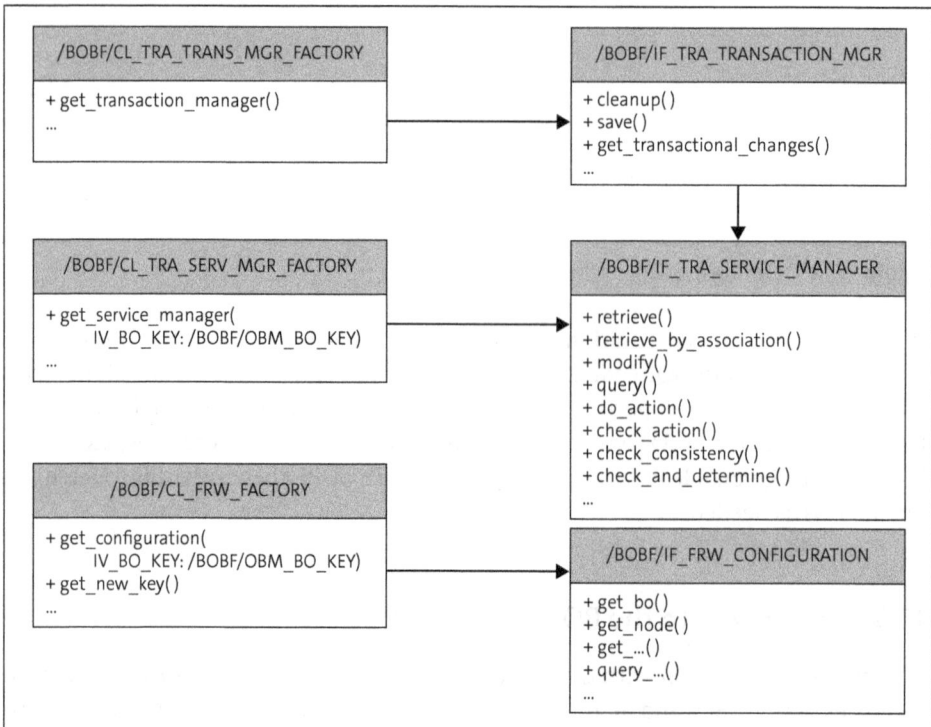

Figure 10.13 Main Elements of the BOPF Client API

As you can see in Figure 10.13, the internals of the BOPF runtime environment are abstracted away behind the three main interfaces we described. To get to the objects that implement these interfaces at runtime, you must go through the factory classes highlighted in Figure 10.13. Then, once you have your hands on these instances, you're off and running.

Business Object Keys and Constants Interfaces

If you look closely at the signature of the factory methods of the /BOBF/CL_TRA_SERV_ MGR_FACTORY and /BOBF/CL_FRW_FACTORY classes shown in Figure 10.13, you'll notice that these methods expect to receive a BO key. What is a BO key? Well, it's a key that's used to look up BO metadata from the BOPF configuration store. The BOPF service manager and configuration managers need to have this information in context to understand how to respond to requests such as "perform this action" or "tell me the data type associated with this particular node." From this, you can glean an important truth about BOPF service managers and configuration managers: If you're working with multiple BOs, you'll need separate service and configuration manager instances for each BO type that you're working with.

Since BO keys are auto-generated by the BOPF design-time configuration tools, you're not expected to simply know the BO key offhand. Instead, each BO has a generated constants interface associated with it that makes it easy to address the BO key as well as various BO node elements. For a given BO type, you can find the constants interface in Transaction /BOBF/CONF_UI by double-clicking on the BO and looking in the **Constants Interface** field in the **Business Object Settings** panel of the default **Business Object** tab (see Figure 10.14).

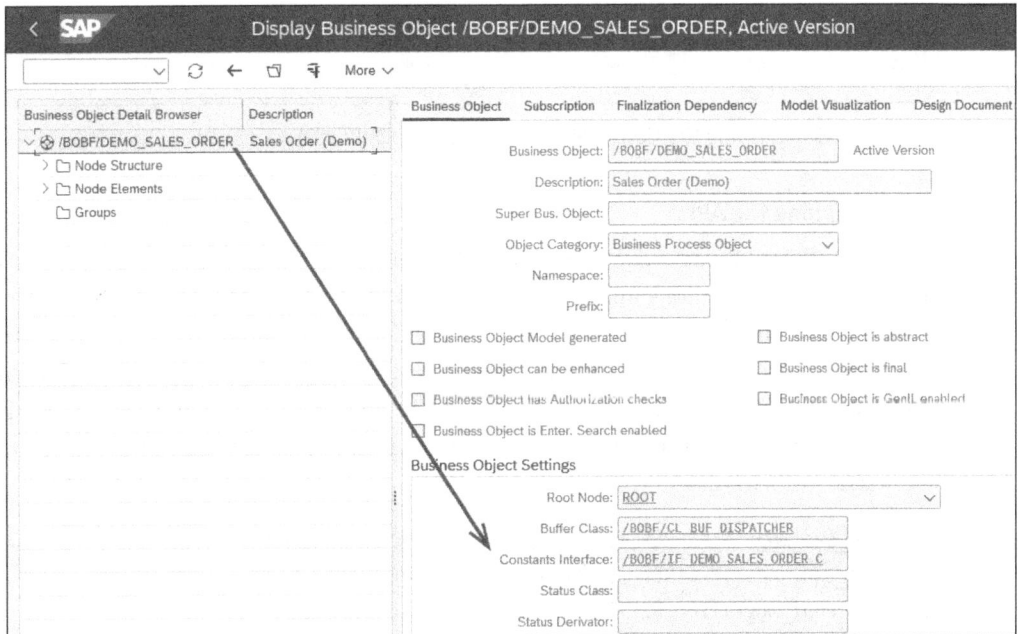

Figure 10.14 Accessing the Constants Interface for a BO (Part 1)

Within Transaction /BOBF/CONF_UI, the **Constants Interface** field is context-sensitive, so you can double-click on it to open the interface in the Class Builder tool. Figure 10.15 shows what the constants interface looks like for a given BO. Here, you can see the target BO key field exposed via the SC_BO_KEY constant. In addition to the key, there are other constants provided to address defined actions, queries, associations, and so forth. These constants are constantly updated to reflect the current state of the BO. For example, if a new query is created in a BO node, the SC_QUERY constant would be updated accordingly. In the upcoming sections, you'll see how these other constants are used in API calls.

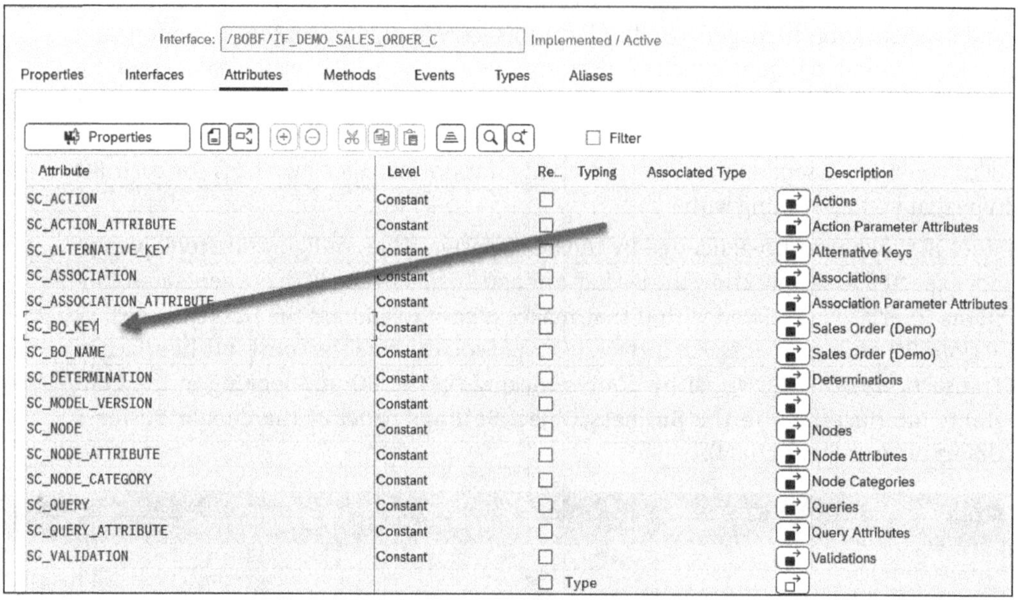

Figure 10.15 Accessing the Constants Interface for a BO (Part 2)

Bootstrapping the API

Having reviewed the basic architecture of the BOPF API, let's now look at what it takes to bootstrap the API using ABAP code. For the purposes of this demonstration, we'll be working with the /BOBF/DEMO_SALES_ORDER BO, which is generally available in most SAP Business Suite systems. In Listing 10.1, you can see how we're preparing to work with this BO using the factory classes illustrated in Figure 10.13. Though there's a fair amount of work going on behind the scenes, the client-side API calls themselves are very straightforward.

```
DATA lo_svc_mngr TYPE REF TO /bobf/if_tra_service_manager.
DATA lo_txn_mngr TYPE REF TO /bobf/if_tra_transaction_mgr.
DATA lo_conf_mngr TYPE REF TO /bobf/if_frw_configuration.
```

```
TRY.
  lo_txn_mngr =
    /bobf/cl_tra_trans_mgr_factory=>get_transaction_manager( ).

  lo_svc_mngr =
    /bobf/cl_tra_serv_mgr_factory=>get_service_manager(
      /BOBF/IF_DEMO_SALES_ORDER_C=>SC_BO_KEY ).

  lo_conf_mngr =
    /bobf/cl_frw_factory=>get_configuration(
      /BOBF/IF_DEMO_SALES_ORDER_C=>SC_BO_KEY ).
CATCH /bobf/cx_frw.
  "TODO: Error handling...
ENDTRY.
```

Listing 10.1 Bootstrapping the BOPF API

Note

You can find a fully developed version of this demonstration in the book's source code bundle at *www.sap-press.com/6093*.

In the upcoming sections, we'll put the API to work performing basic CRUD operations on the /BOBF/DEMO_SALES_ORDER BO.

10.3.2   Creating Business Object Instances and Node Rows

BOPF BOs are managed objects that must be created using the BOPF service manager. Most of the heavy lifting is carried out by the modify() method of the /BOBF/IF_TRA_SERVICE_MANAGER interface. This method receives as its input an internal table called IT_MODIFICATION that contains the nodes that you want to create. Create the right nodes, and you create a BO instance.

The code excerpt in Listing 10.2 demonstrates how you can use the modify() method to create an instance of the /BOBF/DEMO_SALES_ORDER BO. As you can see, the node creation process consists of creating a data reference, filling in the node data, and then adding the node as a record in the aforementioned IT_MODIFICATION table. Once this table is filled out, you can call the modify() method to apply the changes. Internally, the BOPF will process the request and carry out any validations or determinations configured to ensure that the integrity of the BO instance remains intact.

```
DATA lo_svc_mngr TYPE REF TO /bobf/if_tra_service_manager.
DATA lr_s_root TYPE REF TO /bobf/s_demo_sales_order_hdr_k.
DATA lr_s_root_text TYPE REF TO /bobf/s_demo_short_text_k.
```

```
DATA lt_mod TYPE /bobf/t_frw_modification.
FIELD-SYMBOLS <ls_mod> LIKE LINE OF lt_mod.
DATA lo_message TYPE REF TO /bobf/if_frw_message.
DATA lo_change TYPE REF TO /bobf/if_tra_change.

TRY.
  "Initialize the service manager:
  lo_svc_mngr =
    /bobf/cl_tra_serv_mgr_factory=>get_service_manager(
      /BOBF/IF_DEMO_SALES_ORDER_C=>SC_BO_KEY ).

  "Create the order ROOT node:
  CREATE DATA lr_s_root.
  lr_s_root->key = /bobf/cl_frw_factory=>get_new_key( ).
  lr_s_root->order_id = '1234567890'.
  ...
  lr_s_root->sales_org = 'AMER'.
  lr_s_root->amount = '250.00'.
  lr_s_root->amount_curr = 'USD'.

  APPEND INITIAL LINE TO lt_mod ASSIGNING <ls_mod>.
  <ls_mod>-node = /bobf/if_demo_sales_order_c=>sc_node-root.
  <ls_mod>-change_mode = /bobf/if_frw_c=>sc_modify_create.
  <ls_mod>-key = lr_s_root->key.
  <ls_mod>-data = lr_s_root.

  "Create the order description:
  CREATE DATA lr_s_root_text.
  lr_s_root_text->key = /bobf/cl_frw_factory=>get_new_key( ).
  lr_s_root_text->language = sy-langu.
  lr_s_root_text->text = |Order # { lr_s_root->order_id }|.

  APPEND INITIAL LINE TO lt_mod ASSIGNING <ls_mod>.
  <ls_mod>-node = /bobf/if_demo_sales_order_c=>sc_node-root_text.
  <ls_mod>-change_mode = /bobf/if_frw_c=>sc_modify_create.
  <ls_mod>-source_node =
    /bobf/if_demo_sales_order_c=>sc_node-root.
  <ls_mod>-association =
    /bobf/if_demo_sales_order_c=>sc_association-root-root_text.
  <ls_mod>-source_key = lr_s_root->key.
  <ls_mod>-key = lr_s_root_text->key.
  <ls_mod>-data = lr_s_root_text.
```

```
"Apply the changes:
lo_svc_mngr->modify(
  EXPORTING
    it_modification = lt_mod
  IMPORTING
    eo_change       = lo_change
    eo_message      = lo_message ).

"Check the results:
IF lo_message IS BOUND AND lo_message->check( ) EQabap_true.
  "TODO: Error handling...
ENDIF.
CATCH /bobf/cx_frw.
"TODO: Error handling...
ENDTRY.
```

Listing 10.2 Creating a Sales Order Instance Using the BOPF API

Note

You can find a fully developed version of this demonstration in the book's source code bundle at *www.sap-press.com/6093*.

After the modify() method is finished, you can check the results using the following exporting parameters from the method signature:

- **EO_CHANGE**
  This object reference parameter provides you with a handle that you can use to query the status of the updates and check for any failures that might have occurred. For more details about the types of operations you can perform, check out the documentation for interface /BOBF/IF_TRA_CHANGE.

- **EO_MESSAGE**
  This object reference contains any human-readable messages generated during the update process. This includes messages issued from validations or determinations defined against the nodes being created. For more information about how to consume these messages, check out the documentation for interface /BOBF/IF_FRW_MESSAGE.

Assuming the modification process goes off without a hitch, the code excerpt in Listing 10.2 will generate a new sales order BO instance. However, it's worth noting that this instance only exists in shared memory. To persist this instance to the database, you must commit the in-flight transaction using the BOPF transaction manager. You'll see how to accomplish this in Section 10.3.6.

10

10.3.3   Searching for Business Object Instances

As you learned in Section 10.2.1, BOPF node rows are keyed by a UUID field called KEY. Though it's certainly efficient to access these rows by their key, many times you may not know the key offhand. In these situations, you can use BOPF queries to locate the node rows that you want to operate on.

The code excerpt in Listing 10.3 demonstrates how to use a BOPF query called SELECT_ BY_ELEMENTS for a sales order instance using its ID. Using this same query, you could search for sales orders based on the sold-to customer, delivery status, and so on. You simply have to specify the search criteria by filling in a table of type /BOBF/T_FRW_QUERY_ SELPARAM. This table type has a similar look and feel to ABAP range tables.

```
DATA lo_svc_mngr TYPE REF TO /bobf/if_tra_service_manager.
DATA lt_params TYPE /bobf/t_frw_query_selparam.
FIELD-SYMBOLS <ls_param> LIKE LINE OF lt_params.
DATA lt_key TYPE /bobf/t_frw_key.
FIELD-SYMBOLS <ls_key> LIKE LINE OF lt_key.
DATA lt_root TYPE /bobf/t_demo_sales_order_hdr_k.

APPEND INITIAL LINE TO lt_params ASSIGNING <ls_param>.
<ls_param>-attribute_name = 'ORDER_ID'.
<ls_param>-sign = 'I'.
<ls_param>-option = 'EQ'.
<ls_param>-low = '1234567890'.

lo_svc_mngr->query(
  EXPORTING
    iv_query_key =
  /bobf/if_demo_sales_order_c=>sc_query-root-select_by_elements
    it_selection_parameters = lt_params
    iv_fill_data = abap_true
  IMPORTING
    et_data = lt_root
    et_key = lt_key ).

READ TABLE lt_root...
```

Listing 10.3 Searching for BO Instances Using BOPF Queries

As you can see in Listing 10.3, there are two output parameters defined by the query() method:

- **ET_DATA**
  If the IV_FILL_DATA Boolean input parameter is set to true, then the actual node row data for any found rows will be returned via the ET_DATA exporting parameter.

- ET_KEY

  This table parameter passes back the keys of any found node row matching the selection criteria.

Depending on your use case, you can either fetch the keys of records that match some selection criteria, or you can retrieve the actual node row data. In the next section, you'll see what you can do with this data once you retrieve it using the BOPF service manager.

10.3.4   Updating and Deleting Business Object Node Rows

Once you get your hands on the BO node rows, the process of applying updates to these nodes is almost identical to the one you used to create the node rows in the first place (refer to Section 10.3.2). This is demonstrated in the code excerpt in Listing 10.4. Here, we're updating the sales order ROOT node by adjusting the overall amount. Then, we tell the BOPF we want to apply the update by mapping the appropriate value in the CHANGE_MODE attribute highlighted in Listing 10.4. Had we wanted to delete this node row, we would have assigned the value /BOBF/IF_FRW_C=>SC_MODIFY_DELETE. In either case, the BOPF uses the CHANGE_MODE attribute in conjunction with the rest of the data contained in the modification table to identify the target node row (via the KEY attribute) and process the update.

```
DATA lo_svc_mngr TYPE REF TO /bobf/if_tra_service_manager.
DATA lr_s_root TYPE REF TO /bobf/s_demo_sales_order_hdr_k.
DATA lt_mod TYPE /bobf/t_frw_modification.
FIELD-SYMBOLS <ls_mod> LIKE LINE OF lt_mod.
DATA lo_message TYPE REF TO /bobf/if_frw_message.

...
READ TABLE lt_root INDEX 1 REFERENCE INTO lr_s_root.
IF sy-subrc EQ 0.
  lr_s_root->amount = lr_s_root->amount + '20.00'.

  APPEND INITIAL LINE TO lt_mod ASSIGNING <ls_mod>.
  <ls_mod>-node = /bobf/if_demo_sales_order_c=>sc_node-root.
  <ls_mod>-change_mode = /bobf/if_frw_c=>sc_modify_update.
  <ls_mod>-key = lr_s_root->key.
  <ls_mod>-data = lr_s_root.

  APPEND 'AMOUNT' TO <ls_mod>-changed_fields.
ENDIF.

IF lines( lt_mod ) GT 0.
  lo_svc_mngr->modify(
```

```
    EXPORTING
      it_modification = lt_mod
    IMPORTING
      eo_message      = lo_message ).

  "TODO: Error handling...
ENDIF.
```

Listing 10.4 Updating BO Node Rows Using the BOPF API

As is the case with the creation operation, all of the updates/deletes you perform using the modify() method are merely staged in shared memory until you either commit the transaction or roll it back. This feature comes in handy when building user interface (UI) apps since it makes it easy to manage long running stateful sessions with users.

10.3.5   Executing Actions

To invoke a BOPF action, you simply need to call the do_action() method of the /BOBF/IF_TRA_SERVICE_MANAGER interface. This is demonstrated in the code excerpt in Listing 10.5. Here, we're calling the DELIVER action defined for the ROOT node of the /BOBF/DEMO_SALES_ORDER BO. To invoke this action, we simply need to pass in the action key, the key(s) of the node rows we want to process, and an optional parameter structure.

```
DATA lo_svc_mngr TYPE REF TO /bobf/if_tra_service_manager.
DATA lt_key TYPE /bobf/t_frw_key.
FIELD-SYMBOLS <ls_key> LIKE LINE OF lt_key.
DATA lo_message TYPE REF TO /bobf/if_frw_message.
DATA lt_failed_key TYPE /bobf/t_frw_key.

"Call the DELIVER action:
lo_svc_mngr->do_action(
  EXPORTING
    iv_act_key =
      /bobf/if_demo_sales_order_c=>sc_action-root-deliver
    it_key = lt_key
  IMPORTING
    eo_message = lo_message
    et_failed_key = lt_failed_key ).

"Check the results:
IF lines( et_failed_key ) GT 0.
  "TODO: Error handling...
ENDIF.
```

Listing 10.5 Calling an Action Using the BOPF API

If any errors occur during the action processing, the corresponding node row keys will be added to the ET_FAILED_KEY exporting parameter. We can use this information in conjunction with the EO_MESSAGE exporting parameter to determine what might have gone wrong with the action call.

10.3.6   Working with the Transaction Manager

As we noted previously, all updates that you perform via the BOPF service manager are staged in shared memory. To commit these changes, you must invoke the save() method of the BOPF transaction manager. The code excerpt in Listing 10.6 demonstrates how this works.

```
DATA lo_txn_mngr TYPE REF TO /bobf/if_tra_transaction_mgr.
DATA lo_svc_mngr TYPE REF TO /bobf/if_tra_service_manager.
DATA lo_message TYPE REF TO /bobf/if_frw_message.
DATA lv_rejected TYPE boole_d.

TRY.
  "Initialize the BOPF API:
  lo_txn_mngr =
    /bobf/cl_tra_trans_mgr_factory=>get_transaction_manager( ).

  lo_svc_mngr =
    /bobf/cl_tra_serv_mgr_factory=>get_service_manager(
      /BOBF/IF_DEMO_SALES_ORDER_C=>SC_BO_KEY ).

  "Perform various updates using the BOPF service manager:
  ...

  "Commit the changes:
  lo_txn_mngr->save(
    IMPORTING
      eo_message = lo_message
      ev_rejected = lv_rejected ).

  IF lv_rejected EQ abap_true.
    "TODO: Error handling...
  ENDIF.
CATCH /bobf/cx_frw.
  "TODO: Error handling...
ENDTRY.
```

Listing 10.6 Working with the BOPF Transaction Manager

As you can see in Listing 10.6, the transaction manager mostly sits off to the side while you're performing your basic updates. It's keeping track of things behind the scenes, but from an API perspective, you don't really notice it. It only really comes into play when you determine that you either need to commit or rollback a set of changes. For commits, you call the save() method; for rollbacks, the cleanup() method. Note that in either case, you don't have to chase the call with a COMMIT WORK statement—this is handled implicitly within the BOPF internal framework layer.

If a save or commit request is rejected, the EV_REJECTED flag will be set to true. Assuming that's the case, you can use the EO_MESSAGE exporting parameter to determine what went wrong. Most of the time, the messages contained in this parameter will be generated via consistency checks or validations performed within the BOPF nodes staged for updates. This last-minute check ensures that nothing slips through the cracks.

10.4   Where to Go from Here

Now that you have a general sense of what the BOPF is all about, take a moment to digest what you've learned and see how it relates to object-oriented development in ABAP.

10.4.1   Looking at the Big Picture

If you look back on the code excerpts in Section 10.3, you might be inclined to think that BOPF development is rather tedious. While this is true to a point, the BOPF is rarely consumed directly, as we demonstrated in this chapter. Instead, BOPF BOs are normally consumed through high-level frameworks that abstract away the wearisome aspects of the API. These frameworks make it easy to consume BOs using SAP Gateway and OData services as part of the ABAP programming model for SAP Fiori and through Web Dynpro ABAP UI apps based on the floorplan manager.

Since this book is focused on object-oriented programming (OOP) concepts, we're not concerned with the specifics of any one-off consumption framework. Instead, our focus is on reusable API designs. When you think about the BOPF API design, it's completely generic. Whether you're dealing with a sales order BO, or a business partner BO, or a custom BO of your choosing, the API call sequence remains the same. You simply supply the API with a BO key, and you're up and running.

So, what can you take away from all this? In terms of API design, you can see that the BOPF certainly employs many of the best practices described throughout this book:

- **Encapsulation**
  The core business logic for a BOPF BO is encapsulated within its internal node elements (e.g., as actions, determinations, and validations). This is encapsulation on a macro scale, encompassing every aspect of a particular BO or entity.

- **Designing to interfaces**
  If you look at the UML class diagram depicting the BOPF API in Figure 10.13, you can see that clients work with generic interfaces that provide a consistent interface for working with BOPF BOs. With a little bit of up-front RTTI/introspection development, you can build agents that can consume and interact with any BO type.

- **Cohesiveness**
  Because of its strong object model, BOPF BOs tend to have very high *cohesion*. This means that node elements focus on implementing modeling the business logic and not on how they might be consumed via clients and/or UIs.

- **Introspection/discoverability**
  Using the `/BOBF/IF_FRW_CONFIGURATION` interface, clients can discover any aspect of a given BO just by using its key. When combined with the ABAP RTTI API, this makes it possible for clients to dynamically consume BO data or adapt it for use in data binding protocols.

At the end of the day, you're still employing object-oriented concepts; you're just doing it on a larger scale with the BOPF. Since your primary objective is to build a better abstraction, you can think of the BOPF as a means of applying your object-oriented artifacts toward the development of reusable BOs and entities.

10.4.2   Enhancing Business Objects

Over the years, SAP has leveraged the BOPF in core modules such as SAP Transportation Management (SAP TM), SAP Extended Warehouse Management (SAP EWM), and SAP Environment, Health, and Safety Management (SAP EHS Management). While SAP has shifted toward the newer *ABAP RESTful application programming model*—a topic we'll be covering in Chapter 11—the BOPF is still used extensively in SAP S/4HANA systems.

While you might not be building custom BOs, it's still a common requirement to enhance SAP standard business objects. For this task, you can use the *BOPF Enhancement Workbench* provided in Transaction /BOBF/CUST_UI or the more modern toolset included in the ABAP development tools. For more information about extension options, read over the BOPF Enhancement Workbench help documentation.

10.5   UML Tutorial: Advanced Sequence Diagrams

In this section, we'll look at some advanced features of UML sequence diagrams. As a frame of reference for this discussion, we've revised the sequence diagram example from Chapter 3 in Figure 10.16 to include some of the more advanced features that will be discussed in the upcoming subsections.

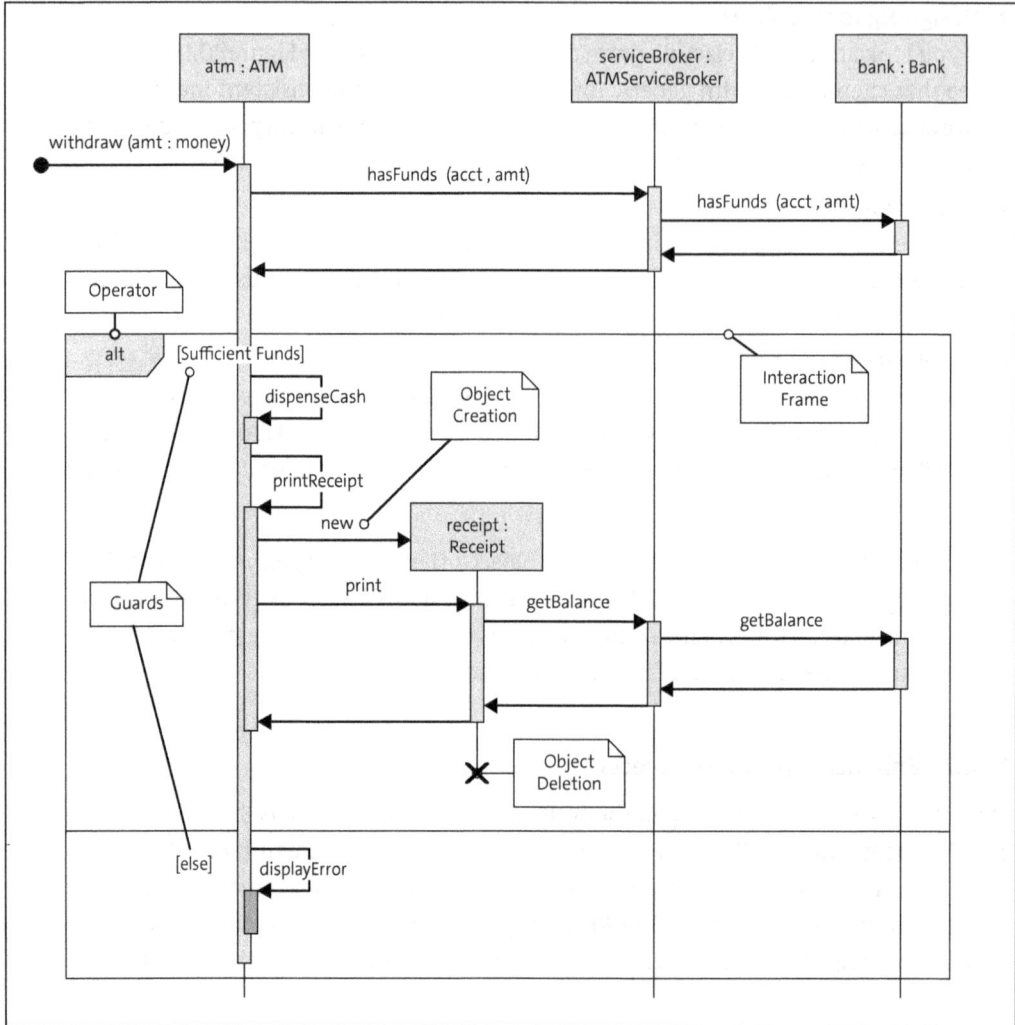

Figure 10.16 Sequence Diagram for Withdrawing Cash from an ATM

10.5.1   Creating and Deleting Objects

Within a given activation, it's not uncommon for a method to need to dynamically create another object to carry out a particular task. As you can see in Figure 10.16, the creation of an object is initiated by a special new message. The message name is optional, but the general convention is to name the message new. Notice that the object box for the receipt object is aligned with the creation message. This notation helps to clarify the fact that the object did not exist whenever the interaction began. Once an object is created, you can send messages to it just like any of the other objects in the sequence diagram.

If the created object is a temporary object (e.g., a local variable inside a method), then you can depict the deletion of the object by terminating the object lifeline with an X (see Figure 10.16). It's also possible for one object to explicitly delete another object by mapping a message from the requesting object to an X on the target object's lifeline.

10.5.2   Depicting Control Logic with Interaction Frames

You typically don't depict a lot of control logic in a sequence diagram. However, it's sometimes helpful to include high-level logic to make the interaction between the objects clear. In UML 2.0, this control flow is depicted using *interaction frames*.

An interaction frame partitions off a portion of the sequence diagram inside of a rectangular boundary. The functionality depicted in an interaction frame is described by an *operator* listed in the upper left-hand corner of the frame. For example, the sequence diagram in Figure 10.16 shows an interaction frame that is using the alt operator. The alt operator is used to depict conditional logic, such as an IF...ELSE or CASE statement. The branches of this conditional logic are divided by a horizontal dashed line. Each branch of the logic also contains a conditional expression called a *guard*. As you would expect, guards determine whether the control flows to a particular branch of the conditional logic. For instance, in the sequence diagram shown in Figure 10.16, the atm object will only dispense cash if there are sufficient funds in the account. Otherwise, an error message will be displayed on the console.

Table 10.2 describes some of the basic operators that you can use with interaction frames. Again, interaction frames should be used very sparingly. If you need to depict more complex logic, consider using an activity diagram or even some basic pseudocode.

Operator	Usage Type
alt	Depicts conditional logic such as an IF...ELSE or CASE statement.
opt	Depicts an optional piece of logic such as a basic IF statement.
par	Depicts parallel behavior. In this case, each fragment in the interaction frame runs in parallel.
loop	Depicts various types of looping structures (e.g., LOOP and DO).
ref	References an interaction defined on another sequence diagram.
sd	Surrounds an embedded sequence diagram within the current sequence diagram.

Table 10.2 Interaction Frame Operators

10.6   Summary

This concludes your introduction to the BOPF. Whether or not you plan to utilize the BOPF in your own custom developments, we hope that this introduction helped you understand how to apply object-oriented concepts on a wider scale in a reusable framework. This is important to consider as your object-oriented designs mature beyond a handful of classes. As your designs increase in scope, it's useful to have a framework like the BOPF to organize development objects and provide an abstraction around common tasks such as transaction handling and persistence.

In the next chapter, we'll show you how you can use your ABAP Objects skills to model behavior with the ABAP RESTful application programming model.

Chapter 11
ABAP RESTful Application Programming Model

A key topic for modern ABAPers is the ABAP RESTful application programming model, which has transformed ABAP development for SAP S/4HANA systems. But how does it work, and how does object-oriented programming come into the picture?

In this chapter, we'll look at SAP's ABAP RESTful application programming model through an object-oriented programming (OOP) lens. While we won't be able to cover every detail and function of this powerful model, we will discuss it at a foundational level so that you can understand how the ABAP RESTful application programming model fits into the SAP landscape. Also, more importantly for this book, we'll explore its relationship to OOP.

There are two main development objects that we'll focus on to give you an idea of how the ABAP RESTful application programming model relates to object-oriented design: classes and core data services (CDS) views. After reviewing how the ABAP RESTful application programming model fits into your SAP system, you'll learn how these two development objects relate to OOP, and you'll gain a new perspective and better understanding of the ABAP RESTful application programming model.

11.1 Introduction

In this section, we'll introduce the basics of the ABAP RESTful application programming model. You'll learn what the model is, how it fits into your SAP system, how it processes during runtime, and the technical objects you'll use to begin developing with the ABAP RESTful application programming model.

11.1.1 What Is the ABAP RESTful Application Programming Model?

SAP's ABAP RESTful application programming model is a model of dynamically implemented ABAP classes and automations that work in conjunction with *behavior definitions*. Behavior definitions are an extension of CDS views that allow you to perform operations (such as create, update, and delete) on the data model defined in the CDS view. Depending on the declarations in the behavior definition, the ABAP RESTful application programming model provides you with different levels of custom implementations where you can insert your own implementation logic. The model provides this function by generating subclasses of certain specially designed classes. This inheritance provides a structure in which you can insert your own implementations, and then the ABAP RESTful application programming model decides when to call each of the methods.

To get a visual of how behavior definitions and the ABAP RESTful application programming model relate to CDS views, SAP Gateway services, and SAP Fiori apps, think of behavior definitions and the ABAP RESTful application programming model as more of an extension alongside of CDS views, as opposed to trying to place them above or below CDS views. Figure 11.1 shows how each of the different development objects relate to each other to create the exposed service.

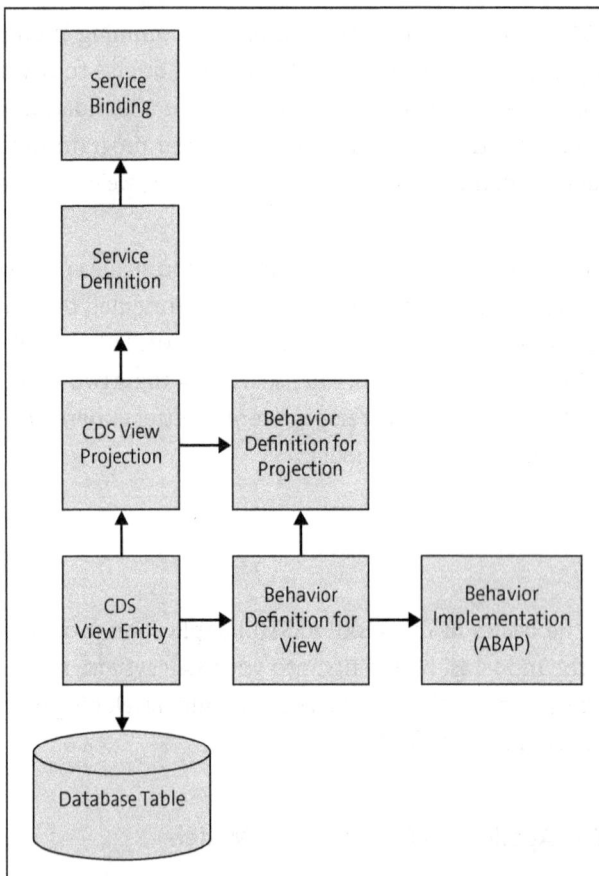

Figure 11.1 Behavior Definition Created as an Extension to the CDS View

In addition to providing standard create, read, update, delete (CRUD) functionality, the ABAP RESTful application programming model also provides powerful features such as draft capabilities, automatic numbering, authorization techniques, validations, functionality related to specific events, and the ability to perform operations on associations. Depending on which of these features are declared in the behavior definition, the model provides you with specific generated methods within the implementation class that you can use to implement your own logic as it relates to the specified feature. This is all determined and generated dynamically using a little bit of inheritance and some help from the kernel layer.

11.1.2 Relationship to SAP Gateway

Depending on your previous experience, you may be familiar with the Internet Communication Framework (ICF) (Transaction SICF), where you may have created a new endpoint and manually implemented a handler class. This implemented the REST interface, where you applied some CRUD logic for another app to consume. Or, perhaps you built CDS views for your data model, and you created a gateway service via Transaction SEGW, where you either let SADL take care of the majority of the implementation or you provided custom implementation within the _DPC_EXT classes. If you've ever activated a new service created in Transaction SEGW yourself, you'll know this is done by going to Transaction SICF.

SAP continues to build on top of its existing solutions to enhance and streamline its developer experience. The ABAP RESTful application programming model is no exception. After you complete the behavior definition and any custom implementation, when you expose the CDS view via a service definition and service binding, and then activate the new node created in Transaction SICF, the _DPC, _DPC_EXT, _MPC, and _MPC_EXT classes are utilized by the framework to call the underlying ABAP RESTful application programming model classes and methods.

To speak in object-oriented terms: The _DPC class uses composition to create objects related to the ABAP RESTful application programming model so that you can call methods on the instances of those ABAP RESTful application programming model objects.

In Figure 11.2, you can see the relationship between the ABAP RESTful application programming model's classes, SAP Gateway, and the ICF. The ICF provides the exposed access point for an external app to call, and SAP Gateway provides the platform on which you perform HTTP methods on various entities in that access point. When you perform an HTTP method on an entity with SAP Gateway, and that entity is extended with a behavior definition and an ABAP RESTful application programming model implementation, the ABAP RESTful application programming model's classes are called to execute the associated logic.

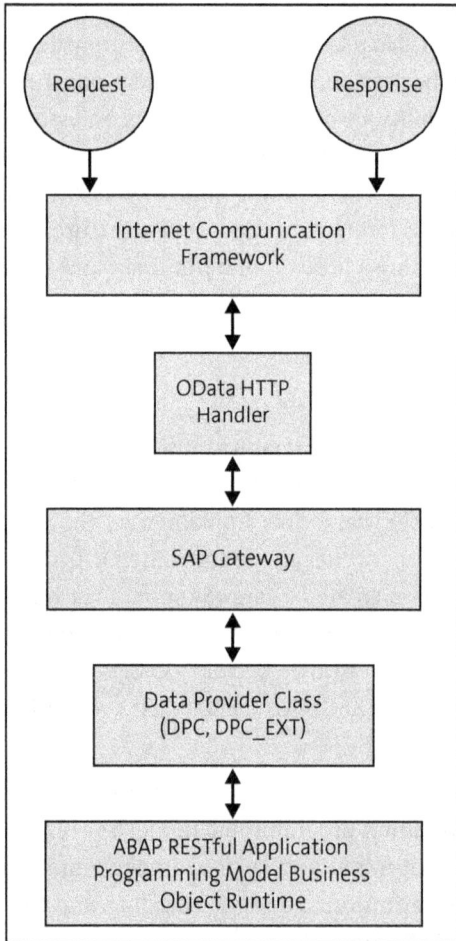

Figure 11.2 ABAP RESTful Application Programming Model Framework in Relation to SAP Gateway

11.1.3   Business Object Runtime

When a request is made from the consuming service or app, after the SAP Gateway service passes control to the ABAP RESTful application programming model framework, the *interaction phase* begins. The HTTP method of POST, PUT, DELETE, etc. is translated into ABAP and the ABAP RESTful application programming model determines whether this method is a CREATE, UPDATE, DELETE, or another type of action or determination. As the request comes in from the consuming service or app, the ABAP RESTful application programming model transactional engine takes over to determine which methods to call and in which order. Assuming all methods and features are defined for the behavior definition (which is not required), the sequence of the method calls would occur as follows:

1. Method FOR GLOBAL AUTHRIZATION

2. Method FOR GLOBAL FEATURES

3. Method FOR LOCK (not called on CREATE)

4. Method FOR READ (not called on CREATE)

5. Method FOR PRECHECK

6. Method FOR MODIFY

7. Method FOR PRECHECK

8. Method FOR INSTANCE AUTHORIZATION (not called on CREATE)

9. Method FOR INSTANCE FEATURES (not called on CREATE)

10. MODIFY method FOR [CREATE, UPDATE, DELETE, etc.]

11. Method FOR NUMBERING

12. Method FOR DETERMINATION on modify [CREATE, UPDATE, DELETE]/field

During the processing of the interaction phase, data that will be committed to the database is added to the *transactional buffer*. This is the in-memory holding area where data is staged so that it can be persisted to the database. In Section 11.1.5, you'll learn when the transactional buffer is automatically managed for you and when you must create your own transactional buffer to persist the data yourself.

After the interaction phase is completed, the *save sequence* is processed. The save sequence calls several methods to access the transaction buffer to get the data in memory to persist to the database, as well as determine any numbering and perform preparation or clean up actions. When the save sequence starts, it performs the following method calls:

1. Method FINALIZE

2. Method FOR DETERMINATION ON SAVE operation/field

3. Method CHECK_BEFORE_SAVE

4. Method FOR VALIDATION on the save field

5. Method ADJUST_NUMBERS

6. Method SAVE_MODIFIED (for managed with additional save scenarios)

7. Method SAVE (for unmanaged scenarios)

8. Method CLEANUP

9. Method CLEANUP_FINALIZE

This process of steps is shown in Figure 11.3.

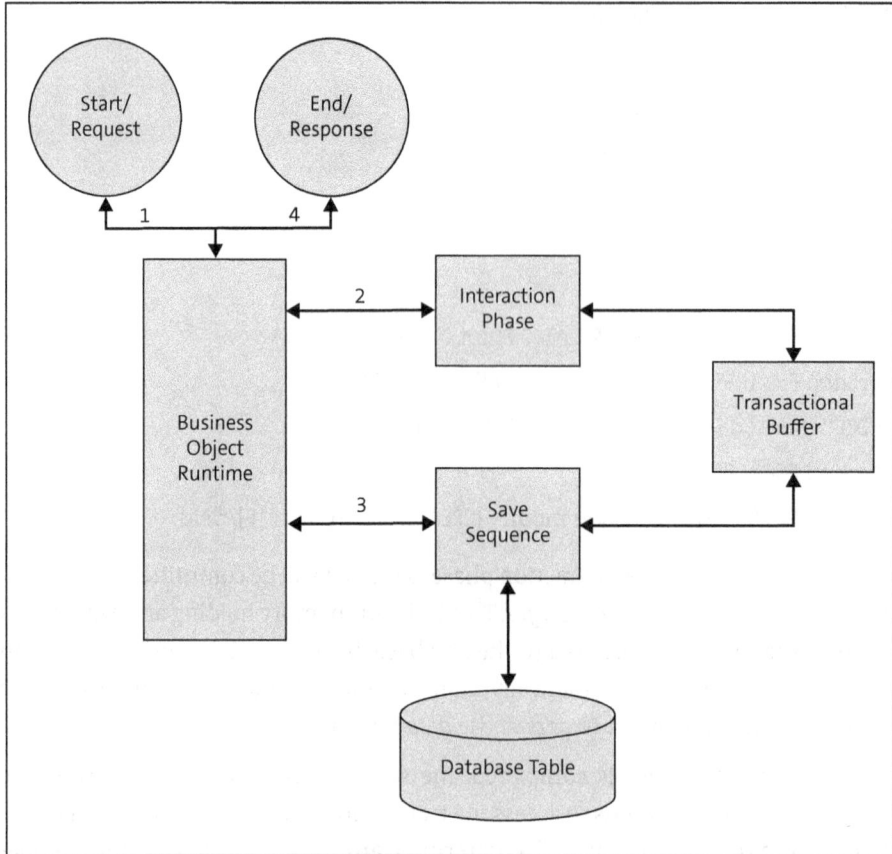

Figure 11.3 Business Object Runtime Sequence

11.1.4   Behavior Definitions

Behavior definitions are extensions of a CDS view that enable the ABAP RESTful application programming model. As their name suggests, they define the behaviors that you want to allow for a specific CDS view.

Behavior definitions must be created for both the projection view (also known as the *consumption view*) and the interface view. The behavior definition implemented on the projection view is the behavior definition that exposes functionalities implemented on the behavior definition of the interface view. The behavior definition on the interface view specifies most of the capabilities being enabled for the view, such as validation, authorization, create, update, delete, and so on.

There are a variety of levels of control when it comes to behavior definitions. Managed behavior definitions allow you the least amount of control and the highest degree of implementation, while unmanaged behavior definitions give you the maximum amount of control within the ABAP RESTful application programming model to implement within your designated pattern. Additionally, there are a few in-between options that provide automatic implementation from the managed definitions while still allowing you to implement your own saving methods. Let's take a look at each type of behavior definition:

- **Managed**

  Managed behavior definitions automatically create the necessary underlying logic for you to be able to perform all the standard operations on the CDS view (CRUD). When the behavior definition is created, it declares which database table the behavior definition should be linked to. Then, additional mapping is declared to explicitly confirm the connections between the fields in the CDS view and the fields in the database table.

  With the managed behavior definition, all the transactional buffering and standard operations are handled for you, and you implement your business logic using actions, determinations, and validations. Managed behavior definitions are most often used when you're starting from scratch and no preexisting function modules or classes have been created for your business logic.

  In instances where you may need a more customized saving technique, you have the option of managed behavior definitions with an unmanaged save, as well as managed behavior definitions with an additional save. We'll explore these options next.

- **Managed with unmanaged save**

  Like fully managed behavior definitions, the managed with unmanaged save strategy handles all of the implementation for standard operations, and you implement your business logic in actions, validations, and determinations. However, it allows you to determine exactly how the data should be saved.

  With an unmanaged save, a database table is not declared in the behavior definition because you are telling the model not to handle the saving aspect for you. You must specify the entire saving implementation yourself. The ABAP RESTful application programming model provides you with a template method to implement your saving technique, however simple or complex it may be.

- **Managed with additional save**

  With an additional save, you declare a database table in the behavior definition so that the model can handle the saving to the database for you, exactly the same as the traditional managed strategy. Additionally, the model provides you with a template method where you can implement additional saving to other tables that you may need to associate with your data. This method is called after the normal save routine of the ABAP RESTful application programming model runtime.

- **Unmanaged**

  Unmanaged behavior definitions do not allow you to specify a database table in the behavior definition, as you are expected to implement all logic yourself within the provided ABAP RESTful application programming model classes and methods. You must provide the implementation for all standard operations, explicitly stating how and when data is created, updated, and deleted. While the ABAP RESTful application programming model does perform the read operation based on the CDS view, it also provides an option for you to supplement the read operation, which is primarily leveraged when utilizing the business object in Entity Manipulation Language (EML) in other ABAP programs.

Unmanaged behavior definitions are often utilized when you have preexisting functional modules and classes that already have a business log implemented. The unmanaged strategy simply gives you the template to insert that existing logic to work with the ABAP RESTful application programming model.

You should implement a managed behavior definition strategy whenever possible. It allows the ABAP RESTful application programming model to do most of the heavy lifting, and SAP Fiori elements apps can more easily consume the service. If your needs exceed the scope of a managed strategy, see if you can use a managed with unmanaged save or managed with additional save strategy instead, as these are the second best options. If it's not possible to fit your needs into a managed scenario, you'll have to use an unmanaged strategy, which gives you the most flexibility and control over implementing what you need within the boundaries of the template and the ABAP RESTful application programming model's classes.

11.1.5   Model Classes

At this point, you should have a basic understanding of how the ABAP RESTful application programming model works. Now, we'll zoom in to look at the classes that the ABAP RESTful application programming model provides. We'll explore their method capabilities briefly, explain them through an object-oriented lens, and discuss the arbitrary lcl_buffer class that you'll inevitably come across.

Model classes are automatically generated for you based on what you defined in a behavior definition. When you create a behavior definition, you'll see a syntax warning at the top of the behavior pool class name alerting you that you need to create the behavior pool class before proceeding. You can generate this class using the code assistant ([Ctrl]+[F1]), which automatically generates two local classes in the **Local Types** tab of that class: a *handler* class and a *saver* class. These local classes are templated for you, so the framework has defined the methods, and it's up to you to provide the implementation for those methods.

Class Hierarchy

Before diving into the handler and saver classes, we'll first discuss the root class that the handler and saver classes inherit from so that you understand where the provided methods come from and the polymorphism created from the subclassing.

Both generated classes ultimately inherit from `cl_abap_behv`. Think of this as the root class for any custom implementation that you do. It provides you with necessities that make it easier for you to work with the ABAP RESTful application programming model. The most commonly used aspect is related to messages. The `cl_abap_behv` class gives you two helper methods for creating messages: `new_message`, and `new_message_with_text`. Each of these communicates various things from the code to the consuming service or app.

The subclasses `cl_abap_behavior_handler` and `cl_abap_behavior_saver`, as shown in Figure 11.4, are explained in the next section.

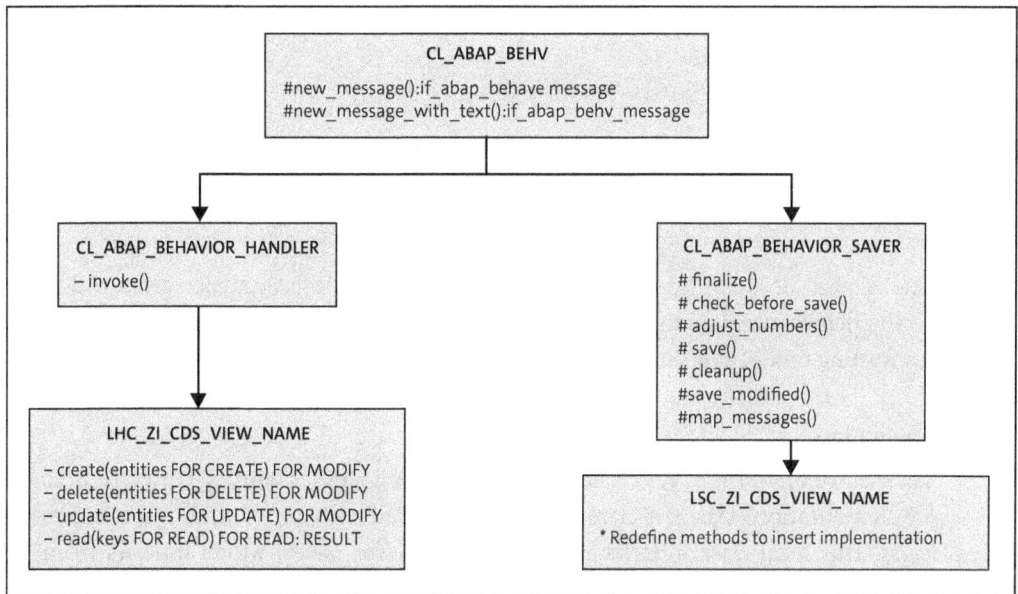

Figure 11.4 UML Diagram Showing How the ABAP RESTful Application Programming Model Behavior Classes Are Arranged

Handler Class

As mentioned in Section 11.1.1, the generated implementation class from the behavior definition creates two local classes in the **Local Types** tab. The first is the handler class, which will be named using the same convention as the implementation class, with `lhc_` prepended to the beginning of the class name, which stands for *local handler class*. This class inherits from `cl_abap_behavior_handler`, which inherits from `cl_abap_behv`.

Inheriting from the `cl_abap_behavior_handler` class implicitly allows you to specify your own methods using keywords FOR MODIFY after the method name. This tells the compiler that this method must be looked at when performing an operation (POST, PUT, DELETE, etc.). You can create a method with many importing parameters for each operation, or you can create multiple methods, each designated to a specific operation. You define which parameter should be looked at during runtime in the IMPORTING parameters sections, where you can specify your operation-specific parameters by adding on the related keywords after the parameter name FOR CREATE, FOR UPDATE, or FOR DELETE as shown in Listing 11.1.

```
CLASS lhc_zi_todo_cdh DEFINITION INHERITING FROM cl_abap_behavior_handler.
PRIVATE SECTION.
METHODS create FOR MODIFY
IMPORTING entities FOR CREATE zi_todo_cdh.
METHODS delete FOR MODIFY
IMPORTING keys FOR DELETE zi_todo_cdh.
METHODS update FOR MODIFY
IMPORTING entities FOR UPDATE zi_todo_cdh.
METHODS read FOR READ
IMPORTING keys FOR READ zi_todo_cdh RESULT result.
ENDCLASS.
```

Listing 11.1 Class Definition for the ABAP RESTful Application Programming Model Handler Class

The cl_abap_behavior_handler class inherits from cl_abap_behv, which gives you access to the helper methods that create messages within the ABAP RESTful application programming model and can be used to return messages to the consumer.

Saver Class

Let's move on to the saver class, which will follow a similar naming convention to the implementation class. The class name is prepended with lsc_, standing for *local saver class*. This local class inherits from cl_abap_behavior_saver, which inherits from cl_abap_behv.

The cl_abap_behavior_saver class is more explicit. If you look at the class, you can see it gives you access to methods, as shown in Listing 11.2: finalize, check_before_save, adjust_numbers, save, cleanup, save_modified, and map_messages. This gives you a template to work from when implementing your logic in the necessary places.

```
CLASS lsc_zi_todo_cdh DEFINITION INHERITING FROM cl_abap_behavior_saver.
PROTECTED SECTION.
METHODS check_before_save REDEFINITION.
METHODS finalize REDEFINITION.
METHODS save REDEFINITION.
ENDCLASS.
CLASS lsc_zi_todo_cdh IMPLEMENTATION.
METHOD check_before_save.
ENDMETHOD.
METHOD finalize.
ENDMETHOD.
```

```
METHOD save.
ENDMETHOD.
ENDCLASS.
```

Listing 11.2 Saver Class Definition and Implementation

Class `cl_abap_behavior_saver` also inherits from the `cl_abap_behv` class, giving you access to the helper methods that create messages in the ABAP RESTful application programming model.

Local Buffer Class

When a behavior definition is *managed*, the transactional buffer is managed for you. However, when a behavior definition is *unmanaged*, you must provide and handle the transactional buffer. In this case, you must define your own local class alongside the generated handler and saver classes in the **Local Types** tab. You can call this anything you'd like, but for the sake of this book, we'll name it `lcl_buffer` and refer to it as the *buffer class*.

Simply put, the buffer class is a holding place to store data during runtime when data is passed between the handler class and the saver class. There are a few ways you can set this up to adhere to different object-oriented patterns, but considering this is a local class, it's not accessible to the global environment. It isn't technically necessary to implement the usual protections, but you may want to do so to adhere to clean code practices or your team's development guidelines.

Considering you don't need any logic surrounding the holding of the data, the simplest form of the buffer class is to create static attributes to hold the data to be passed between the handler and saver class, as shown in Listing 11.3. An example of its use is shown in Listing 11.4.

```
CLASS lcl_buffer DEFINITION.
PUBLIC SECTION.
CLASS-DATA: m_data TYPE ztodo.
ENDCLASS.
CLASS lcl_buffer IMPLEMNTATION.
ENDCLASS.
```

Listing 11.3 Buffer Class Using Public Static Attributes to Store Data in the Buffer Class

```
DATA(ls_data) = lcl_buffer=>m_data.
```

Listing 11.4 Buffer Class and Static Attribute Accessing the Data in the Buffer Class

A more advanced form is to leverage a *singleton pattern*, where only one instance of the buffer class can be instantiated during the runtime. This ensures that both the handler and saver class reference the same instance of the buffer and pass the same data to each other. Listing 11.5 shows an example where instance creation is protected inside of the get_instance method and a getter and setter method is implemented to control access to the internal attributes.

```abap
CLASS lcl_buffer DEFINITION.
PUBLIC SECTION.
CLASS-METHODS get_instance
RETURNING VALUE(ro_buffer) TYPE REF TO lcl_buffer.
METHODS set_data
IMPORTING is_data TYPE ztodo.
METHODS get_data
RETURNING VALUE(rs_data) TYPE ztodo.
PRIVATE SECTION.
CLASS-DATA: mo_buffer TYPE REF TO lcl_buffer.
DATA m_data type ztodo.

ENDCLASS.

CLASS lcl_buffer IMPLEMENTATION.
METHOD get_instance.
IF mo_buffer IS INITIAL.
mo_buffer = NEW lcl_buffer( ).
ENDIF.
ro_buffer = mo_buffer.
ENDMETHOD.

METHOD set_data.
m_data = is_data.
ENDMETHOD.

METHOD get_data.
rs_data = m_data.
ENDMETHOD.

ENDCLASS.
```

Listing 11.5 Example of a Buffer Class Using the Singleton Pattern

Listing 11.6 shows how the buffer class is instantiated and utilized to set and get data from the object. Normally, the set and get happen in different places in the context of your implementation. For the sake of this example, they are shown one after the other.

```abap
DATA(lo_buffer) = lcl_buffer=>get_instance( ).
lo_buffer->set_data( VALUE #( item = 'to do item 1' ) ).
DATA(ls_data) = lo_buffer->get_data( ).
```

Listing 11.6 Using the Buffer Singleton Class to Set and Get Data

These are just a few ways you can implement the buffer class to hold data at runtime to communicate between the handler and saver class. If you want to learn more about the singleton pattern, refer to Chapter 13, Section 13.2.2.

11.2   CDS and SAP Gateway Service Bindings

By now, you should have a basic understanding of how the ABAP RESTful application programming model's classes relate to each other and use inheritance to give you a template for implementing your own logic. Let's explore the slightly more advanced area of CDS views and their relation to the ABAP RESTful application programming model.

In this section, you'll learn how CDS views form the foundation of the ABAP RESTful application programming model and how associations provide relational features. You'll build a simple ABAP RESTful application programming model-based service, from Z table and CDS creation to publication in an SAP Gateway service.

11.2.1   CDS Views

CDS views are the foundation of working with the ABAP RESTful application programming model. You can't have the ABAP RESTful application programming model without CDS views! Creating a CDS view involves defining the type of structure for the business object entity you want to expose in a web service or SAP Fiori app. This can be as simple as mirroring the fields in a single table or as complex as creating a CDS view with a structure that combines fields from multiple tables via joins or stacking similar sets of data via unions.

Depending on how and when you've implemented your existing CDS views, you may need to make minor adjustments to them to use them with the ABAP RESTful application programming model and create the behavior definitions you need to associate with them. If you're creating entirely new CDS views, keep the following rules in mind:

- Behavior definitions can only be created on *root views*.
- Views should be *view entities*.
- The consumption view should be a *projection view*.

Listing 11.7 shows an example of a root view. This tells the system that this is the main base view.

Declaring CDS views as view entities eliminates the requirement to specify the SQL view name in the top annotations of the CDS view. You can perform SQL directly on the CDS view name, as opposed to the SQL view name in ABAP, for the version of ABAP that ABAP RESTful application programming model runs on. You can see the declaration of the view entity in Listing 11.7.

```
...
Define root view entity zi_my_cds_view as select from z_my_table
...
```

Listing 11.7 An Example Definition of a Root View Entity

The consumption view should be declared with as projection on (see Listing 11.8) as opposed to as select from, since you shouldn't perform an additional select on the interface view to expose the necessary fields. You simply need to project and/or expose the necessary fields. It helps the ABAP RESTful application programming model's projection behavior definition to only have to project the specific behaviors needed for the service, as opposed to requiring the behavior implementation class that the interface view-level behavior definition requires.

```
...
Define root view entity zc_my_cds_view as projection on zi_my_cds_view
...
```

Listing 11.8 An Example Definition of a Projection View

11.2.2   Associations

If you've dealt with CDS views in the past, you've probably encountered or created *associations*. These allow you to relate one or more CDS views to another without having to duplicate fields or introduce code smells by including duplicate or unrelated fields in certain views.

When dealing with the ABAP RESTful application programming model, it's important to make sure your CDS views are associated in a specific way. The ABAP RESTful application programming model framework has a CREATE_BY_ASSOCIATION declaration, which allows the entities to be related in an expected way. This means that the child entity is only accessible through the parent entity. You cannot make a request directly on a /child endpoint. It must go through the parent, as in the following example: /parent('id_001')/child.

To use CREATE BY ASSOCIATION, you must set up the associations on your CDS views appropriately, using composition and association to the parent. Composition should be a familiar word from what you've learned in earlier chapters. It means that two objects have a *has-a* relationship, as opposed to inheritance, where objects have an *is-a* relationship. For example, in the entity-relationship diagram shown in Figure 11.5, Order has a *has-a* relationship to Item, meaning that an order can have one or more items, and an item belongs to one order.

Listing 11.9 shows how you can create this with CDS views by creating an order view and an item view and setting up the association. Here, the ZI_Item view is the child, so the association is declared as association to parent. You define the the the ZI_Order entity as the parent entity, declaring the association as composition of. Using these declarations as opposed to just association to [view] holds the entity relationship to a parent-child relationship and restricts the child entity from trying to be a parent entity to the already existing parent entity. In other words, the ZI_Item entity cannot be a parent to a ZI_Order entity.

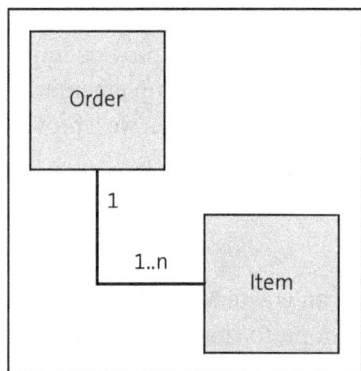

Figure 11.5 Entity-Relationship Diagram Displaying the Relationship Between Order and Item

```
define view entity ZI_Item
  as select from zorder
  association to parent zi_order on $projection.ordernum = ordernum
{
  . . .,
_items
}

define root view entity ZI_Order
  as select from zorder
  composition of zitem as _items
{
  . . .,
_items
}
```

Listing 11.9 Parent-Child Relationship Between Item and Order CDS View Entities

When CDS views are created with an association like this, and when the behavior definition is generated, the system automatically prepopulates with the CREATE BY ASSOCIATION declaration. Likewise, when the behavior implementation is generated, the corresponding CREATE BY ASSOCIATION methods will be automatically generated and available for you to implement your own code into.

When a consumer wants to create one of the composing entities, they must do it through the parent entity, as follows: /service/order(100001)/_item.

This type of structured modeling allows the ABAP RESTful application programming model to work well with a managed behavior definition strategy and SAP Fiori elements apps. It creates a predictable process to reduce the amount of customized code.

11.3   Modeling Behavior Using Object-Oriented Techniques

In this section, we'll explore how you can leverage your OOP skills to provide implementations for ABAP RESTful application programming model implementation classes. First, we'll walk through the process of creating a service, and then we'll provide a few implementation examples using an object-oriented approach.

11.3.1   Creating the Service

We'll use a chore list app as our example, where you can create a new list and mark tasks as complete. As we'll discuss in the following sections, to create the service, you must first create the following:

- Custom tables for lists and tasks
- CDS views
- Behavior definitions
- Service definition
- Service binding

Create the Custom Tables

First off, you need to create two custom tables that will persist the data for your ABAP RESTful application programming model app. The first of two tables will be the Chore List table. This will hold the header-level data of the different chore lists you'll create. Listing 11.10 shows the declaration of the table and its fields.

```
@EndUserText.label : 'Chore List'
@AbapCatalog.enhancementCategory : #NOT_EXTENSIBLE
@AbapCatalog.tableCategory : #TRANSPARENT
@AbapCatalog.deliveryClass : #A
@AbapCatalog.dataMaintenance : #RESTRICTED
define table zchorelist {
  key client : abap.clnt not null;
  key listid : /bobf/conf_key not null;
  title     : char255;
  createdby : uname;
  createdat : timestampl;
}
```

Listing 11.10 Definition of the Chore List Table

Next, create the Chore List Task table. This will hold the chore lists' item-level data. Listing 11.11 shows the declaration of the task table and its fields.

```
@EndUserText.label : 'Chore List Task'
@AbapCatalog.enhancementCategory : #NOT_EXTENSIBLE
@AbapCatalog.tableCategory : #TRANSPARENT
@AbapCatalog.deliveryClass : #A
@AbapCatalog.dataMaintenance : #RESTRICTED
define table zchoretask {
  key client   : abap.clnt not null;
  key listid   : /bobf/conf_key not null;
  key taskid   : /bobf/conf_key not null;
  description  : char255;
  complete     : boolean_flg;
  createdby    : uname;
  createdat    : timestampl;
  updatedat    : timestampl;
}
```

Listing 11.11 Definition of the Chore Task Table

After you save and activate these tables, your database is ready to persist data from your app.

Create CDS Views

You'll need to create several CDS views: two interface views that correspond to your database tables, which you created in the previous section, and one consumption view that will be a projection of your header level view of the chore list.

First, create the Chore List (header level) view, as shown in Listing 11.12.

```
@AccessControl.authorizationCheck: #NOT_REQUIRED
@EndUserText.label: 'Chore List'
@Metadata.ignorePropagatedAnnotations: true
@Metadata.allowExtensions: true
define root view entity ZI_ChoreList
  as select from zchorelist as List
  composition [0..*] of ZI_ChoreTask as _Task
{
  key listid    as ListId,
      title     as Title,
      createdby as CreatedBy,
      createdat as CreatedAt,
      _Task
}
```

Listing 11.12 Definition of the Chore List CDS View Entity

Next, create the ZI_ChoreTask (item level) view, as shown in Listing 11.13.

```
@AccessControl.authorizationCheck: #NOT_REQUIRED
@EndUserText.label: 'Chore List Task'
@Metadata.ignorePropagatedAnnotations: true
@Metadata.allowExtensions: true
define view entity ZI_ChoreTask
  as select from zchoretask as Task
  association to parent ZI_ChoreList as _List on $projection.ListId = _
List.ListId
{
  key listid      as ListId,
  key taskid      as TaskId,
      description as Description,
      complete    as Complete,
      createdby   as CreatedBy,
      createdat   as CreatedAt,
      updatedat   as UpdatedAt,
      _List
}
```

Listing 11.13 Definition of the Chore Task CDS View Entity

These views are mostly straightforward, as you're including the fields from the database table and defining the appropriate alias names for the field names. You might be curious about how the Chore List view relates to the Chore Task view. Here, you're using the concept of composition, as opposed to using associations for both views. By using composition in the Chore List view, it tells the ABAP RESTful application programming model framework that Chore List is the parent entity, and it contains tasks. It restricts the Chore List Task from being a parent to any Chore Lists. Likewise, in your Chore List Task view, you define the relationship as association to parent, declaring the Chore List Task entity as the child. When you generate the behavior handler class for your unmanaged behavior definition, you'll see how setting up a parent-child relationship with composition helps the ABAP RESTful application programming model framework know which methods to generate and call.

Now that you've created two interface views for your database tables, you can create your consumption view (Chore List Projection) to project your Chore List entity, as shown in Listing 11.14.

```
@AccessControl.authorizationCheck: #NOT_REQUIRED
@EndUserText.label: 'Chore List Projection'
@Metadata.ignorePropagatedAnnotations: true
define root view entity ZC_ChoreList
  as projection on ZI_ChoreList
```

```
{
  key ListId,
      Title,
      CreatedBy,
      CreatedAt,
      /* Associations */
      _Task
}
```

Listing 11.14 Definition of the CDS View Projection for Consumption

Now that you've created the necessary CDS views, you can move on to creating the behavior definitions to unlock your entities' CRUD capabilities.

Create Behavior Definitions

To allow your CDS views to be used by the ABAP RESTful application programming model framework, you must first create behavior definitions for them. You need to create a behavior definition for your root interface view as well as your projection consumption view.

Listing 11.15 displays the behavior definition for the root interface view of ZI_ChoreList. Notice that we created it as an unmanaged implementation type so that we can implement object-oriented techniques into the ABAP RESTful application programming model framework's methods. The behavior definition for ZI_ChoreList defines the methods of access for the entity: create, update, delete, and its association with the child entity, _Task. Because of that association, you also define the behavior definition for the ZI_ChoreTask and its methods of access: create, update, delete.

```
unmanaged implementation in class zbp_i_chorelist unique;

define behavior for ZI_ChoreList
//late numbering
//lock master
//etag master <field_name>
{
  create;
  update;
  delete;
  association _Task { create; }
}

define behavior for ZI_ChoreTask
//late numbering
//lock dependent by <association>
//etag master <field_name>
```

```
{
  create;
  update;
  delete;
}
```

Listing 11.15 Definition of the Behavior Definition for the CDS View

The behavior pool implementation class for the behavior definition is generated based on the declarations made in the behavior definitions you defined in Listing 11.15. In Listing 11.16, you can see the behavior handler classes that were generated from the behavior definition. The implementation of these handler classes occurs in the **Local Types** area generated within the zbp_i_chorelist behavior pool implementation class object. The generated implementation includes all of the methods defined from the behavior definition, such as create, update, and delete. Because of the composition relationship originally set up on the CDS view entities, the ABAP RESTful application programming model framework automatically adds the method definitions for cba_task (create by association) and rba_task (read by association) for the task entity.

```
CLASS lhc_zi_chorelist DEFINITION INHERITING FROM cl_abap_behavior_handler.
  PRIVATE SECTION.
    METHODS create FOR MODIFY
      IMPORTING entities FOR CREATE zi_chorelist.
    METHODS delete FOR MODIFY
      IMPORTING keys FOR DELETE zi_chorelist.
    METHODS update FOR MODIFY
      IMPORTING entities FOR UPDATE zi_chorelist.
    METHODS read FOR READ
      IMPORTING keys FOR READ zi_chorelist RESULT result.
    METHODS cba_task FOR MODIFY
      IMPORTING entities_cba FOR CREATE zi_chorelist\_task.
    METHODS rba_task FOR READ
      IMPORTING keys_rba FOR READ zi_chorelist\_task FULL result_requested
RESULT result LINK association_links.
ENDCLASS.

CLASS lhc_zi_chorelist IMPLEMENTATION.
  METHOD create.
  ENDMETHOD.

  METHOD delete.
  ENDMETHOD.

  METHOD update.
  ENDMETHOD.
```

```
    METHOD read.
    ENDMETHOD.

    METHOD cba_task.
    ENDMETHOD.

    METHOD rba_task.
    ENDMETHOD.
ENDCLASS.

CLASS lhc_zi_choretask DEFINITION INHERITING FROM cl_abap_behavior_handler.
  PRIVATE SECTION.
    METHODS create FOR MODIFY
      IMPORTING entities FOR CREATE zi_choretask.
    METHODS delete FOR MODIFY
      IMPORTING keys FOR DELETE zi_choretask.
    METHODS update FOR MODIFY
      IMPORTING entities FOR UPDATE zi_choretask.
    METHODS read FOR READ
      IMPORTING keys FOR READ zi_choretask RESULT result.
ENDCLASS.

CLASS lhc_zi_choretask IMPLEMENTATION.
  METHOD create.
  ENDMETHOD.

  METHOD delete.
  ENDMETHOD.

  METHOD update.
  ENDMETHOD.

  METHOD read.
  ENDMETHOD.
ENDCLASS.

CLASS lsc_zi_chorelist DEFINITION INHERITING FROM cl_abap_behavior_saver.
  PROTECTED SECTION.
    METHODS check_before_save REDEFINITION.
    METHODS finalize         REDEFINITION.
    METHODS save             REDEFINITION.
ENDCLASS.
```

```
CLASS lsc_zi_chorelist IMPLEMENTATION.
  METHOD check_before_save.
  ENDMETHOD.

  METHOD finalize.
  ENDMETHOD.

  METHOD save.
  ENDMETHOD.
ENDCLASS.
```

Listing 11.16 Behavior Pool Class Implementation Generated by the ABAP RESTful Application Programming Model Framework Based on the Behavior Definitions

The final step in creating your behavior definitions is to create the behavior definition for your projection view. Listing 11.17 shows that the behavior definition for the projection view is defined with projection at the top. It's also high level and admits access to the available behaviors defined in the interface view-level behavior definition, as opposed to actually defining the types of access that will be implemented. You can spot this because of the use declaration on each of the access methods. For example, if your interface view-level behavior definition didn't define create as an access method, you wouldn't have the option to declare use create on your projection level behavior definition. As a contrary example, you may have defined create as an access method on your interface level behavior defnition, but you could choose to omit use create on your projection level behavior definition. You would choose this option if you needed to restrict your app or service from creating entities for some reason.

```
projection;

define behavior for ZC_ChoreList //alias <alias_name>
{
  use create;
  use update;
  use delete;
  use association _Task { create; }
}
```

Listing 11.17 Definition of the Projection Behavior Definition

Create Service Definition and Service Binding

The last step in building out the foundation of your ABAP RESTful application programming model app is to expose the entities you've created as a web service or UI app. In our case, we'll use a UI app, and we'll add some annotations to leverage an SAP Fiori elements app to help us with our implementation.

Start by creating the service definition for the chore list, as shown in Listing 11.18. As shown in this example, you should expose the projection view entity ZC_ChoreList and the ZI_ChoreTask view so that the app or service knows it can access these specific entities.

```
@EndUserText.label: 'Chore List Service Definition'
define service ZSD_ChoreList {
  expose ZC_ChoreList;
  expose ZI_ChoreTask;
}
```

Listing 11.18 Service Definition for Our Chore List

Next, create the service binding, as shown in Figure 11.6. In this example, our service binding is created with a **Binding Type** of **OData V2 - UI**. Make sure you publish it and activate the binding.

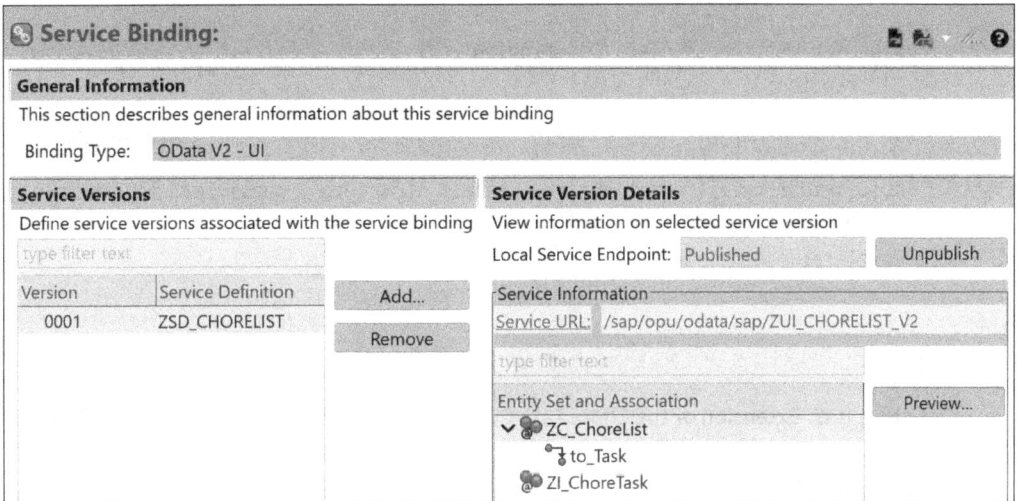

Figure 11.6 Service Binding for the Chore List

To verify that it's working, click on the **ZC_ChoreList** entity set and click **Preview.** This should load a blank SAP Fiori app. The app will be blank because you have yet to add any annotations. You can add some basic annotations in the extension view so that you can view the data and the available functions.

Listing 11.19 shows the extension annotations for the chore task view. These annotations define a UI facet so that you can view the ZI_ChoreTask entity in the UI. Note that you'll need to add the fields Task Description and Complete to view them on the UI screen within that facet.

```
@Metadata.layer: #CORE
annotate entity ZI_ChoreTask with
{

    @UI.facet: [{
    id: 'TaskMain',
    purpose: #STANDARD,
    type: #FIELDGROUP_REFERENCE,
    label: 'Chore Task',
    targetQualifier: 'TaskMain',
    position: 10
  }]


  @UI.lineItem: [{ position: 20 }]
  @UI.fieldGroup: [{ position: 20,  label: 'Description', qualifier: 'TaskMain'
}]
  @EndUserText.label: 'Task Description'
  Description;

  @UI.lineItem: [{ position: 30 }]
  @UI.fieldGroup: [{ position: 30,  label: 'Complete', qualifier: 'TaskMain' }]
  @EndUserText.label: 'Complete'
  Complete;


}
```

Listing 11.19 Extension of the Chore Task CDS View to Implement Necessary Annotations

Listing 11.20 shows the extension annotations for the chore list view. These annotations define a UI facet so that you can view the ZC_ChoreList entity in the UI. Note that you'll need to add the Title field to view it on the UI screen within that facet.

```
@Metadata.layer: #CORE
annotate entity ZC_ChoreList
    with
{

@UI.facet: [{
    id: 'Main',
    purpose: #STANDARD,
    type: #FIELDGROUP_REFERENCE,
    label: 'Chore List',
    targetQualifier: 'List',
```

```
      position: 10
  },
  {
    id: 'Tasks',
    purpose: #STANDARD,
    type: #LINEITEM_REFERENCE,
    label: 'Tasks',
    targetElement: '_Task',
    position: 20
  }]

  @UI.fieldGroup: [{ position: 10, qualifier: 'List' }]
  @UI.lineItem: [{ position: 10 }]
  @UI.identification: [{ position: 10 }]
  @EndUserText.label: 'Title'
  Title;
}
```

Listing 11.20 Extension for Chore List Consumption View for Annotations

Listing 11.21 shows the extension annotations for the list view. These annotations define a UI facet so that you can view the ZI_ChoreList entity in the UI. Note that you'll need to add the Title field to view it on the UI screen within that facet.

```
@Metadata.layer: #CORE
annotate entity ZI_ChoreList
    with
{

@UI.facet: [{
    id: 'Main',
    purpose: #STANDARD,
    type: #FIELDGROUP_REFERENCE,
    label: 'Chore List',
    targetQualifier: 'List',
    position: 10
  },
  {
    id: 'Tasks',
    purpose: #STANDARD,
    type: #LINEITEM_REFERENCE,
    label: 'Tasks',
    targetElement: '_Task',
    position: 20
  }]
```

```
@UI.fieldGroup: [{ position: 10, qualifier: 'List' }]
@UI.lineItem: [{ position: 10 }]
@UI.identification: [{ position: 10 }]
@EndUserText.label: 'Title'
Title;
}
```

Listing 11.21 Extension View for Chore List Annotations

After you add these extensions, refresh the preview app to see the **Title** field. After creating an initial list, you should be able to see where you can create tasks within the list. However, no data will persist, and if you try saving a task to the list, it will error out. This is because you don't have an implementation for your unmanaged scenario yet. We'll cover this in the next section.

11.3.2   Object-Oriented Implementation

Building out the implementation of your behavior pool can ultimately be done however you see fit. In our case, we'll focus on using object-oriented principles to accomplish this task. Since the ABAP RESTful application programming model framework automatically gives you a template of methods to work with, you'll need to build your objects to work with those templates. If you already have existing classes or function modules, you might need to refactor them to work with the ABAP RESTful application programming model framework. At the very least, you can build a façade or adapter class to act as a wrapper to make the ABAP RESTful application programming model framework fit with your existing objects.

In our example, we're starting from scratch, so we'll walk you through building some simple classes that work well on their own and with the ABAP RESTful application programming model framework. Because the ABAP RESTful application programming model framework decouples the save method from the CRUD methods, it's common to use a transactional buffer class to temporarily hold the data that will be used in the save method. You'll create the following classes for each of your entities, making sure that persistence can occur separately when called from the save method:

- ZBP_I_CHORELIST
  - LCL_CHORELIST_BUFFER
  - LHC_ZI_CHORELIST
  - LHC_ZI_CHORETASK
  - LSC_ZI_CHORELIST
- ZCL_CHORELIST
- ZCL_CHORETASK

Start by creating the ZCL_CHORELIST and ZCL_CHORETASK classes. Similar to how your CDS views are related, where the chore list view is the parent and the chore task view is the child, you'll also use composition so that your chore list class contains one-to-many chore task objects within it. Listing 11.22 and Listing 11.23 show the complete class implementation of the chore list class ZCL_CHORELIST and the chore task class ZCL_CHORE_TASK, respectively.

```abap
CLASS zcl_chorelist DEFINITION
  PUBLIC
  FINAL
  CREATE PUBLIC .

  PUBLIC SECTION.
    METHODS:
      constructor
        IMPORTING
          iv_listid TYPE /bobf/conf_key OPTIONAL,

      set_title
        IMPORTING
          iv_title TYPE char255,

      delete,
      read,
      save,

      add_task
        IMPORTING
          iv_description TYPE char255
          iv_new         TYPE abap_bool DEFAULT abap_false,
      delete_task
        IMPORTING
          iv_taskid TYPE zi_choretask-taskid,
      get_task
        IMPORTING
          iv_taskid     TYPE zi_choretask-taskid
        RETURNING
          VALUE(r_task) TYPE REF TO zcl_choretask.

  PROTECTED SECTION.
  PRIVATE SECTION.
    TYPES: BEGIN OF ty_task_ref,
             taskid TYPE /bobf/conf_key,
             task   TYPE  REF TO zcl_choretask,
           END OF ty_task_ref.
```

```
    DATA:
        lt_tasks TYPE STANDARD TABLE OF ty_task_ref.
    DATA:
        lm_title TYPE char255.
    DATA: lm_listid TYPE /bobf/conf_key.
ENDCLASS.
CLASS zcl_chorelist IMPLEMENTATION.

  METHOD constructor.
    lm_listid = iv_listid.
  ENDMETHOD.

  METHOD set_title.
    lm_title = iv_title.
  ENDMETHOD.

  METHOD delete.
    DELETE FROM zchorelist WHERE listid = lm_listid.
    DELETE FROM zchoretask WHERE listid = lm_listid.
  ENDMETHOD.

  METHOD save.

    DATA ls_chorelist TYPE zchorelist.

    IF lm_listid IS INITIAL.

      DATA lv_timestamp TYPE timestampl.
      GET TIME STAMP FIELD lv_timestamp.

      ls_chorelist = VALUE #( listid = /bobf/cl_frw_factory=>get_new_key( )
                              createdat = lv_timestamp
                              createdby = sy-uname
                              title = lm_title ).

      MODIFY zchorelist FROM ls_chorelist.
    ENDIF.

    LOOP AT lt_tasks ASSIGNING FIELD-SYMBOL(<ls_task>).
      <ls_task>-task->save( ).
    ENDLOOP.

  ENDMETHOD.
```

```
METHOD add_task.
  DATA(lo_task) = NEW zcl_choretask( iv_listid = lm_listid  ).
  lo_task->set_description( iv_description ).
  IF iv_new = abap_true.
    lo_task->set_new_ind( ).
  ENDIF.
  APPEND INITIAL LINE TO lt_tasks ASSIGNING FIELD-SYMBOL(<task>).
  <task>-task = lo_task.
  <task>-taskid = ''.
ENDMETHOD.

METHOD delete_task.
  DELETE FROM zchoretask WHERE taskid = iv_taskid.
ENDMETHOD.

METHOD get_task.
  DATA(ls_task_ref) = VALUE #( lt_tasks[  taskid = iv_taskid ] OPTIONAL ).
  DATA(lo_task) = ls_task_ref-task.
  IF lo_task IS BOUND.
    r_task = lo_task.
    RETURN.
  ENDIF.

  SELECT SINGLE *
  FROM zchoretask
  WHERE taskid = @iv_taskid
  INTO @DATA(ls_choretask).

  lo_task = NEW zcl_choretask( iv_listid = ls_choretask-listid
                               iv_taskid = ls_choretask-taskid ).
  lo_task->set_description( ls_choretask-description ).
  lo_task->set_complete( ls_choretask-complete ).

  r_task = lo_task.
  ENDMETHOD.
ENDCLASS.
```

Listing 11.22 The Complete Chore List Class Definition and Implementation

In the Chore List class, we implemented a setter and getter for the List Title. We also implemented basic logic to integrate with the database for the create, update, and delete methods of the Chore List. The Chore List class contains an attribute of lt_tasks, which is a table of the Chore Task objects. This way, we can easily store tasks associated to the Chore List in memory.

```
CLASS zcl_choretask DEFINITION
  PUBLIC
  FINAL
  CREATE PUBLIC .
  PUBLIC SECTION.
    METHODS:
      constructor
        IMPORTING
          iv_listid TYPE /bobf/conf_key
          iv_taskid TYPE /bobf/conf_key OPTIONAL,

      set_description
        IMPORTING
          iv_description TYPE char255,

      delete,
      save,
      set_new_ind,
      set_complete
        IMPORTING
          iv_complete TYPE abap_bool.

  PROTECTED SECTION.
  PRIVATE SECTION.
    DATA: lm_description TYPE char255.
    DATA: lm_listid   TYPE /bobf/conf_key,
          lm_taskid   TYPE /bobf/conf_key,
          lm_new      TYPE abap_bool,
          lm_complete TYPE abap_bool.
ENDCLASS.

CLASS zcl_choretask IMPLEMENTATION.

  METHOD constructor.
    lm_listid = iv_listid.
    lm_taskid = iv_taskid.
  ENDMETHOD.

  METHOD set_description.
    lm_description = iv_description.
  ENDMETHOD.

  METHOD save.
    DATA ls_task TYPE zchoretask.
    DATA lv_timestamp TYPE timestampl.
```

```abap
    IF lm_new = abap_false.
      GET TIME STAMP FIELD lv_timestamp.
      ls_task = VALUE #(
          listid  = lm_listid
          taskid  = lm_taskid
          description = lm_description
          complete = lm_complete
          createdby = sy-uname
          createdat = lv_timestamp
          updatedat = lv_timestamp
      ).
      MODIFY zchoretask FROM ls_task.
      RETURN.
    ENDIF.

    GET TIME STAMP FIELD lv_timestamp.
    ls_task = VALUE #(
        listid  = lm_listid
        taskid  = /bobf/cl_frw_factory=>get_new_key( )
        description = lm_description
        complete = abap_false
        createdby = sy-uname
        createdat = lv_timestamp
        updatedat = lv_timestamp
    ).

    MODIFY zchoretask FROM ls_task.

  ENDMETHOD.

  METHOD set_new_ind.
    lm_new = abap_true.
  ENDMETHOD.

  METHOD set_complete.
    lm_complete = iv_complete.
  ENDMETHOD.
ENDCLASS.
```

Listing 11.23 The Complete Chore Task Class Definition and Implementation

Note that we implemented setter methods for the Description and Complete fields in the Chore Task class, as well as a basic implementation for database access for the create, update, and delete methods.

You can now utilize these classes in the behavior pool class that the ABAP RESTful application programming model framework generated for you. In the behavior pool class ZBP_I_CHORELIST, you can find existing templated classes and methods under the **Local Types** tab. Listing 11.24 shows the implementation of the chore list and chore task classes into the ABAP RESTful application programming model behavior pool class, along with the transactional buffer class, in which we named LCL_CHORELIST_BUFFER to help store data in memory between the behavior handler classes and the saver classes.

```
CLASS lcl_chorelist_buffer DEFINITION.
  PUBLIC SECTION.
    CLASS-METHODS:
      get_list
        IMPORTING
          iv_listid      TYPE /bobf/conf_key OPTIONAL
        RETURNING
          VALUE(r_list) TYPE REF TO zcl_chorelist.
    CLASS-METHODS:
      set_list
        IMPORTING
          ir_list TYPE REF TO zcl_chorelist.

  PRIVATE SECTION.
    CLASS-DATA: lr_chorelist TYPE REF TO zcl_chorelist.
ENDCLASS.

CLASS lcl_chorelist_buffer IMPLEMENTATION.
  METHOD get_list.
    IF lr_chorelist IS BOUND.
      r_list = lr_chorelist.
      RETURN.
    ENDIF.

    IF iv_listid IS NOT INITIAL.

      SELECT SINGLE *
      FROM zchorelist
      WHERE listid = @iv_listid
      INTO @DATA(chorelist).

      lr_chorelist = NEW zcl_chorelist( chorelist-listid ).
      r_list = lr_chorelist.
      lr_chorelist->set_title( chorelist-title ).

      SELECT *
```

```
      FROM zchoretask
      WHERE listid = @iv_listid
      INTO TABLE @DATA(choretasks).

      LOOP AT choretasks ASSIGNING FIELD-SYMBOL(<task>).
        lr_chorelist->add_task( <task>-description ).
      ENDLOOP.

    ENDIF.

  ENDMETHOD.

  METHOD set_list.
    lr_chorelist = ir_list.
  ENDMETHOD.
ENDCLASS.

CLASS lhc_zi_chorelist DEFINITION INHERITING FROM cl_abap_behavior_handler.
  PRIVATE SECTION.
    METHODS create FOR MODIFY
      IMPORTING entities FOR CREATE zi_chorelist.

    METHODS delete FOR MODIFY
      IMPORTING keys FOR DELETE zi_chorelist.

    METHODS update FOR MODIFY
      IMPORTING entities FOR UPDATE zi_chorelist.

    METHODS read FOR READ
      IMPORTING keys FOR READ zi_chorelist RESULT result.

    METHODS cba_task FOR MODIFY
      IMPORTING entities_cba FOR CREATE zi_chorelist\_task.

    METHODS rba_task FOR READ
      IMPORTING keys_rba FOR READ zi_chorelist\_task FULL result_requested
RESULT result LINK association_links.

ENDCLASS.

CLASS lhc_zi_chorelist IMPLEMENTATION.
  METHOD create.
    LOOP AT entities ASSIGNING FIELD-SYMBOL(<chorelist>).
      DATA(lo_chorelist) = NEW zcl_chorelist( ).
```

```
          lcl_chorelist_buffer=>set_list( lo_chorelist ).
          lo_chorelist->set_title( <chorelist>-title ).
      ENDLOOP.
    ENDMETHOD.

    METHOD delete.
      BREAK-POINT.
      DATA(lo_chorelist) = NEW zcl_chorelist( keys[ 1 ]-listid ).
      lo_chorelist->delete( ).
    ENDMETHOD.

    METHOD update.
    ENDMETHOD.

    METHOD read.
    ENDMETHOD.

    METHOD cba_task.
      LOOP AT entities_cba ASSIGNING FIELD-SYMBOL(<entities>).
        LOOP AT <entities>-%target ASSIGNING FIELD-SYMBOL(<task>).
          DATA(lo_chorelist) = lcl_chorelist_buffer=>get_list( <entities>-listid
).
          lo_chorelist->add_task(
              iv_description = <task>-description
              iv_new = abap_true ).
        ENDLOOP.
      ENDLOOP.

    ENDMETHOD.

    METHOD rba_task.
    ENDMETHOD.

ENDCLASS.

CLASS lhc_zi_choretask DEFINITION INHERITING FROM cl_abap_behavior_handler.
  PRIVATE SECTION.

    METHODS create FOR MODIFY
      IMPORTING entities FOR CREATE zi_choretask.

    METHODS delete FOR MODIFY
      IMPORTING keys FOR DELETE zi_choretask.
```

```
    METHODS update FOR MODIFY
       IMPORTING entities FOR UPDATE zi_choretask.

    METHODS read FOR READ
       IMPORTING keys FOR READ zi_choretask RESULT result.

ENDCLASS.

CLASS lhc_zi_choretask IMPLEMENTATION.

  METHOD create.
  ENDMETHOD.

  METHOD delete.
    DATA(lo_chorelist) = lcl_chorelist_buffer=>get_list( keys[ 1 ]-listid ).
    lo_chorelist->delete_task( keys[ 1 ]-taskid ).
  ENDMETHOD.

  METHOD update.
    data(ls_entity) = entities[ 1 ].

    DATA(lo_chorelist) = lcl_chorelist_buffer=>get_list( ls_entity-listid ).
    DATA(lo_task) = lo_chorelist->get_task( ls_entity-taskid ).

    if ls_entity-%control-Description = cl_abap_behv_ctrl=>flag_provided.
        lo_task->set_description( ls_entity-Description ).
    endif.

    if ls_entity-%control-Complete = cl_abap_behv_ctrl=>flag_provided.
        lo_task->set_complete( ls_entity-complete ).
    endif.
  ENDMETHOD.

  METHOD read.
  ENDMETHOD.

ENDCLASS.

CLASS lsc_zi_chorelist DEFINITION INHERITING FROM cl_abap_behavior_saver.
  PROTECTED SECTION.
    METHODS check_before_save REDEFINITION.
    METHODS finalize          REDEFINITION.
    METHODS save              REDEFINITION.
ENDCLASS.
```

```
CLASS lsc_zi_chorelist IMPLEMENTATION.

  METHOD check_before_save.
  ENDMETHOD.

  METHOD finalize.
  ENDMETHOD.

  METHOD save.
    DATA(lo_chorelist) = lcl_chorelist_buffer=>get_list( ).
    IF lo_chorelist IS BOUND.
      lo_chorelist->save( ).
    ENDIF.
  ENDMETHOD.

ENDCLASS.
```

Listing 11.24 ABAP RESTful Application Programming Model Behavior Pool Class Local Types Tab, Including the Buffer Class and the Generated Behavior Handler and Saver Classes

11.4   Summary

In this chapter, you learned about the ABAP RESTful application programming model and how it was built as a successor of SAP Gateway to provide a structured service and app-building model. You learned how the ABAP RESTful application programming model has a set of classes that aid you in implementation and how those classes relate to object orientation. Additionally, you reviewed CDS views and associations, including how they relate to both the ABAP RESTful application programming model and object orientation in general. You now understand behavior definitions as they relate to CDS views and the ABAP RESTful application programming model and how they enable you to implement varying degrees of control between managed and unmanaged strategies.

In the next chapter, we'll expand upon the ABAP RESTful application programming model concepts and see how ABAP works in the cloud.

Chapter 12
ABAP Cloud

This chapter provides a surface-level introduction to ABAP Cloud and the benefits of transitioning to cloud-native ABAP development. This chapter will cover how the ABAP layer within SAP Business Technology Platform (SAP BTP) is the bridge between traditional SAP development and cloud-based app development for the modern era.

ABAP Cloud empowers businesses to modernize their development practices without abandoning the customizations and valuable skillsets they've accumulated over the years. SAP BTP has evolved into SAP's go-to cloud service platform, offering a comprehensive suite of tools and technologies that equip ABAP developers to build future-ready solutions using their existing expertise.

By leveraging cloud-native capabilities, businesses can seamlessly transition ABAP enhancements to the cloud while taking advantage of enhanced scalability, flexibility, and integration opportunities. Whether organizations want to extend current solutions, build new apps from scratch, or ensure a seamless coexistence between cloud and traditional systems, SAP BTP provides the essential foundation. For ABAP developers, this transformation is not about abandoning a familiar language; it's about evolving how you work.

In this chapter, we'll explore the significance of cloud-native development in SAP BTP and how the SAP BTP ABAP environment acts as a bridge between traditional SAP programming and modern, cloud-based app strategies. We'll trace SAP BTP's evolution from its early beginnings to the robust, comprehensive platform it is today, highlighting the key architectural components that make cloud-based ABAP development possible.

You'll gain insights into SAP BTP's core capabilities, including its runtime environments, data management tools, analytics, and planning solutions. We'll also discuss how ABAP Cloud enables in-app extensibility and touch on the integration frameworks and intelligent technologies that fuel business innovation. Whether you're extending SAP S/4HANA or building entirely new apps, we'll unpack the differences between working in the SAP BTP ABAP environment and using developer extensibility so you can make informed development choices.

We'll start by walking through setting up your ABAP development environment, covering prerequisites, tools, and best practices to get started with cloud-based development. We'll close with a case study showing how to implement key elements of the

ABAP RESTful programming model in the cloud, building on concepts from Chapter 11. By the end of this chapter, you'll have a clear understanding of how ABAP Cloud fits within SAP BTP. You'll be equipped to design scalable, maintainable, and enterprise-ready ABAP solutions aligned with today's business needs.

12.1   Introduction to SAP Business Technology Platform

Before we dive into the specifics of ABAP Cloud, it's essential to understand the evolution of SAP BTP and how it has become the cornerstone of SAP's cloud strategy. It brings together a suite of services—ranging from app development and integration to data management and artificial intelligence (AI)—all within a unified platform as a service (PaaS) model.

SAP BTP's journey started with a focus on cloud-based data management and app development and grew from there, with SAP continuously refining it to become more comprehensive over the years and align with the ever-changing demands of enterprise technology. For ABAP developers, SAP BTP opens a new path: the ability to build and extend enterprise-grade solutions using ABAP, while fully embracing cloud-native design principles.

12.1.1   What Is SAP Business Technology Platform?

SAP BTP serves as the foundation for SAP's enterprise cloud solutions. Originally introduced in 2012 as SAP HANA Cloud Platform, it was later rebranded in 2017 to SAP Cloud Platform. Over time, the platform expanded in scope and eventually consolidated under the SAP BTP name in 2021. Today, it serves as the digital foundation for SAP's entire cloud ecosystem.

In a nutshell, SAP BTP presents a unified PaaS offering that integrates data management, analytics, app development, and intelligent technologies to support businesses in their digital transformation efforts for the modern era. With SAP BTP, businesses can innovate and adapt more quickly to achieve business goals more efficiently. To help you visualize where development activities are managed, Figure 12.1 shows the SAP BTP cockpit, which is the entry point for developers and administrators into the platform's services and runtimes.

SAP BTP serves as the backbone for creating, extending, and maintaining enterprise apps in the cloud. It provides an environment where developers can seamlessly design solutions using prebuilt connectors and advanced analytics tools, all leveraging a robust integration framework. SAP BTP supports several runtime environments to cater to diverse development needs, including Cloud Foundry, Kyma, SAP HANA, and most recently, ABAP.

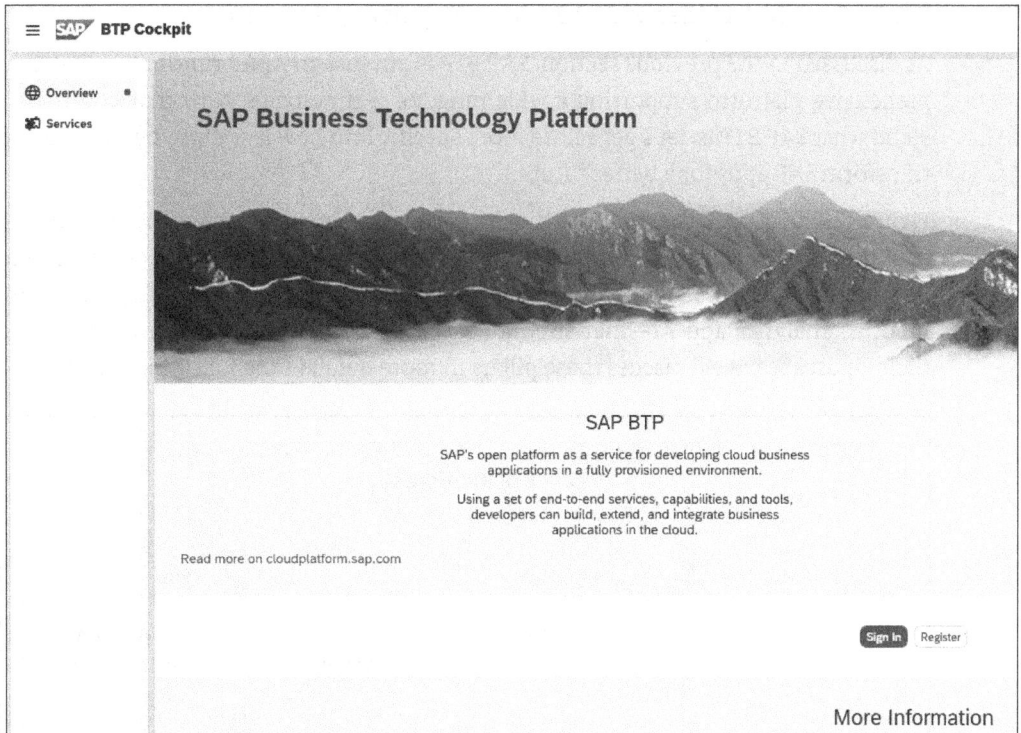

Figure 12.1 SAP BTP Cockpit Dashboard

Initially, SAP introduced Cloud Foundry, an open-source cloud platform providing developers with a flexible environment for building and scaling apps using popular languages like Java, Node.js, and Python. As the need for containerization grew, SAP incorporated Kyma, an open-source Kubernetes-based runtime enabling developers to extend apps and build cloud-native apps with event-driven capabilities. Additionally, SAP BTP now includes the SAP HANA Cloud database, which enables developers to leverage the powerful in-memory database capabilities of SAP HANA to quickly and efficiently handle large-scale data processing and transactional workloads.

In recognition of the true importance of SAP's traditional developer community, SAP recently introduced the ABAP runtime, which brings the familiar and widely used ABAP programming language to the cloud. The addition of this ABAP layer within SAP BTP allows enterprises to modernize their existing ABAP-based apps while maximizing the possibilities of cloud scalability, integration capabilities, lifecycle management tools, and more. It's an exciting time for ABAP developers—you're empowered to build both traditionally and in the cloud, so you can maximize the strengths of each option, while also looking ahead toward progress to come from more future-ready SAP solutions.

With these versatile runtime environments, SAP BTP provides you with the flexibility to choose the best approach for your cloud transformation journey, whether you're building from scratch, extending existing solutions, or modernizing legacy apps.

12.1.2   Core Capabilities of SAP Business Technology Platform

As discussed in the previous section, SAP BTP is not just an ABAP runtime—it's a comprehensive platform supporting a wide range of technologies. Now that you understand what SAP BTP is, let's get a little more specific into how it equips you to build and transform SAP apps for a better future.

Figure 12.2 illustrates the multiple runtime environments supported within SAP BTP, including Cloud Foundry, Kyma, SAP HANA, and the SAP BTP ABAP environment. Beneath this runtime layer are SAP BTP's core capability pillars—from app development to analytics and AI—that support developers and enterprises in modernizing their landscapes. We'll discuss these pillars in more detail in the following sections.

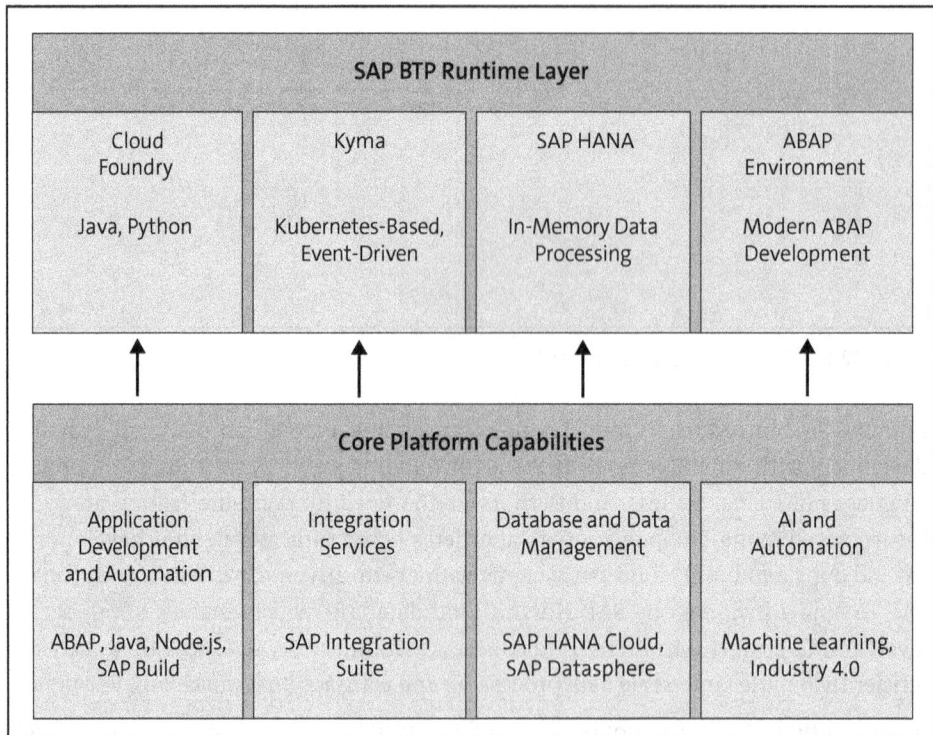

SAP BTP Runtime Layer			
Cloud Foundry	Kyma	SAP HANA	ABAP Environment
Java, Python	Kubernetes-Based, Event-Driven	In-Memory Data Processing	Modern ABAP Development

Core Platform Capabilities			
Application Development and Automation	Integration Services	Database and Data Management	AI and Automation
ABAP, Java, Node.js, SAP Build	SAP Integration Suite	SAP HANA Cloud, SAP Datasphere	Machine Learning, Industry 4.0

Figure 12.2 Overview of SAP BTP Runtimes and Core Capabilities

Database and Data Management

Robust database and data management capabilities are the central offerings of SAP BTP. SAP HANA is the in-memory database at the core of SAP BTP, providing unmatched real-time performance. SAP BTP supports the use of both structured and unstructured data, enabling businesses to more efficiently handle large-scale datasets. Using SAP BTP ensures that enterprises can gain actionable insights from their data while also maintaining high levels of performance.

Analytics and Planning

SAP BTP provides powerful analytical tools for reporting, visualizing trends, predictive forecasting, and operational planning. This suite of tools empowers organizations to make more informed and data-driven strategic decisions. Specific products like SAP Datasphere and SAP Analytics Cloud play crucial roles in enabling advanced analytics and planning capabilities in SAP BTP.

SAP Datasphere, formerly known as SAP Data Warehouse Cloud, is a comprehensive data management solution providing a unified data foundation that enables businesses to connect and analyze data across various sources with built-in security.

SAP Analytics Cloud is an all-in-one analytics platform that combines business intelligence and predictive analytics to allow organizations to visualize data, generate insights, and create strategic plans through interactive dashboards and AI-driven capabilities.

Leveraging the full power of analytics and planning tools available to developers within SAP BTP will empower business to improve operational efficiency and gain a competitive edge in today's data-driven landscape.

Development and Integration

SAP BTP's comprehensive suite of tools allows developers to not only develop but also to extend and integrate apps for businesses. SAP BTP allows developers to capitalize on their existing ABAP skills within the SAP BTP ABAP environment to create cloud-native apps while ensuring seamless connectivity between on-premise and cloud systems, thus guaranteeing business processes will remain uninterrupted across different environments.

SAP extension capabilities enable businesses to enhance and customize SAP apps with great ease, with one of its key advantages being its support for low-code and no-code development. This allows both professional developers and business users to build apps quickly with minimal coding effort.

Intelligent Technologies

SAP BTP keeps your business innovating by incorporating cutting-edge technologies like AI, machine learning, and Industry 4.0 into your processes. Let's explore these key technologies:

- SAP BTP's intelligence tools empower enterprises to better automate processes and predict outcomes to enhance operational efficiencies. An example of this would be leveraging Internet of Things (IoT) devices for real-time data from sensors and other connected sources, while AI optimizes supply chain operations or identifies opportunities to enhance customer experiences.

- SAP BTP also offers robust capabilities that support modern cloud-native development across multiple domains. For example, in app development and automation, developers can leverage a broad range of tools including ABAP, Node.js, Java, and low-code environments such as SAP Build. These tools enable rapid app creation with minimal setup, supporting both professional developers and business users.

- Integration services are powered by SAP Integration Suite, which provides a rich set of prepackaged connectors, workflows, and application programming interfaces (APIs). This simplifies the integration of SAP and non-SAP systems and allows for seamless data and process flow across diverse landscapes.

- On the data side, SAP BTP delivers powerful database and data management capabilities. SAP HANA Cloud serves as the core in-memory engine, with support for data lakes, federated sources, and large-scale analytical workloads.

- For analytics and planning, SAP BTP includes native integration with SAP Analytics Cloud and data warehousing tools. These allow for real-time dashboards, business intelligence, and predictive planning in a centralized interface.

- Finally, SAP BTP's AI and automation services enable innovation by offering tools for machine learning, Industry 4.0 scenarios, robotic process automation (RPA), and AI-driven decision support. These intelligent technologies help organizations continuously optimize operations and respond proactively to business needs.

These capabilities empower organizations to move faster, reduce technical debt, and innovate continuously. For ABAP developers, the key takeaway is that you can now build apps that operate within this broader landscape—while still writing ABAP.

12.1.3   ABAP and SAP Business Technology Platform

The SAP BTP ABAP environment is the bridge between traditional SAP development and cloud-based app development for the modern era. Using SAP BTP enables developers to extend and build apps within a scalable and secure yet familiar environment. The SAP BTP ABAP environment provides efficient tools and help for developers who are building entirely new cloud-native solutions, extending existing systems, or a mix of the two.

ABAP's integration into SAP BTP empowers businesses to modernize their development practices without abandoning legacy systems or expertise. This makes transitioning to the cloud a whole lot smoother while retaining the robust capabilities native to ABAP. It also provides the following business advantages:

- **Flexibility**
  Companies innovate at their own pace and leverage scalable resources as needed.

- **Scalability**
  Since it's a cloud-native platform, SAP BTP dynamically scales to meet business demands, ensuring consistently maximized performance.

- **Unified platform**
  Combining development, analytics, and integration tools into one environment allows SAP BTP to reduce complexities and accelerate development cycles.

- **Ease of integration**
  SAP BTP offers ready-to-use prebuilt APIs and connectors to simplify integration with legacy systems and third-party apps.

SAP BTP is all about scalability and flexibility. It provides an on-ramp to a broader set of technologies and services not typically available to ABAP developers who are building on top of on-premise SAP NetWeaver stacks.

Figure 12.3 illustrates how the SAP BTP ABAP environment bridges SAP BTP and SAP S/4HANA Cloud, showing how developers can extend systems with cloud-native ABAP apps.

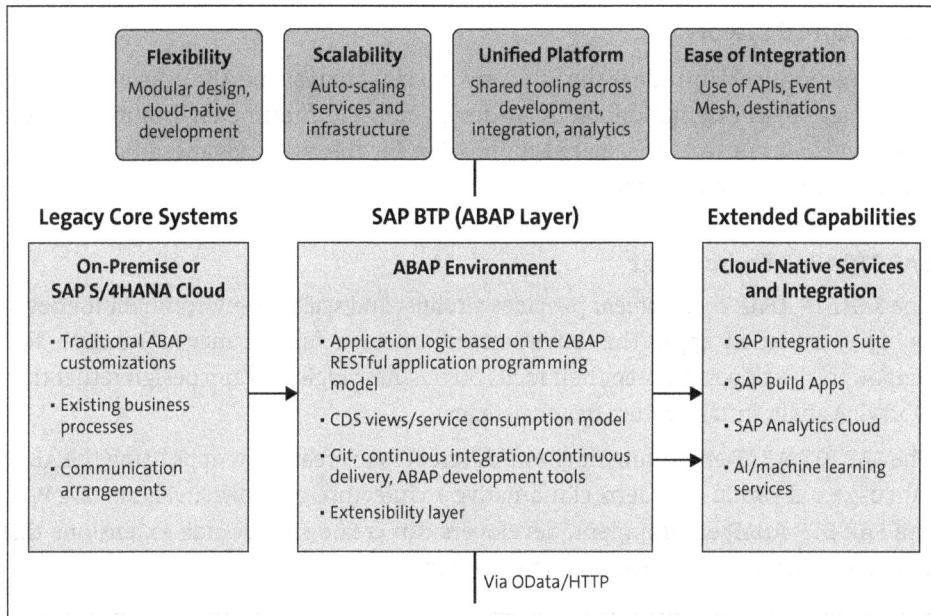

Figure 12.3 ABAP Cloud Development for SAP BTP and SAP S/4HANA Cloud

SAP BTP is much more than just a technology platform. It's a catalyst for innovation that enables businesses to transform their operations and marketplace value-add like never before.

12.2 ABAP Environment Overview

ABAP has been the cornerstone of SAP's enterprise solutions for decades. With the modern shift towards cloud computing, ABAP is evolving to meet current and future needs for app development. ABAP Cloud is not just a mere extension of traditional ABAP, but rather a transformation emphasizing scalability, security, and modernization of development standards.

At the heart of cloud-based ABAP development is SAP BTP ABAP environment, previously referred to by its internal codename: *Steampunk*. This is a fully managed cloud runtime solution for ABAP developers, and this environment enables developers to build and run ABAP apps in the cloud—decoupled from any specific SAP S/4HANA system and aligned with clean core principles. Basically, it allows developers to build apps for their businesses that are tightly integrated with SAP systems while adhering to the more modern best practices and principles of cloud-native development.

Let's explore the key features and in-app extensibility benefits available through ABAP Cloud development.

12.2.1   Core Concepts

In this section, we'll introduce the fundamental ideas that shape ABAP development in the cloud: object orientation, modular design, and the runtime environment. These building blocks provide the shared foundation for the more advanced patterns and techniques we'll explore later.

SAP BTP ABAP Environment

The *SAP BTP ABAP environment* provides a robust and scalable environment for developing cloud-based apps. This environment offers a fully managed runtime that empowers developers to keep their focus on business logic and app design rather than worrying about infrastructure or consistency.

The SAP BTP ABAP environment allows developers to create new apps using the ABAP language, but within a modern, cloud-native architecture, as shown in Figure 12.4. With the SAP BTP ABAP environment, developers can create side-by-side extensions that seamlessly integrate with SAP and other systems.

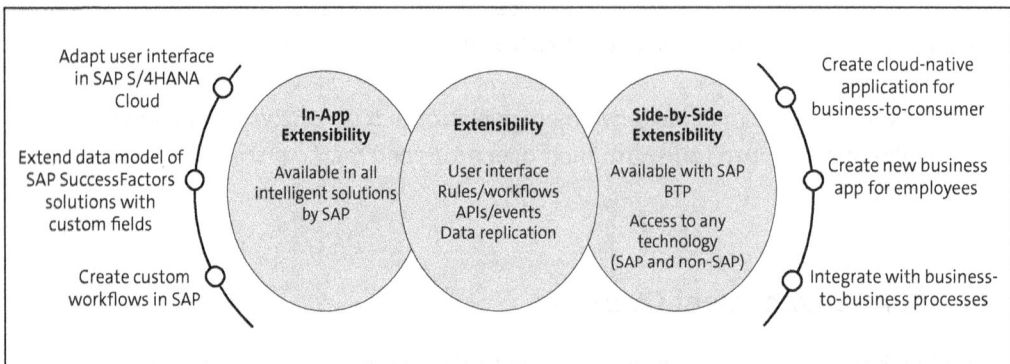

Figure 12.4 SAP BTP ABAP Environment

One of the key features of working within the ABAP layer on SAP BTP is its support for *in-app extensibility*, which refers to capability for users to customize and/or enhance SAP apps directly within the target system without requiring deep technical skills or modifications to the core codebase. If you want to learn more about in-app extensibility or start to practice it, SAP provides guided tools such as the *key user extensibility framework*.

Organizations can achieve a high level of flexibility while maintaining system stability, and modifications like adding custom fields or adjusting user experience interfaces can be made directly within the SAP-provided framework rather than needing to alter the core app. This allows developers and business users to extend SAP cloud apps while staying within a clean and upgrade-safe environment.

The following is a summary of key features within the SAP BTP ABAP environment:

- **Cloud-native tooling**
  Development is done using ABAP development tools, integrated with Git-based version control.

- **Decoupled architecture**
  Apps run independently of backend systems like SAP S/4HANA but can connect to them via APIs.

- **Strict clean core enforcement**
  Only released objects and APIs are accessible; there is no access to internal tables or modification of SAP standard code.

- **Multitenancy support**
  SAP BTP ABAP environment supports multitenant software as a service (SaaS) scenarios and lifecycle management via lifecycle management APIs.

Figure 12.5 illustrates the architecture of the SAP BTP ABAP environment. Apps are developed using cloud-native solutions like ABAP development tools and Git for version control. The environment supports RESTful services and the ABAP RESTful application programming model while remaining decoupled from SAP S/4HANA backend systems. Integration is achieved via external APIs, cloud connectors, and managed platform services such as IoT, analytics, and user management.

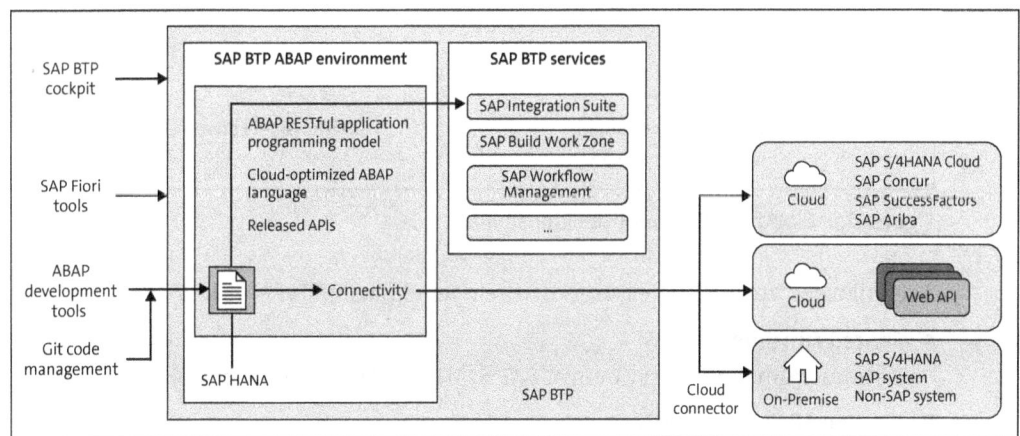

Figure 12.5 SAP BTP Integration Model and SAP BTP ABAP Environment Architecture

SAP S/4HANA Cloud ABAP Environment

In addition to the SAP BTP ABAP environment, SAP also offers a variant embedded directly into SAP S/4HANA Cloud called SAP S/4HANA Cloud ABAP environment (previously referred to as Embedded Steampunk). While both share the same runtime and tooling, they serve different purposes and offer distinct architectural options.

Developer extensibility is an SAP-introduced concept that brings the capabilities of the SAP BTP ABAP environment directly into SAP S/4HANA Cloud. It allows you to build custom business logic directly within your SAP S/4HANA Cloud system while still adhering to the same modern ABAP principles enforced in the cloud.

It's important to note that developer extensibility is available only within SAP S/4HANA Cloud ABAP environment. It's not available in the traditional on-premise SAP S/4HANA environment, even though they share similar concepts. Figure 12.6 introduces the concept of developer extensibility, where the ABAP environment is integrated directly into the SAP S/4HANA Cloud landscape.

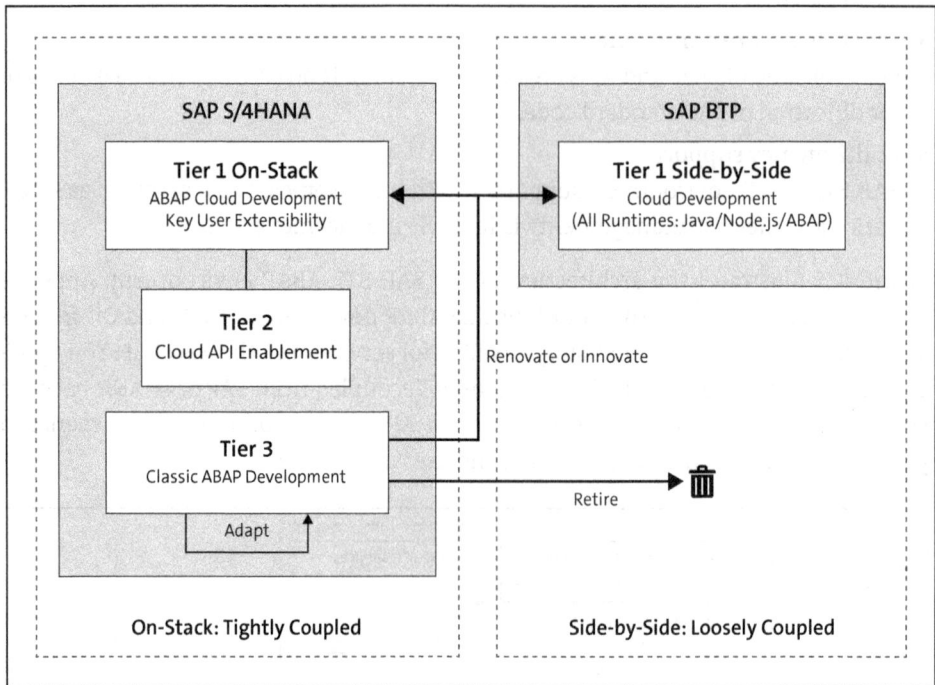

Figure 12.6 Cloud and On-Premise Extensibility

The following are the key features of developer extensibility:

- **Shared runtime**
  It runs within the same system as SAP S/4HANA Cloud but in an isolated layer.

- **Released API access only**
  Developers must use whitelisted objects, ensuring upgrade compatibility.

- **Tightly coupled extensibility**
This is ideal for building extensions that directly impact SAP S/4HANA processes without external dependencies.

- **In-app extensibility**
SAP S/4HANA Cloud provides a controlled environment for developer extensibility to operate where extensions are developed directly inside the system. This ensures tight integration with core business processes without compromising stability. With in-app extensibility governed by SAP's extensibility framework, organizations can achieve a high level of flexibility while maintaining system stability and compliance with SAP updates.

- **SAP BTP ABAP environment similarities**
Developer extensibility shares the following features with the SAP BTP ABAP environment:

  - Development tools: Developers use the same ABAP development tools as they would for SAP BTP.

  - Modern ABAP standards: Extensions follow consistent syntax rules and optimization standards to ensure security, performance, and scalability.

  - ABAP RESTful application programming model: The model is supported for developer use in building modern service-based apps and APIs.

  - Core data services (CDS): Developers can define and utilize familiar data models and CDS views.

Figure 12.7 shows SAP's three-tier extensibility model, illustrating how in-app, key user, and developer extensibility options complement one another.

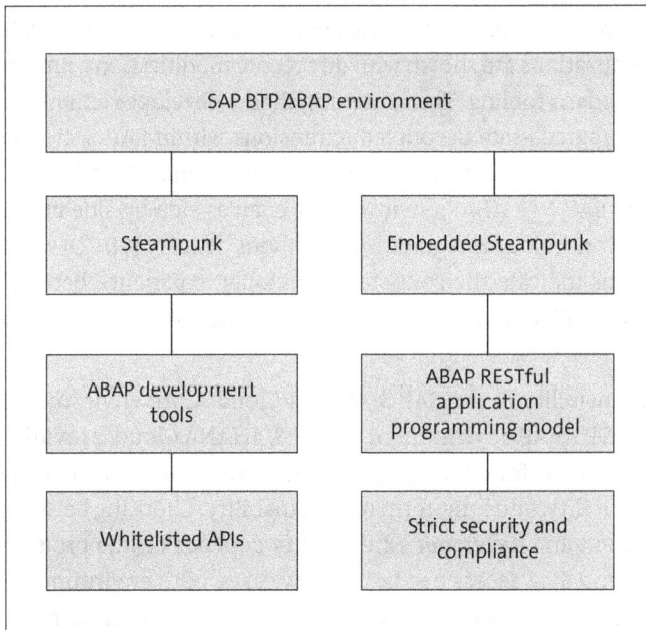

Figure 12.7 SAP BTP ABAP Environment Similarities

- **Ease of use**
  Unlike on-premise extensions that modify the core system, developer extensibility ensures that extensions are decoupled from standard SAP S/4HANA code. This enables smoother upgrades and prevents disruptions due to custom code or system updates.

- **Extension management**
  Developers can utilize packages, version control (such as Git), and pipelines just as they would in the SAP BTP ABAP environment.

- **Service-oriented architecture**
  The ABAP RESTful application programming model allows developers to expose custom logic as OData services to allow for integration with other apps or systems.

- **Deployment and testing**
  Extensions can be tested and deployed within the controlled environment of SAP S/4HANA Cloud, always ensuring compliance with SAP best practices.

- **Flexibility**
  Businesses can adapt their SAP S/4HANA Cloud systems to meet specific requirements without altering core functionalities. Other options for building cloud-native extensions exist as well such as Java, Node.js, and even the bring your own language (BYOL) approach.

- **Sustainability**
  Extensions developed with developer extensibility remain compatible and essentially future proof, keeping up with the latest SAP S/4HANA Cloud upgrades to ensure long-term stability.

Figure 12.8 illustrates three different approaches to extending SAP systems. On the left, traditional on-premise customizations are shown with direct core modifications, limited upgrade safety, and lack of modern tooling. The center highlights developer extensibility, where developers create upgrade-safe decoupled extensions within SAP S/4HANA Cloud using the ABAP RESTful application programming model, Git, and ABAP development tools. On the right, the SAP BTP ABAP environment enables side-by-side cloud-native development with full continuous integration/continuous delivery (CI/CD) workflows. Arrows and shared icons indicate the consistent developer experience between developer extensibility and SAP BTP, emphasizing extensibility, sustainability, and modernization.

To sum it up, developer extensibility in the SAP S/4HANA Cloud ABAP environment brings the power of the SAP BTP ABAP environment to SAP S/4HANA Cloud, providing a modern yet familiar environment for developing custom extensions with superior performance, compatible flexibility, and long-term maintainability. Choosing between the SAP BTP ABAP environment and developer extensibility depends on the project's architecture, business goals, and data access needs. In many cases, both environments are used together: Core extensions are built in developer extensibility, while complementary services and UIs are delivered via the SAP BTP ABAP environment.

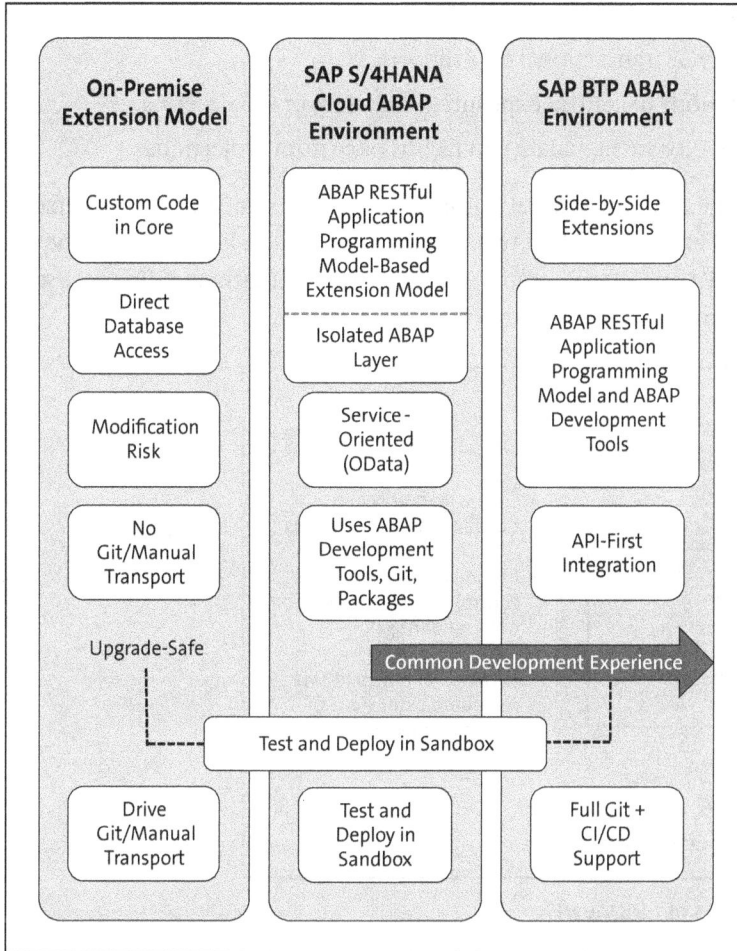

Figure 12.8 SAP Extensibility Approaches

12.2.2   Key Architectural Components

Now that we've gone over some key core concepts, let's take a moment to walk through some of the architectural components that comprise SAP BTP.

Core Data Services

CDS views, which we introduced in Chapter 11, are a central component of ABAP Cloud, allowing developers to define and consume data models in a standardized way for more efficient data access and manipulation. CDS views are foundational to ABAP Cloud. They allow developers to model data structures declaratively while also defining relationships, aggregations, and semantic annotations that drive both backend and frontend behavior. This ensures enhanced consistency and performance in cloud-based apps.

CDS views are capable of the following tasks:

- Defining read-only or transactional data models.

- Including annotations to control exposure, validation, or UI rendering.

- Allowing extensibility via metadata extensions or custom projections.

CDS views replace the need for manual SQL logic in many scenarios and are often consumed by other CDS views, behaviors, or analytical tools. Figure 12.9 illustrates the role of CDS views in ABAP Cloud development, showing how annotations define data modeling, UI behavior, and service exposure.

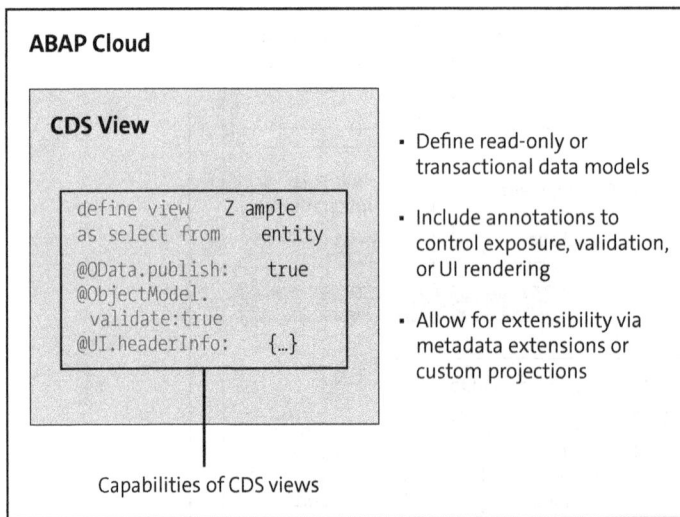

Figure 12.9 Capabilities of CDS Views

ABAP RESTful Application Programming Model

The ABAP RESTful application programming model, which we discussed in detail in Chapter 11, is the central framework for developing modern ABAP apps in both SAP BTP ABAP environment and developer extensibility. It's the framework for developing service-oriented apps in ABAP, which simplifies the creation of APIs and services to make building transactional and analytical apps easier.

The ABAP RESTful application programming model enables developers to design scalable, maintainable, and upgrade-safe apps using a combination of declarative modeling, behavior definitions, and service exposure layers. The ABAP RESTful application programming model prioritizes modularity and reusability, which aligns with modern development principles for cloud-native builds. Unlike classical ABAP development, which typically relies on procedural logic and tightly coupled customizations, the ABAP RESTful application programming model promotes a loosely coupled, service-oriented architecture that supports clean core principles.

ABAP RESTful application programming model apps are built around three key layers:

- **Data modeling layer**
  The data modeling layer defines the structure of business entities using CDS views. These views represent the schema and provide a clean interface for the underlying data.

- **Behavior layer**
  The behavior layer describes the permissible actions (e.g., create, update, and delete) on those business entities via behavior definitions and implementations. These rules are enforced consistently across the UI and backend.

- **Service exposure layer**
  The service exposure layer exposes the behavior of business objects as RESTful OData services, which can be consumed by SAP Fiori apps or external systems.

Figure 12.10 depicts the ABAP RESTful application programming model's layered architecture (data modeling, behavior definitions, and service exposure), demonstrating how ABAP Cloud apps are structured.

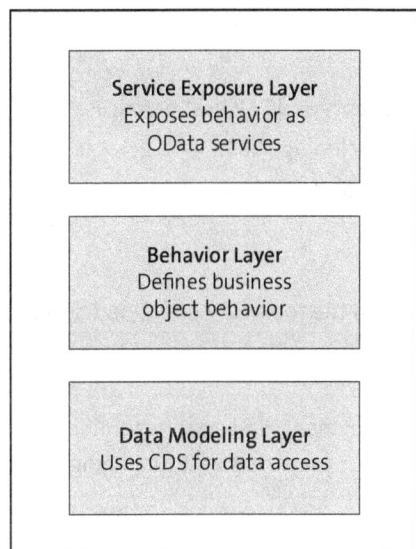

Figure 12.10 ABAP RESTful Application Programming Model Layered Architecture

The goal of the ABAP RESTful application programming model is to separate business logic from technical implementation, enabling modular development that is easier to maintain and test over time.

ABAP Development Tools

ABAP development tools is the primary integrated development environment (IDE) for cloud-native ABAP development. Built on Eclipse, ABAP development tools provides a rich set of features for code completion, syntax highlighting, debugging, and seamless

Git integration. These tools enhance developer productivity and streamline the development process in the cloud.

ABAP Cloud empowers developers to create ABAP solutions that are scalable and agile yet deeply integrated within SAP's ecosystem. By leveraging the power of the cloud, the SAP BTP ABAP environment helps turn businesses into intelligent enterprises that can respond to changing market demands more effectively and innovate more efficiently.

12.3   Setting Up Your Cloud Development Environment

Now, you're ready to get your cloud development environment up and running. In the following sections, we'll discuss considerations for your transition to the cloud and then walk through the process to set up the SAP BTP ABAP environment.

Prerequisites

Before diving fully into ABAP development in the cloud, you should review the following prerequisite knowledge, tools, and resources:

- **Required knowledge and skills**
  - Familiarity with basic ABAP programming concepts and SAP best practices.
  - A general understanding of cloud computing, including PaaS.
  - Experience with version control systems like Git and CI/CD workflows.
- **SAP BTP subscription setup**
  - Set up an SAP BTP account.
  - Subscribe to the ABAP environment service with the tools for ABAP Cloud development.
- **User role configuration**
  - Assign appropriate roles for development, deployment, and system administration to your SAP BTP user. This is necessary for accessing and managing the ABAP environment.
- **Cloud development tools**
  - Install the Eclipse IDE with the ABAP development tools plugin.
  - Ensure internet access to connect to the SAP BTP ABAP system.

SAP Discovery Center is a great resource for tutorials and missions if you need any help. Check out this one, which walks you through a trial environment setup step by step: *https://discovery-center.cloud.sap/missiondetail/3061/3116/*.

12.3.1   Transitioning from On-Premise to Cloud Development

Moving ABAP development from an on-premise SAP NetWeaver system to the cloud is not a simple lift-and-shift activity. It requires adaptation to new syntax standards, security constraints, and adherence to modern development principles such as the ABAP RESTful application programming model. The following sections highlight the main challenges you might face and strategies to ensure a smooth transition.

Common Challenges and Strategies for Success

When adapting legacy ABAP to the cloud, developers often face the following recurring issues:

- Adapting legacy ABAP code to cloud standards.
- Refactoring to comply with stricter syntax and security requirements in ABAP for the cloud.
- Addressing gaps in performance optimization for the cloud environment.

To overcome these challenges, SAP recommends combining code modernization with developer enablement:

- **Code refactoring**
  Use modern ABAP constructs like CDS for data modeling. Replace deprecated legacy code with cloud-optimized alternatives.

- **Iterative development**
  Break development into smaller, more manageable tasks. Test frequently to ensure functionality at every stage.

- **Dependency management**
  Identify dependencies between development objects as early as possible. Leverage ABAP development tools to manage and resolve any dependencies.

- **System compatibility**
  Verify compatibility with other existing SAP systems, such as SAP S/4HANA or on-premise SAP Business Suite systems, during development to ensure smooth integration.

- **Cloud-specific training**
  Invest in cloud training so that developers are familiar with the ABAP RESTful application programming model and best practices for cloud-native development.

- **Leverage tools and resources**
  Use SAP-provided migration tools and templates to accelerate the transition process. Engage with SAP's community forums and support for more guidance.

Version Control with Git

A major shift in cloud ABAP is adopting Git for lifecycle management. Rather than relying on traditional transport-based workflows, Git provides modern collaboration and CI/CD enablement.

To initialize Git integration in Eclipse, navigate to the Git perspective by selecting **Window** • **Perspective** • **Open Perspective** • **Other** • **Git**. You can then set up your project repository by linking it to a remote Git server.

You should regularly commit your development changes to maintain a version history and push updates to the remote repository for collaboration and backup.

Benefits of Git integration include the following:

- Git enables efficient and seamless collaboration among team members.
- Integration with CI/CD pipelines ensures streamlined deployments and automated testing.

> **Note**
>
> If you want to learn more about best practices for and the benefits of Git version control, refer to the E-Bite *Hands On with abapGIT* (SAP PRESS, 2024; available at *https://www.sap-press.com/5874/*) for more detailed information.

Security, Performance, and Development Guidelines

ABAP Cloud enforces a stricter set of development principles than traditional environments. These constraints are intentional: they ensure apps are secure, maintainable, and upgrade-safe across multi-tenant cloud landscapes.

Security best practices include the following:

- Only use released APIs and CDS views; access to unreleased internal objects is blocked by design.
- Enforce data isolation where required (especially in multi-tenant apps).
- Use OAuth and principal propagation for secure authentication flows.
- Respect authorization scopes and user roles when designing service layers.

Also, be aware of the following performance considerations:

- Minimize unnecessary joins and data loads by structuring CDS views efficiently.
- Use annotations to push logic to the database (e.g., aggregations or filters).
- Avoid SELECT * and always project only the required fields.
- Regularly analyze service execution using SAP-provided performance tools on SAP BTP.

These constraints help developers avoid many of the common pitfalls of on-premise development while improving long-term app health and adaptability. Adhering to best practices when building in the cloud will ensure you are well-prepared to transition into modern, scalable, and efficient cloud-native app development that propels your business forward.

12.3.2   Creating Your SAP BTP ABAP Environment

Before you start creating your SAP BTP ABAP environment, make sure you have access to an SAP BTP account (trial or shared). Figure 12.11 shows an SAP BTP trial account access screen, the starting point for exploring ABAP Cloud.

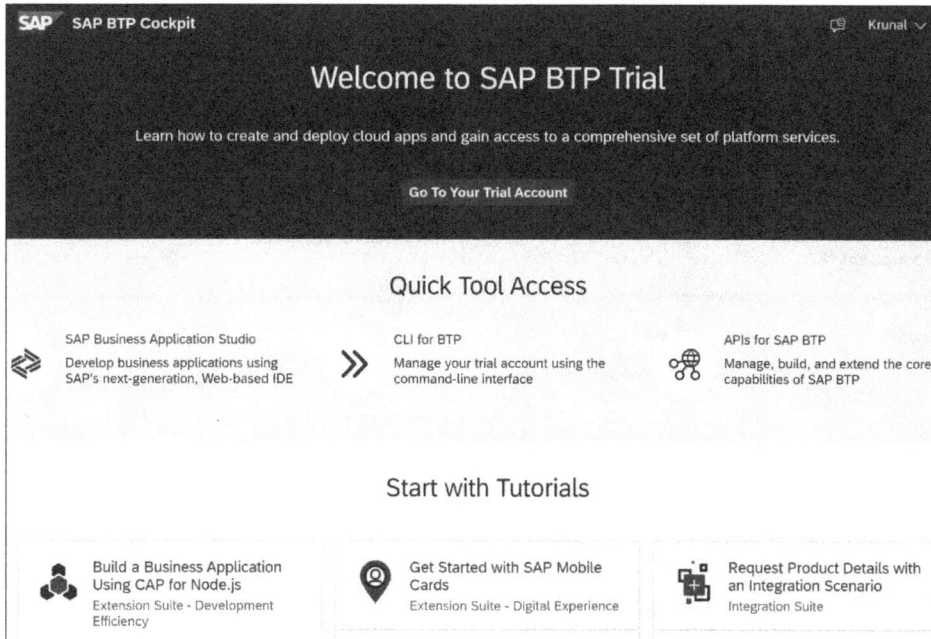

Figure 12.11 SAP BTP Trial Account Access Screen

As an additional prerequisite to setting up your SAP BTP ABAP environment, you need to be able to log in to an up-to-date version of ABAP development tools. With the prerequisites in place, you can follow these steps to set up your cloud environment:

1. Navigate from the SAP BTP cockpit to the **Instances and Subscriptions** tab, as shown in Figure 12.12.

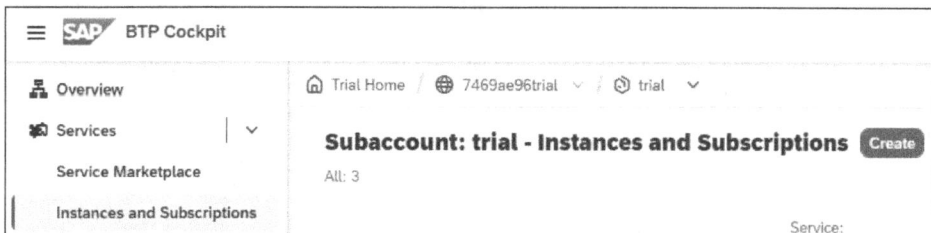

Figure 12.12 SAP BTP Cockpit: Instances and Subscriptions

2. Search for "ABAP" in the search bar and select **ABAP Environment** to create a new ABAP environment instance.

3. As shown in Figure 12.13, in the **New Instance or Subscription** popup, you'll need to enter **Basic Info**. Choose **ABAP environment** as your **Service**, **Cloud Foundry** as your **Runtime Environment**, **dev** as your **Space**, and provide an **Instance Name**.

4. On the next **Parameters** step, as shown in Figure 12.14, you can enter your necessary parameters in JavaScript Object Notation (JSON) format or choose to upload a file.

5. Finally, on the **Review** step (see Figure 12.15), you can click the **Create** button.

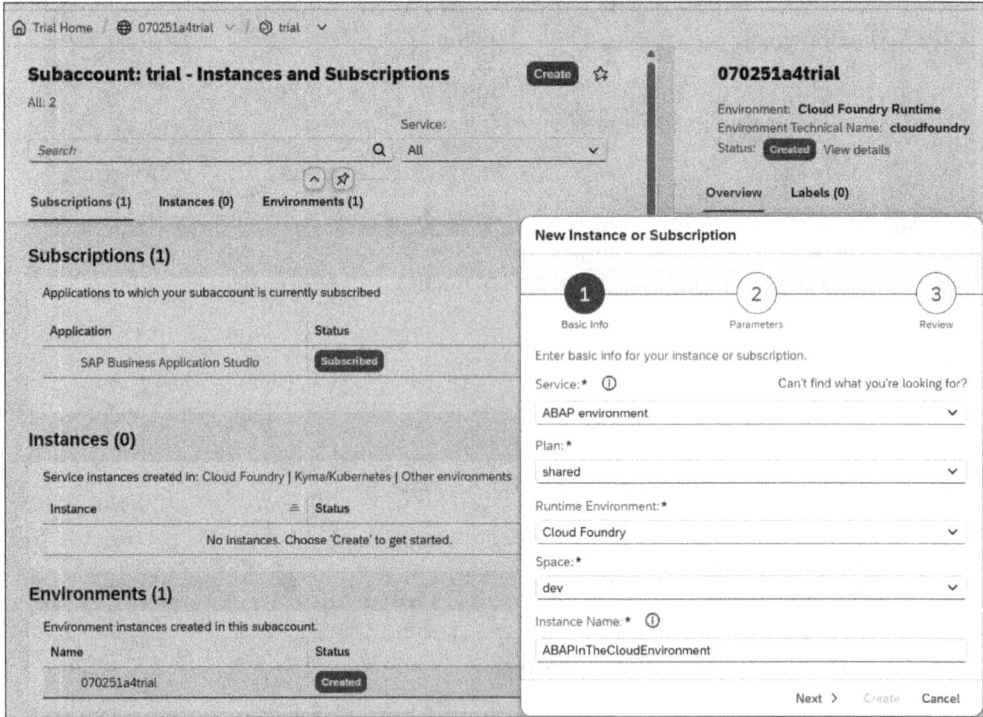

Figure 12.13 Name New SAP BTP ABAP Environment Instance

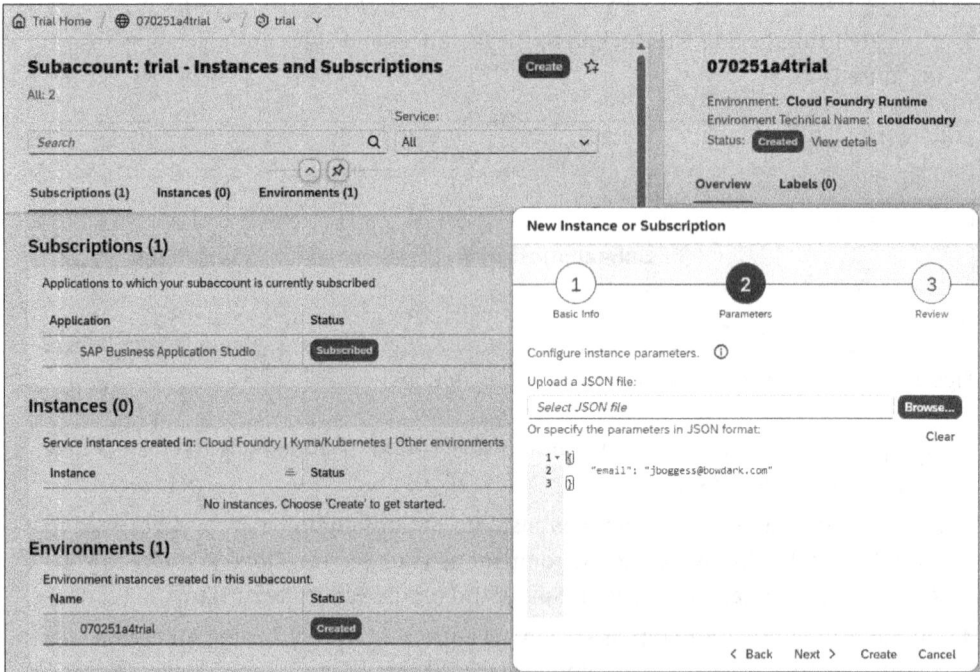

Figure 12.14 Configure ABAP Instance Parameters

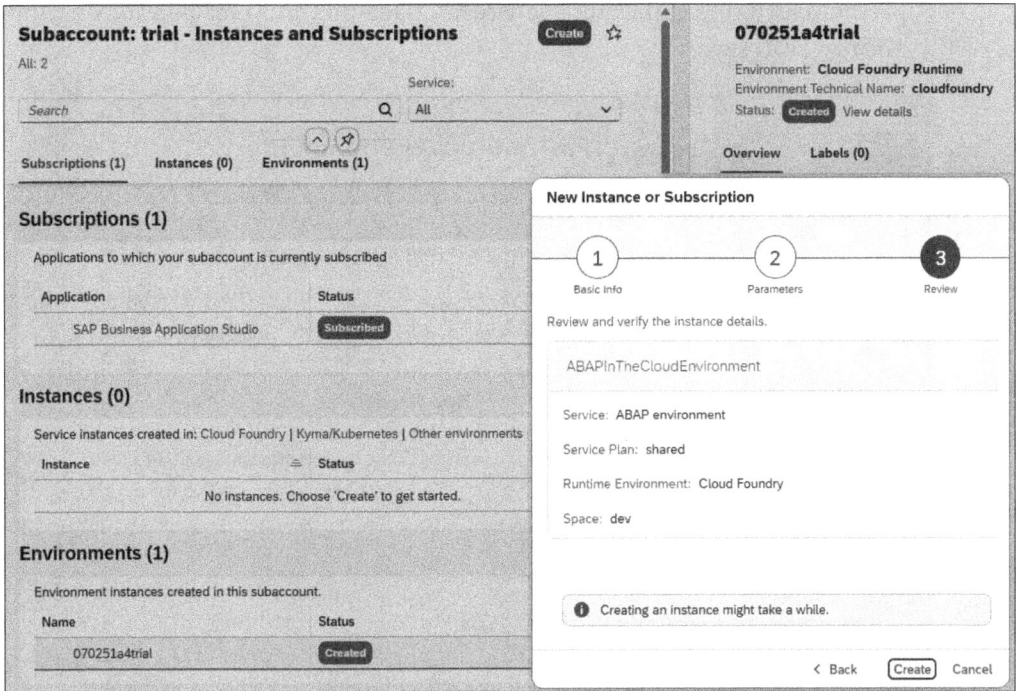

Figure 12.15 Create New SAP BTP ABAP Environment Instance

6. Once the new instance status says **Created**, open your SAP BTP ABAP environment instance, click the ellipses icon in the top right section of the screen, and select **Create Service Key**, as shown in Figure 12.16.

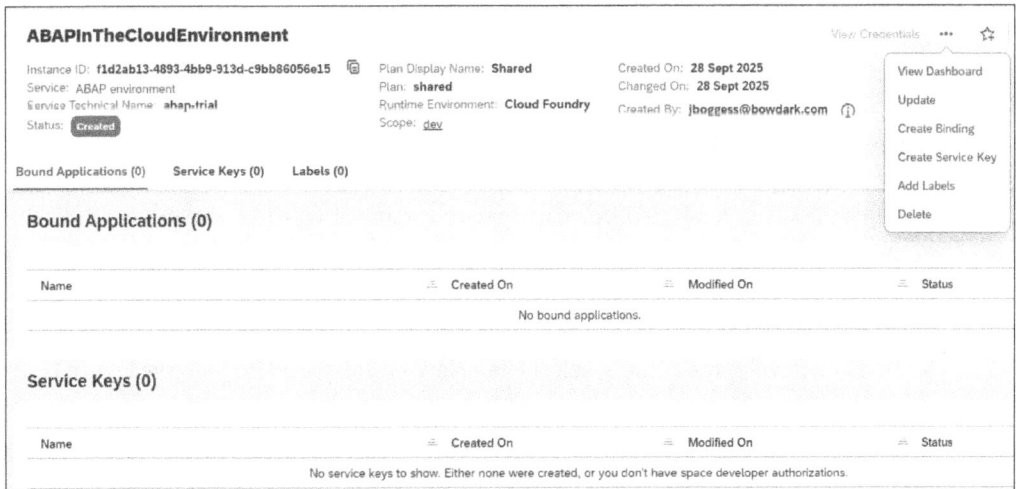

Figure 12.16 Create New Service Key Once Status Says Created

7. As shown in Figure 12.17, input a **Service Key Name** and click **Create** to auto-generate the new JSON key.

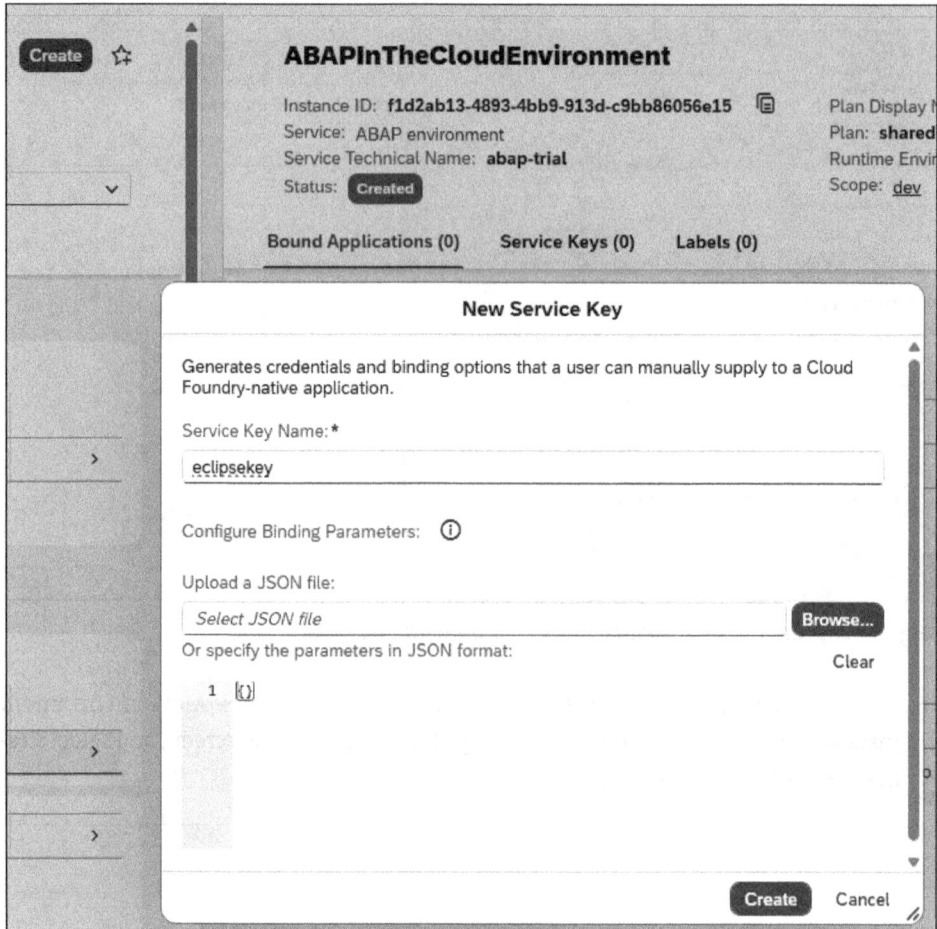

Figure 12.17 Create SAP BTP ABAP Environment Service Key

8. Once the key is generated, you can open it by navigating to **Service Keys** and clicking on your key. Copy the JSON credentials (url, clientid, and clientsecret), as shown in Figure 12.18.

9. Now, open Eclipse and connect ABAP development tools to your SAP BTP instance. In ABAP development tools, select **File • New • ABAP Cloud Project** (see Figure 12.19).

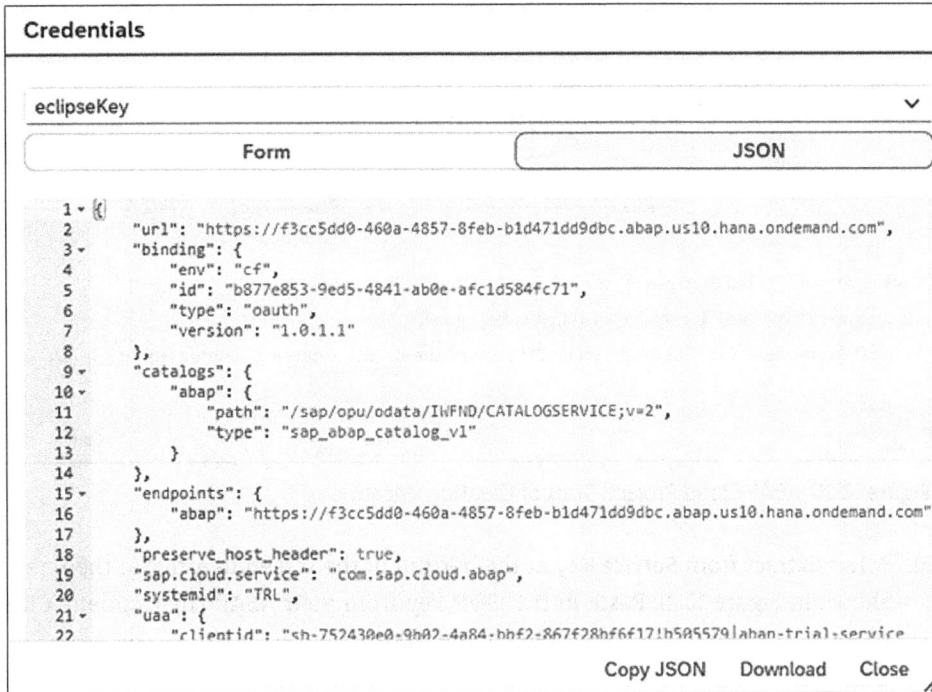

Figure 12.18 Service Key JSON Credentials

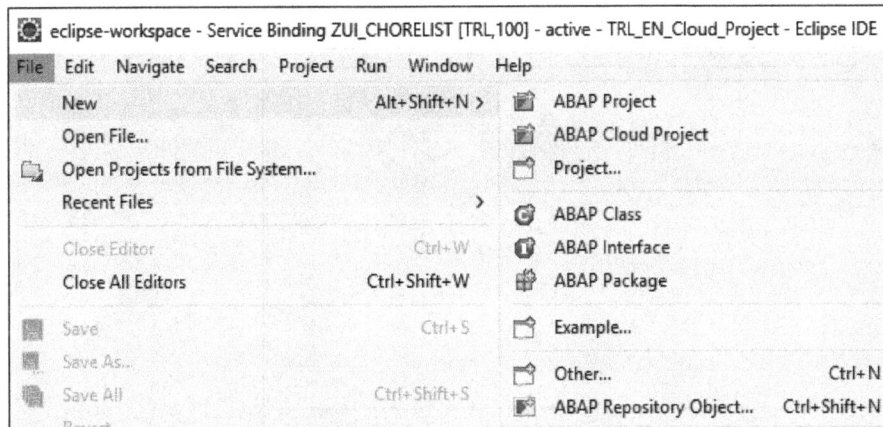

Figure 12.19 Create New ABAP Cloud Project

10. You'll arrive at the screen shown in Figure 12.20, which is the start of the creation wizard. Click **Next**.

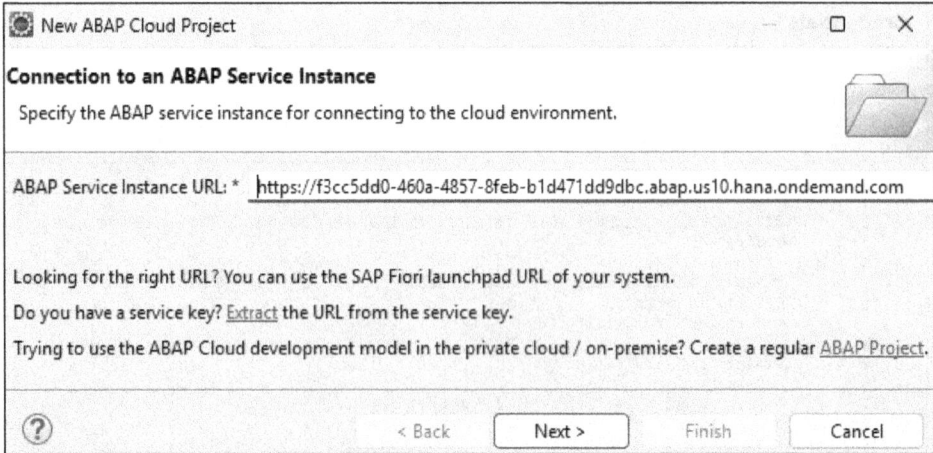

Figure 12.20 ABAP Cloud Project: Start of Creation Wizard

11. Select **Extract from Service Key** at the bottom of the screen to arrive at the screen shown in Figure 12.21. Paste in the JSON key from your ABAP environment. Click **Close** to be taken to the next screen of the wizard.

Figure 12.21 Paste Service Key JSON from SAP BTP ABAP Environment into ABAP Development Tools

12. Click the **Finish** button to complete the wizard to connect your ABAP package to the cloud system instance, as shown in Figure 12.22.

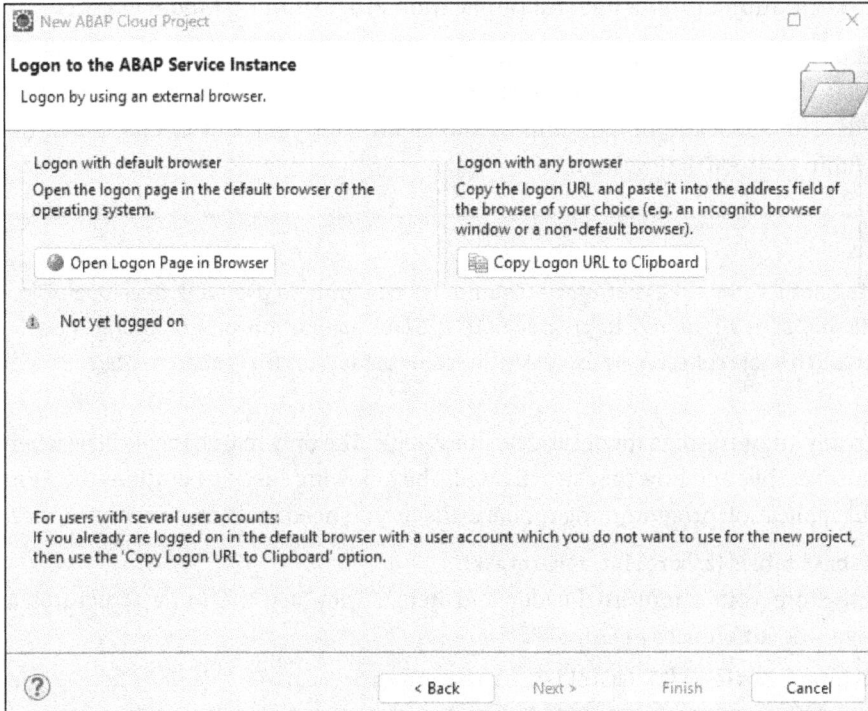

Figure 12.22 Click Finish to Complete the Project Creation Wizard

12.4   Case Study: Implementing ABAP RESTful Application Programming Model Elements via ABAP Cloud

Now that we've gone over most of the basics, let's walk through a hands-on implementation example using the ABAP RESTful application programming model in the SAP BTP ABAP environment. This case study will demonstrate how the concepts covered in this chapter—such as CDS modeling, behavior definitions, and service exposure—work together in a real-world business scenario.

In Chapter 11, we built a chore list and chore task ABAP RESTful application programming model service in a traditional ABAP system. That implementation taught the fundamentals of defining CDS entities, modeling behaviors, and exposing a service.

In this example, we'll take the same domain model and redeploy it in the SAP BTP ABAP environment. The ABAP RESTful application programming model building blocks remain the same—CDS entities, behavior definitions, behavior implementations, service definitions, and service bindings—but the runtime platform and activation process are different. This gives you a chance to see how everything you learned for on-premise ABAP systems applies to SAP's modern cloud offering. By the end of this case study, you'll have the same chore list ABAP RESTful application programming model service running and accessible on SAP BTP.

12.4.1   Recreation of ABAP RESTful Application Programming Model Objects

For the purposes of this implementation, we'll assume your cloud project is connected to the SAP BTP ABAP environment (Section 12.3.2). The next step is to reimplement the ABAP RESTful application programming model artifacts you built in Chapter 11, this time within your SAP BTP system.

> **Note**
>
> For this step, verify that your user account has been granted the ABAP developer role collection, as it's required to create ABAP RESTful application programming model objects. (This role collection is usually included by default for trial subaccounts.)

You already know the data model and business logic. The only real change will be where these artifacts live and how they're activated. The following checklist outlines the ABAP RESTful application programming model artifacts you need to create for your service:

- **Database tables** (zchorelist, zchoretask)
  These store your chore list header and items. They use the same structures as before—no differences in SAP BTP.

- **CDS interface views** (ZI_ChoreList, ZI_ChoreTask)
  These define the root entity and its composition with tasks. Composition ensures that create-by-association (CBA) and read-by-association (RBA) methods are generated.

- **CDS projection view** (ZC_ChoreList)
  This controls which fields and associations are exposed externally. It works the same in the cloud as on-premise.

- **Behavior definitions**
  These define how entities behave: create, update, delete, and association handling. You'll create both the interface behavior definitions (for ZI_ views) and the projection behavior definition (for ZC_).

- **Behavior implementation** (behavior implementation class, e.g., ZBP_I_CHORELIST)
  This provides handler methods for validations, determinations, and save logic. In the cloud, ABAP development tools generates the same skeleton classes for you.

- **Service definition** (ZSD_ChoreList)
  This decides which CDS views are exposed to the outside world.

- **Service binding**
  This publishes the service via OData V4 (either UI preview or web API).

We won't go through the modeling steps again here. Simply recreate the same objects you built in Chapter 11, using the exact same names and code. ABAP RESTful application programming model artifacts behave identically in the cloud; the only nuance is in the publishing and consumption steps, which we'll cover next.

12.4.2 Publishing and Consuming the Service

Now that you've rebuilt the ABAP RESTful application programming model artifacts from Chapter 11 (tables, CDS, behaviors, and service definition), those artifacts will behave the same on SAP BTP. The main differences for the cloud will be in how you publish them (service binding in ABAP development tools instead of Transaction /IWFND/ MAINT_SERVICE) and how you consume them (OAuth2 token from Extended Services – User Account and Authentication [XSUAA] and cross-site request forgery [CSRF] for writes, instead of basic authentication).

Start by opening your service binding in ABAP development tools (**Project Explorer • Business Services • Service Bindings**) or create a new service binding if you haven't already (**New • Service Binding • ODATA V4**).

For your **Binding Type** selection, **OData V4 - UI** is best for quick SAP Fiori previews. Use **OData V4 - Web API** when you need a pure API with no UI preview.

Then, right-click on your service binding and select **Publish**.

Once you click **Publish**, ABAP development tools generates and activates the runtime artifacts and Internet Communication Framework (ICF) handler automatically. Copy the service URL for the service binding or click **Preview** to open it in your browser. You can now send external requests to your service using this URL.

12.5 Summary

ABAP Cloud is more than just a technology shift—it represents a strategic realignment of SAP's enterprise development philosophy. By adopting clean core principles, decoupled architecture, and service-oriented design, developers can build future-proof apps that are easier to maintain, test, and scale.

This chapter introduced the role of SAP BTP and its capabilities, explained the difference between SAP BTP ABAP environment and developer extensibility, and provided a foundational overview of the ABAP RESTful application programming model and its core components. We also explored:

- How development in the cloud differs from traditional ABAP
- Where performance and security guidelines must be enforced
- Why the ABAP RESTful application programming model offers a scalable and maintainable path forward

For ABAP developers, this is a moment of transformation. ABAP isn't going anywhere— it's growing up and moving to the cloud. The syntax may be familiar, but your mindset must evolve. Cloud-native ABAP emphasizes declarative design, lifecycle-aware extension strategies, and full integration with SAP's broader platform capabilities.

In the next chapter, we'll build on what you learned about bringing your traditional ABAP skills to the cloud by sharing best practices and design patterns that will serve you in any development environment.

Chapter 13
Best Practices and Design Patterns

You've now mastered object-oriented programming with ABAP Objects and explored practical apps for modern landscapes. To conclude your journey, in this chapter, you'll gain a broad understanding of object-oriented best practices as well as design patterns.

Good object-oriented programming (OOP) in ABAP starts with principles such as programming to interfaces, keeping classes focused on a single responsibility, favoring composition over inheritance, and applying dependency injection where appropriate. These best practices, which we'll explore in this chapter, form the foundation of clean, reusable code and make it easier to adapt when requirements inevitably change.

To build on that foundation, we'll then look at design patterns in OOP. Design patterns can be found throughout just about any code library involving OOP, whether the developer intended to use design patterns or not. That's because when you're programming in an object-oriented way, you tend to instantiate objects and organize your logic in certain patterns, even if the objects themselves provide different functionality. Design patterns also allow you to write and think about code in a reusable way, allowing you to refactor monolithic methods into sets of classes that encapsulate specific functionality.

We'll review and provide examples of a few categories of design patterns, including creational patterns, structural patterns, and behavioral patterns. We'll explain each category at a high level, provide a list of some of the patterns included in the category, and go into detail with example patterns in each category. Afterwards, we'll share some insight into how you can deepen your knowledge of design patterns and provide you with helpful resources.

13.1 Object-Oriented Analysis and Design

Before you dive into design patterns, it's important to understand the broader principles of object-oriented analysis and design. Patterns are most effective when they are built on top of solid fundamentals, and without those fundamentals, they can easily be misapplied or lead to overly complex solutions. Object-oriented best practices provide that foundation by giving you guidelines for structuring classes, managing dependencies, and thinking about responsibilities in your code.

Let's explore these core best practices:

- **Program to interfaces**
  One of the most important practices is programming to *interfaces* rather than concrete classes. In ABAP, this means defining interfaces that describe behavior and coding against those interfaces, rather than hard-wiring dependencies to specific implementations. By doing so, you make your code more flexible, more testable, and easier to extend. For example, instead of directly instantiating a class that exports data in JavaScript Object Notation (JSON), you can work with an interface zif_exporter. Later, you can swap in a different implementation for PDF or XML without changing the calling code.

- **Follow the single responsibility principle**
  Another key practice is the *single responsibility principle*, which states that each class should have one clear reason to change. In ABAP development, it's tempting to add more and more logic into a single global class or even a single method, but this quickly leads to code that is difficult to maintain or reuse. By keeping classes focused, you make your code easier to test and easier for other developers to understand.

- **Choose composition over inheritance**
  Closely related is the idea of favoring *composition* over inheritance. While inheritance is a powerful feature in ABAP, relying too heavily on it can create rigid hierarchies that are hard to change later. Composition, on the other hand, allows you to build complex behavior by assembling smaller, reusable parts. A sales order service, for example, might compose separate components for validation, pricing, and persistence, rather than inheriting all of those responsibilities from a single base class.

- **Use dependency injections**
  Finally, dependency injection is an essential practice for building maintainable ABAP code. Instead of having a class create its own dependencies with CREATE OBJECT, you pass those dependencies in from outside—either through a constructor or a setter method. This makes your classes more adaptable and easier to test since you can provide different implementations in different contexts (for example, injecting a mock service in ABAP Unit tests).

Together, these practices form the groundwork for clean, object-oriented ABAP. They ensure that your code is modular, testable, and ready to evolve as your business needs change. Once these foundations are in place, you can layer design patterns on top to provide proven solutions to recurring problems. Patterns and best practices go hand in hand: best practices guide you to think about classes and responsibilities, and patterns show you how to organize those classes into flexible and reusable designs.

Design patterns are proven solutions to recurring problems in software design. They capture ways that developers across many languages have learned to organize code so that it remains clear, reusable, and adaptable as systems grow. In object-oriented ABAP, patterns are especially valuable because business requirements change constantly, and long-lived systems must absorb new features without becoming brittle. By learning to recognize and apply design patterns, you gain more than just coding techniques; you get a shared vocabulary that makes it easier to discuss architecture with other developers and to reason about the structure of your own programs.

Design patterns can be broken into three general categories: *creational, structural*, and *behavioral*. Each category groups together patterns that are beneficial for certain parts of your development. This chapter is only an introduction to design patterns, but the real value comes when you start recognizing patterns in your own projects and deliberately applying them. A useful next step is to revisit some of your existing ABAP code and ask yourself: How is object creation handled? How are responsibilities divided? How do different objects communicate? You'll often find that patterns are already present, even if they were never named. By naming them, you adopt a common vocabulary for discussing designs with other developers, and you sharpen your ability to refactor.

Diving Deeper

If you want to learn more about design patterns after reading this chapter, consider the following resources. An excellent reference specific to ABAP is *Design Patterns in ABAP Objects* by Kerem Koseoglu (SAP PRESS, 2017), which adapts classical patterns to the SAP ecosystem. It shows how patterns appear not only in custom code but also in SAP frameworks themselves, from ABAP List Viewer (ALV) to Business Object Processing Framework (BOPF) to SAP Gateway runtime. Because ABAP has its own conventions—global classes, ABAP Dictionary integration, event models—seeing patterns expressed in ABAP terms helps bridge the gap between theory and practice.

Beyond ABAP, the foundational text is *Design Patterns: Elements of Reusable Object-Oriented Software* by Eric Gamma, Richard Helm, Ralph Johnson, and John Vlissides (Addison-Wesley Professional, 1994), often referred to as the "Gang of Four" book. This is where the vocabulary of creational, structural, and behavioral patterns was first established. While the examples are in C++ and Smalltalk, the concepts are universal and apply just as well to ABAP.

Finally, for a quick reference, the website *https://sourcemaking.com/design_patterns* provides concise explanations, diagrams, and example code for all major patterns, along with coverage of refactoring and UML. Its brevity is ideal when you need to refresh your understanding of a pattern or see its structure at a glance. It's particularly useful for ABAP developers, because once you grasp the essence of a pattern, it's straightforward to adapt it to ABAP syntax and frameworks.

13.2   Creational Patterns

The focus of creational design patterns involves designing how objects get created. The traditional development practice from ABAP developers involves the CREATE OBJECT statement or NEW statement to make new class instances whenever developers need them. This approach works for small programs or new classes until apps grow beyond direct instantiation methods. This can lead to a number of issues:

- The calling code remains rigidly linked to specific classes.
- The practice of implementing objects becomes challenging when you're testing different solutions.
- The system contains multiple places with duplicate code for creating objects.

With creational patterns, the object creation process becomes abstracted to solve these issues. These patterns introduce standardized mechanisms that handle object creation and help developers avoid ad-hoc decisions in their code. The implementation of these patterns enables you to achieve better maintainability in addition to flexibility and consistency throughout your ABAP code.

The most prevalent creational patterns in practice consist of:

- **Abstract factory**
  The abstract factory pattern provides an interface for creating families of related or dependent objects without specifying their exact classes. In ABAP, you can use it to return different types of output handlers (such as ALV, PDF, or JSON) depending on user or Customizing settings.

- **Builder**
  The builder pattern separates the construction of a complex object from its representation so the same process can create different representations. A common ABAP use case is constructing a complex sales order structure step by step before passing it to a Business Application Programming Interface (BAPI) for saving.

- **Factory method**
  The factory method pattern defines a method that creates objects but allows subclasses to decide which class to instantiate. In ABAP, you might use this when creating different loggers (file logger, app log logger, etc.), depending on your configuration.

- **Prototype**
  The prototype pattern creates new objects by cloning existing ones rather than building from scratch. In ABAP, this could be applied to duplicate a material master template or replicate a sales order draft with only a few modifications.

- **Singleton**
  The singleton pattern ensures a class has only one instance and provides a global point of access to it. In ABAP, this is often used for shared resources such as configuration readers or app loggers where multiple instances would be redundant or harmful.

Of these patterns, we'll pick two commonly used ones to explore in the following sections: singleton and factory method. But first, we'll discuss when you should use creational patterns.

13.2.1   When to Use Creational Patterns in ABAP

In ABAP, object creation carries a few special considerations:

- **Global versus local classes**
  ABAP uses two distinct types of classes: global classes stored in the Class Builder for system-wide use and local classes defined within reports and includes. The creation of global classes benefits from controlled instantiation methods, which prevent developers from using numerous CREATE OBJECT and NEW statements throughout their code.

- **Testability**
  In ABAP Unit, mocking or substituting dependencies is common. When your code directly instantiates classes, it's almost impossible to replace them with test doubles. Creational patterns enable straightforward substitution of objects.

- **SAP framework usage**
  SAP uses creational patterns throughout its framework, as you've seen in BOPF and ALV grid factory classes. Knowing these design patterns allows you to write code that matches SAP's recommended development methods.

ABAP developers should implement creational patterns whenever they need flexible code that also maintains high testability levels and ease of maintenance. The direct use of CREATE OBJECT or NEW to instantiate classes leads to management problems because object creation logic spreads throughout the entire codebase. The need to change object creation methods requires updates in dozens of code locations, thus increasing the chance of introducing bugs and the total maintenance expense. The centralization and abstraction of object creation through creational patterns establishes a single management point for instance creation that allows you to enforce single-instance policies with singleton or dynamic subclass selection through the factory method.

These patterns serve an essential function because they improve testability. ABAP Unit testing frequently requires dependency replacement with mock objects to maintain testing isolation and predictability. Your production code becomes incapable of dependency substitution when it directly creates dependencies. Introducing factories, builders, and similar creational mechanisms enables runtime or test-class injection, which boosts the quality and reliability of your tests.

SAP framework designs align naturally with the patterns found in creational design. The ALV grid control uses factory methods to create suitable display objects, and some parts of BOPF depend on patterns that eliminate direct entity creation. Knowledge of standard SAP code patterns allows developers to integrate with existing frameworks and implement SAP-approved best practices.

You should implement creational patterns in your ABAP codebase to achieve adaptability during changes, enhance testing efficiency, and follow common design principles. The patterns serve three essential functions: creating *clean code* while simultaneously reducing maintenance tasks and following SAP development standards.

13.2.2   Singleton Pattern

The singleton pattern is a basic yet extensively used creational design pattern. The singleton pattern guarantees one instance of a class throughout program execution while offering global access to that instance. Multiple instances of specific objects usually create redundancy and inefficiency in the system. For example, an app should have only one logger instance because multiple instances would be unnecessary and wasteful. Another example is having both a configuration reader and cache object, as they should exist as a single instance that all components can access through reuse. The singleton pattern achieves this by making object creation private and offering a single controlled method to retrieve the instance.

ABAP developers implement the singleton pattern by blocking external access to object creation through CREATE OBJECT and NEW statements. The class offers a static method named get_instance that delivers the same instance during every invocation. The method creates an object and stores it in a class-level attribute during its first execution. The stored reference is returned to the caller during subsequent method calls, ensuring only one active instance remains. Consider the example of a logger implemented as a singleton in ABAP in Listing 13.1.

```
CLASS zcl_logger DEFINITION
  PUBLIC CREATE PRIVATE.
  PUBLIC SECTION.
    CLASS-METHODS get_instance
      RETURNING VALUE(ro_logger) TYPE REF TO zcl_logger.
    METHODS log_message
      IMPORTING iv_msg TYPE string.
  PRIVATE SECTION.
    CLASS-DATA go_instance TYPE REF TO zcl_logger.
ENDCLASS.

CLASS zcl_logger IMPLEMENTATION.
  METHOD get_instance.
    IF go_instance IS INITIAL.
      go_instance = NEW #( ).
    ENDIF.
    ro_logger = go_instance.
  ENDMETHOD.

  METHOD log_message.
    "In a real-world example, this might write to BAL logs or a custom table
    WRITE: / iv_msg.
  ENDMETHOD.
ENDCLASS.
```

Listing 13.1 Implementation of the Singleton Pattern for a Class Used for Logging

The class design follows the fundamental structure of the singleton pattern. The CREATE PRIVATE keyword prevents external code from making logger instances through NEW. The zcl_logger=>get_instance( ) function requires consumers to make requests instead of allowing direct instantiation of the logger through NEW. When get_instance is called for the first time, it creates and stores the object in the go_instance class attribute. After the object is created, get_instance returns the stored object to maintain program-wide logger consistency.

The advantages of this approach are clear. It prevents unneeded duplication of resources by providing a controlled way to manage global resources. Different program sections that use logging functions ensure message consistency through the shared logger instance, which simplifies troubleshooting processes. The singleton configuration reader enables system-wide Customizing to be read once and cached, avoiding multiple database hits.

The singleton pattern provides valuable benefits to users but needs proper handling during implementation. The main weakness of the singleton pattern is that it creates a global state throughout your program. Object sharing between consumers makes dependencies harder to track and complicates testing processes. Using a real database table for your singleton logger makes it difficult to switch to a mock database for ABAP Unit testing. The singleton pattern can be combined with dependency injection to solve this issue by passing references from central program points and using mock implementations in test classes.

The singleton design pattern does not work properly in situations where objects appear to be shared. Each session of your program runs its own version of the singleton instance when parallel processes or user sessions are active. System-wide caches need database tables or shared memory objects instead of singleton instances due to cross-session uniqueness requirements.

The singleton pattern remains one of the most widely used creational patterns when developing ABAP apps despite some potential limitations. The SAP standard framework classes, along with custom development, will present this pattern to you because they require this access method. Your ability to manage shared resources effectively in apps will improve when you understand the singleton pattern, and you'll become better at working with the singletons that SAP provides by default.

13.2.3   Factory Method Pattern

The factory method pattern is another commonly used pattern within the creational pattern family. The factory method pattern functions differently than the singleton because it determines which class will get instantiated at runtime. Direct object creation with CREATE OBJECT or NEW should be replaced by a separate method that resides inside abstract classes or interfaces. Subclasses can modify this method to determine which specific object type should be created. The system becomes more adaptable and easier to maintain because the object usage code remains independent from the object creation code.

ABAP developers find this pattern useful when they need to implement a general process that allows different specific implementations based on configuration settings, environmental factors, or user input. For example, developing a reporting tool requires result export functionality. Different users require their reports to display as ALV grids, PDFs, or JSON files for external system consumption. The direct instantiation of output classes in your code requires modifications when the calling code for each new output format, which produces brittle and error-prone systems. The factory method pattern enables you to store creation logic in one place while adding new output formats through the extension of additional subclasses. Listing 13.2 shows a simplified ABAP implementation of the factory method pattern.

```abap
INTERFACE zif_report_output.
  METHODS display IMPORTING it_data TYPE STANDARD TABLE.
ENDINTERFACE.

CLASS zcl_output_alv DEFINITION.
  PUBLIC SECTION.
    INTERFACES zif_report_output.
ENDCLASS.

CLASS zcl_output_alv IMPLEMENTATION.
  METHOD zif_report_output~display.
    " Example: call ALV grid display
    cl_demo_output=>display( it_data ).
  ENDMETHOD.
ENDCLASS.

CLASS zcl_output_json DEFINITION.
  PUBLIC SECTION.
    INTERFACES zif_report_output.
ENDCLASS.

CLASS zcl_output_json IMPLEMENTATION.
  METHOD zif_report_output~display.
    " Example: simple JSON conversion
    DATA(json) = /ui2/cl_json=>serialize( data = it_data ).
    WRITE: / json.
  ENDMETHOD.
ENDCLASS.

" Factory Method superclass
CLASS zcl_output_factory DEFINITION ABSTRACT.
```

```
  PUBLIC SECTION.
    CLASS-METHODS create_output
      IMPORTING iv_type TYPE string
      RETURNING VALUE(ro_output) TYPE REF TO zif_report_output.
ENDCLASS.

CLASS zcl_output_factory IMPLEMENTATION.
  METHOD create_output.
    CASE iv_type.
      WHEN 'ALV'.
        CREATE OBJECT ro_output TYPE zcl_output_alv.
      WHEN 'JSON'.
        CREATE OBJECT ro_output TYPE zcl_output_json.
      WHEN OTHERS.
        " Default to ALV
        CREATE OBJECT ro_output TYPE zcl_output_alv.
    ENDCASE.
  ENDMETHOD.
ENDCLASS.
```

Listing 13.2 Implementation of the Factory Method Pattern for a Class that Displays Different Outputs

The interface zif_report_output establishes the interface that every output handler must fulfill. Each concrete implementation of the interface through zcl_output_alv and zcl_output_json defines their individual display functionality. The factory class zcl_output_factory features one primary method that decides the creation of a class instance based on input data. The addition of a PDF output requires you to implement zcl_output_pdf alongside factory extension modifications because the consuming code functions through the interface.

This design pattern allows developers to decouple their code. The calling code operates independently of whichever concrete class gets instantiated. The code requires only a zif_report_output interface implementation from its received outputs regardless of the concrete class used. Extensibility becomes possible through this approach because you can introduce new types without affecting current consumers. The pattern makes your code easier to test because you can replace the factory output logic with a mock class that captures function calls instead of displaying actual output.

The factory method pattern functions independently, so it doesn't require a separate factory class to exist. The factory logic is implemented in abstract superclasses before subclasses modify it to return their specific implementation instances. SAP frameworks demonstrate this pattern by letting abstract classes define standard procedures, which subclasses implement with their particular data instances. The ALV grid display classes require you to use factory methods like cl_salv_table=>factory to obtain the correct instance based on data and context conditions.

As with the singleton pattern, the factory method design pattern requires developers to understand its trade-offs. The additional abstraction layer introduces complexity and makes it difficult for beginners to read the code. The initial reading experience reveals a factory method but does not immediately display the resulting object. The slight initial confusion from indirection does not outweigh the long-term benefits that the factory method provides in maintaining and scaling professional ABAP apps. The ability to extend functionality without rewriting extensive code sections proves more valuable than the minimal initial complexity.

Mastering the factory method is a fundamental requirement for any ABAP developer. The SAP standard includes factory method implementations in custom output handling scenarios as well as throughout the standard framework. Your ability to design robust apps that adapt to change will improve when you recognize factory method implementation as SAP uses it.

13.3   Structural Patterns

Creational patterns show how objects get made, but structural patterns demonstrate how objects assemble into a system. The focus of structural patterns lies in providing adaptable methods for object organization while also enabling extension capabilities and connection establishment. The value of structural patterns in ABAP becomes evident during two scenarios: when you need to add new functionality without modifying existing classes, or when you want to hide technical details through interfaces that remain clean and consistent for consumers.

Structural patterns help developers compose classes and objects into arrangements that allow their systems to expand in capability without becoming complicated. Structural patterns direct attention toward object relationships throughout their lifetime rather than during object creation, since they focus on peer collaboration, client exposure, design replacement, or extension methods that preserve overall design stability. ABAP developers must focus on integration boundaries because their work primarily involves BAPI wrapping, remote function call (RFC) isolation, and data transformation for ALV and SAPUI5, as well as protecting app logic from database and legacy API specifics. The problem names and reusable solutions in structural patterns enable you to modify one part of the system, which then produces clean benefits across the entire design.

Some of the common structural patterns you'll encounter include:

- **Adapter**
  The adapter pattern translates one interface into another so that classes with incompatible interfaces can work together. In ABAP, you might build an adapter around a BAPI return structure so that it can be consumed by an SAP Fiori OData service without changing either side.

- **Bridge**

  The bridge pattern decouples an abstraction from its implementation so that the two can vary independently. For example, you could define a document renderer abstraction and provide different implementations for ALV, PDF, or JSON outputs in ABAP.

- **Composite**

  The composite pattern treats individual objects and groups of objects uniformly by giving them a common interface. In ABAP, this is useful for modeling hierarchical data like bills of materials (BOMs) or organizational structures, where a node and a group of nodes can both respond to the same methods.

- **Decorator**

  The decorate pattern attaches additional responsibilities to an object dynamically, without subclassing. In ABAP, you might decorate a data exporter with logging or encryption functionality, layering on behavior without modifying the base class.

- **Façade**

  The façade pattern provides a simplified, high-level interface to a complex subsystem. ABAP developers often create façades around clusters of function modules or service classes so that consumers only need to call a clean, intention-revealing method instead of navigating a tangle of dependencies.

- **Flyweight**

  The flyweight pattern reduces memory consumption by sharing common, immutable data across many small objects. In ABAP, this can be applied to reuse ABAP Dictionary metadata or value help descriptions instead of instantiating new copies for every consumer.

- **Proxy**

  The proxy pattern acts as a stand-in for another object, often controlling access, deferring expensive initialization, or providing local behavior before calling a remote service. This is common in ABAP when working with RFC or web service stubs, which act as proxies for the actual remote implementations.

In the following sections, we'll dive into the benefits of using structural patterns and take a closer look at two of the most important patterns: adapter and decorator.

13.3.1 When to Use Structural Patterns in ABAP

All structural patterns share the common element of indirection as a fundamental theme. The introduction of a minimal interface layer serves as a protective barrier between concrete classes by transforming interfaces, combining multiple collaborators, or delaying work to alternative implementations. The everyday practice of ABAP programming allows you to maintain domain models that generate organized internal tables with type protection, but your apps may require different data transformations for ALV, PDF, and SAPUI5 components. The adapter maintains a position between the model and consumers to transform standardized internal interfaces into consumer-specific formats without altering core logic during new output channel additions. A façade pattern provides a solution to hide repetitive collaboration wiring between an authorization checker, a configuration reader, and a service client by exposing a single unified operation. The client invokes a single method that the façade controls to perform additional operations thus decreasing coupling and mental workload.

Several patterns enable developers to separate functional domains with dedicated areas for each modification axis. The bridge pattern is useful because it divides abstractions from implementations so that developers can modify them independently. In ABAP, you establish the *document renderer* interface at the abstraction level while implementing different renderer classes for ALV, PDF, and JSON. The abstraction serves as the sole dependency for your app code, so implementation changes become possible without affecting the rest of the codebase. The composite pattern presents a different method of operation, since it allows users to handle single objects alongside groups of objects in the same way. The approach works well with hierarchical structures— including BOMs, organizational units, and nested UI components—because a node and a group of nodes share the same interface for traversal and operations. The runtime addition of behavior occurs through the decorator pattern, which wraps objects during execution. You can use decorators to enrich or constrain behavior through layers instead of subclassing a renderer to implement log caching or field-level masking because each concern remains isolated and testable.

Other structural patterns excel at handling infrastructure boundary requirements. Proxy objects function as stand-ins for other objects that are distant or costly to create so that developers can delay access while adding caching mechanisms and managing lifecycles. In an SAP environment, RFC and web service stubs function as proxies because they enable local surrogate access to remote services. When dealing with numerous delicate objects that contain common data elements, the flyweight pattern emerges as an effective solution. You can reduce memory consumption while preserving object-oriented interfaces through externalization of shared immutable parts that can be reused instead of duplicating identical values such as value help metadata, text symbols, and ABAP Dictionary descriptions. Standard SAP libraries use the façade pattern extensively to create a single, well-designed entry point that controls multiple collaborating classes and function modules and delivers a stable external interface while developers perform internal refactoring at their own speed.

The practical benefit of structural patterns in ABAP lies in their ability to evolve interfaces and dependencies without destroying your business logic. A bridge pattern solves the problem of inheritance hierarchy overload, while decorators prevent subclass proliferation by adding cross-cutting concerns and façades provide a clean interface to handle brittle legacy APIs. The safety valves of these patterns protect ABAP systems that operate for years by controlling localized and predictable changes in the system as it integrates new channels. (That could be SAP Fiori today and something else tomorrow.)

Structural patterns provide developers with methods to organize how classes and objects relate to each other. These patterns direct attention away from creation toward established relationships, composition, and reuse. ABAP developers can use these patterns to decrease coupling, conceal complexity, and add functionality to existing code.

13.3.2   Adapter Pattern

The adapter pattern addresses a typical issue that occurs in ABAP projects when two components need to work together using different interfaces. Your domain logic includes methods with neat interface signatures and contemporary ABAP types, yet the consumer requires a BAPI-style table, an OData entity shape, or a legacy structure. A small class functions as an interface adapter to convert two incompatible interfaces without requiring modifications to either system. Both components function independently after the interface implementation. Clients can still use the interface they understand, and the adaptee maintains its original API.

Two essential conceptual variations of the adapter pattern exist in the programming world. An *object adapter* uses instance wrapping to forward calls with parameter and return value translation, whereas a *class adapter* changes behavior through inheritance. ABAP developers should use object adapters to create interface dependencies because they define an interface for client code and implement an adapter class that maintains references to external or legacy classes. Your adapter receives clean ABAP types as input and transforms them to match the adaptee's requirements before calling the adaptee and transforming its response back into your interface's types. The use of interface-based design, combined with composition over inheritance, leads to both idiomatic and testable solutions.

Consider this example: A reporting service generates internal tables of domain rows with standardized field names and types, as well as semantic meaning that follows your app structure. The UI team needs to present this report through an OData service, but the gateway project requires data in a different format with field codes needing translation to text values. Your domain code will stay clean from OData implementation by using an adapter that operates at the system boundary to perform the required translation. The OData project interacts with a basic interface which delivers the needed structure while remaining unaware of HTTP and EDMX and SAP Gateway-specific details.

Let's see this in action. First, define the interface your client will use—what the rest of your codebase wants to call—as shown in Listing 13.3.

```
TYPES: BEGIN OF ty_domain_row,
         matnr TYPE matnr,
         qty   TYPE decfloat34,
         uom   TYPE meins,
```

13

```
          END OF ty_domain_row,
          ty_domain_tab TYPE STANDARD TABLE OF ty_domain_row WITH EMPTY KEY.

INTERFACE zif_report_port
  PUBLIC.
  METHODS get_data
    IMPORTING iv_date_from TYPE d
              iv_date_to   TYPE d
    RETURNING VALUE(rt_data) TYPE ty_domain_tab.
ENDINTERFACE.
```

Listing 13.3 Interface Definition Used for Your Abstract Pattern

Assume you already have a legacy provider that returns data in a different shape, such as a function group or an older global class. We'll model it as a simple class with an incompatible API in Listing 13.4.

```
TYPES: BEGIN OF ty_legacy_row,
         material TYPE matnr,
         menge    TYPE menge_d,
         meins    TYPE meins,
       END OF ty_legacy_row,
       ty_legacy_tab TYPE STANDARD TABLE OF ty_legacy_row WITH EMPTY KEY.

CLASS zcl_legacy_provider DEFINITION PUBLIC.
  PUBLIC SECTION.
    METHODS fetch
      IMPORTING iv_from TYPE d
                iv_to   TYPE d
      EXPORTING et_rows TYPE ty_legacy_tab.
ENDCLASS.
```

Listing 13.4 Implementation Used as Legacy Code That the Adapter Pattern Will Leverage

Now, create the adapter, as shown in Listing 13.5. This implements the *target* interface (zif_report_port) and holds a reference to the *adaptee* (zcl_legacy_provider). The method does the translation, nothing more.

```
CLASS zcl_report_adapter DEFINITION PUBLIC CREATE PUBLIC.
  PUBLIC SECTION.
    INTERFACES zif_report_port.
    METHODS constructor
      IMPORTING io_legacy TYPE REF TO zcl_legacy_provider.
  PRIVATE SECTION.
    DATA mo_legacy TYPE REF TO zcl_legacy_provider.
ENDCLASS.
```

```
CLASS zcl_report_adapter IMPLEMENTATION.
  METHOD constructor.
    mo_legacy = io_legacy.
  ENDMETHOD.

  METHOD zif_report_port~get_data.
    DATA lt_legacy TYPE ty_legacy_tab.
    mo_legacy->fetch(
      EXPORTING
        iv_from = iv_date_from
        iv_to   = iv_date_to
      IMPORTING
        et_rows = lt_legacy ).

    DATA lt_result TYPE ty_domain_tab.
    LOOP AT lt_legacy ASSIGNING FIELD-SYMBOL(<ls_legacy>).
      APPEND VALUE ty_domain_row(
        matnr = <ls_legacy>-material
        qty   = CONV decfloat34( <ls_legacy>-menge )
        uom   = <ls_legacy>-meins ) TO lt_result.
    ENDLOOP.
    rt_data = lt_result.
  ENDMETHOD.
ENDCLASS.
```

Listing 13.5 Implementation of the Adapter Pattern Leveraging the Interface and Adapting to the Legacy Provider

The adapter remains hidden to the caller, who uses zif_report_port for programming and receives ty_domain_tab as output. You can implement zif_report_port to add a new adapter to handle different table requirements, as it sources data from different places or applies different data transformations. Your app code remains unaffected, since it never referenced the legacy provider directly.

The adapter design pattern demonstrates its greatest value with multiple system interfaces. BAPIs require deep export structures and status tables to return data that modern object-oriented interfaces cannot match effectively. You can create an adapter to convert BAPI calls into a minimal interface with domain type return methods and exception-raising capabilities instead of parsing out the BAPIRET2 message return structure across your codebase. The adapter handles all the return code processing and message aggregation activities while keeping your system free from clutter. This approach enables you to maintain a stable reporting port interface while switching between SAPUI5 and ALV by introducing data-shaping adapters that simplify the addition of new presentation channels through adapter implementation rather than core logic modifications.

Keep in mind several important operational factors when implementing this approach. The first performance consideration requires adapters to perform a single pass when processing large internal tables while minimizing unnecessary data duplication. The performance-focused approach requires building the target table once, but you can achieve better results by using field symbols or aligned field names with MOVE-CORRESPONDING statements or streaming conversion logic for reduced memory allocation. The adapter should only handle translations and not collect domain-related business rules. When domain decisions start entering your adapter, move them back to the core system and maintain the adapter as a simple mechanical transformation tool. Interface-based dependency improves testability because you can use fake adaptees in adapters for translation verification while running separate tests on adaptees to check their behavior. ABAP Unit tests use mock legacy providers that return small known tables to verify get_data returns domain rows with the correct structure.

Adapters protect against compatibility issues and allow you to make older ABAP code refactorable. The use of adapters enables you to modernize legacy providers without affecting existing callers. It introduces new UI technologies through fresh adapters that do not modify the domain logic. The thin interface layer becomes profitable through multiple iterations because it enables peripheral changes to reach the system core without compromising the middle sections.

An adapter becomes inappropriate when your current task involves adding cross-cutting features such as logging, caching, and masking but requires no modifications to base classes. This is where the decorator pattern steps in. It enables dynamic behavior enhancement through multiple layers by working with objects that already match the requirements. We'll explore the decorator pattern in the next section.

13.3.3   Decorator Pattern

The decorator pattern enables you to add new behaviors to existing objects through wrappers instead of modifying their original class structure. You can avoid subclassing concrete implementations for variations by using decorator objects that implement the same interface to wrap the original object and add functionality before or after the delegate object is executed. This approach enables runtime flexibility when mixing different components while keeping each one small, testable, and reusable.

The decorator pattern works well for handling *cross-cutting* concerns, which should remain separate from core classes, including logging, masking sensitive data, caching, retry logic, authorization checks, and performance timing. Rather than cluttering a JSON exporter with knowledge of audit logs and personally identifiable information (PII) handling, you keep the exporter focused on formatting and layer those concerns in decorators that you can enable in each scenario.

Let's look at a specific example to help you understand this concept. The exporter interface serves multiple channels (ALV, JSON, and PDF) in your app. The pattern works identically for any implementation, so we'll focus on JSON for simplicity.

First, define a stable interface that your clients already depend on, as shown in Listing 13.6.

```
INTERFACE zif_exporter PUBLIC.
  METHODS export
    IMPORTING it_data TYPE STANDARD TABLE
    RETURNING VALUE(rv_payload) TYPE string
    RAISING   cx_static_check.
ENDINTERFACE.
```

Listing 13.6 Interface Used for the Decorator Pattern

Next, write a straightforward concrete implementation—your baseline component (see Listing 13.7).

```
CLASS zcl_exporter_json DEFINITION PUBLIC CREATE PUBLIC.
  PUBLIC SECTION.
    INTERFACES zif_exporter.
ENDCLASS.

CLASS zcl_exporter_json IMPLEMENTATION.
  METHOD zif_exporter~export.
    " Serialize to JSON (placeholder for real logic)
    rv_payload = /ui2/cl_json=>serialize( data = it_data ).
  ENDMETHOD.
ENDCLASS.
```

Listing 13.7 Base Class Implementation for Use Within the Decorator Pattern

The decorator itself is also a class that implements the same interface and wraps another zif_exporter. It delegates to the wrapped object and adds behavior around that call. An abstract base decorator keeps the forwarding boilerplate in one place, as shown in Listing 13.8.

```
CLASS zcl_exporter_decorator DEFINITION
  PUBLIC ABSTRACT CREATE PUBLIC.
  PUBLIC SECTION.
    INTERFACES zif_exporter.
    METHODS constructor IMPORTING io_inner TYPE REF TO zif_exporter.
  PROTECTED SECTION.
    DATA mo_inner TYPE REF TO zif_exporter.
ENDCLASS.
```

13

```
CLASS zcl_exporter_decorator IMPLEMENTATION.
  METHOD constructor.
    mo_inner = io_inner.
  ENDMETHOD.

  METHOD zif_exporter~export.
    " Default pass-through; concrete decorators will redefine.
    rv_payload = mo_inner->export( it_data = it_data ).
  ENDMETHOD.
ENDCLASS.
```

Listing 13.8 Base Decorator Class From Which Additional Decorators Will Inherit

With that base in place, you can add behaviors one at a time. Listing 13.9 shows a logging decorator that measures duration and writes an app log entry. The specific decoration-related logic occurs before and after the call to the mo_inner export method.

```
CLASS zcl_exporter_logging DEFINITION PUBLIC INHERITING FROM zcl_exporter_deco-
rator.
  PUBLIC SECTION.
    METHODS zif_exporter~export REDEFINITION.
ENDCLASS.

CLASS zcl_exporter_logging IMPLEMENTATION.
  METHOD zif_exporter~export.
    DATA(lv_start) = cl_abap_context_info=>get_system_time( ).
    TRY.
        rv_payload = mo_inner->export( it_data = it_data ).
      CATCH cx_static_check INTO DATA(lx).
        " Log failure and re-raise
        MESSAGE lx->get_text( ) TYPE 'E'. "Replace with real BAL logging
        RAISE EXCEPTION lx.
    ENDTRY.
    DATA(lv_end) = cl_abap_context_info=>get_system_time( ).
    " Log duration (placeholder)
    MESSAGE |Export completed in { lv_end - lv_start } ms| TYPE 'I'.
  ENDMETHOD.
ENDCLASS.
```

Listing 13.9 Implementation for a Decorator Meant for Logging

A masking decorator can redact sensitive fields from the payload, as shown in Listing 13.10. Once again, the masking logic is "decorated" around the call to the base class mo_inner export method.

```
CLASS zcl_exporter_masking DEFINITION PUBLIC INHERITING FROM zcl_exporter_deco-
rator.
  PUBLIC SECTION.
    METHODS zif_exporter~export REDEFINITION.
  PRIVATE SECTION.
    METHODS mask_payload
      IMPORTING iv_json TYPE string
      RETURNING VALUE(rv_json) TYPE string.
ENDCLASS.

CLASS zcl_exporter_masking IMPLEMENTATION.
  METHOD zif_exporter~export.
    DATA(lv_raw) = mo_inner->export( it_data = it_data ).
    rv_payload   = mask_payload( lv_raw ).
  ENDMETHOD.

  METHOD mask_payload.
    " Naive example: replace account numbers (illustrative)
    REPLACE ALL OCCURRENCES OF REGEX '"acct"\s*:\s*"\d{8,}"'
      IN iv_json WITH '"acct":"****"'.
    rv_json = iv_json.
  ENDMETHOD.
ENDCLASS.
```

Listing 13.10 Implementation for a Decorator Meant for Masking

A caching decorator helps you avoid recomputation for identical inputs (though you'd use a better keying function in production), as shown in Listing 13.11. Here, we've created some caching related logic to occur before and after the call to the base class mo_inner export method.

```
CLASS zcl_exporter_cache DEFINITION PUBLIC INHERITING FROM zcl_exporter_decora-
tor.
  PUBLIC SECTION.
    METHODS zif_exporter~export REDEFINITION.
  PRIVATE SECTION.
    TYPES: BEGIN OF ty_entry,
             key     TYPE string,
             payload TYPE string,
           END OF ty_entry,
           ty_tab TYPE STANDARD TABLE OF ty_entry WITH EMPTY KEY.
    CLASS-DATA gt_cache TYPE ty_tab.
    METHODS make_key
      IMPORTING it_data TYPE STANDARD TABLE
      RETURNING VALUE(rv_key) TYPE string.
ENDCLASS.
```

13

```
CLASS zcl_exporter_cache IMPLEMENTATION.
  METHOD make_key.
    " Simple hash surrogate: size + first row as text (illustrative)
    DATA(lv_count) = lines( it_data ).
    rv_key = |CNT:{ lv_count }|.
  ENDMETHOD.

  METHOD zif_exporter~export.
    DATA(lv_key) = make_key( it_data ).
    READ TABLE gt_cache WITH KEY key = lv_key INTO DATA(ls_hit).
    IF sy-subrc = 0.
      rv_payload = ls_hit-payload.
      RETURN.
    ENDIF.

    rv_payload = mo_inner->export( it_data = it_data ).

    APPEND VALUE ty_entry( key = lv_key payload = rv_payload ) TO gt_cache.
  ENDMETHOD.
ENDCLASS.
```

Listing 13.11 Implementation for a Decorator Meant for Caching

The beauty is in the composition. As shown in Listing 13.12, you can stack decorators in any order to achieve the effect you need for a particular scenario—no subclass explosion and no changes to the base component.

```
DATA(lo_core)     = NEW zcl_exporter_json( ).
DATA(lo_masked)   = NEW zcl_exporter_masking( lo_core ).
DATA(lo_logged)   = NEW zcl_exporter_logging( lo_masked ).
DATA(lo_cached)   = NEW zcl_exporter_cache( lo_logged ).

DATA(result) = lo_cached->export( it_data = lt_items ).
```

Listing 13.12 How Instantiation of the Exporter Object Occurs with Its Various Decorators

The wiring exists in a factory environment to provide a link with the factory method section. The factory method section allows your configuration to activate decorators that determine their implementation. The development client receives logging decoration while the production client gets caching and masking enabled. The internal admin tool bypasses masking altogether.

You should keep these practical aspects in mind. Masking must precede logging to prevent sensitive information from appearing in logs. Logging before caching will hide performance metrics that you want to track, while masking must happen first to prevent raw unmasked data from appearing. The decorator pipeline requires deliberate sequence arrangement and proper documentation. When working with decorators, you should avoid using IS BOUND checks on specific concrete classes because the decorated object loses its original reference. The program should operate on interfaces instead of concrete classes. Performance measurement is important, since each wrapper introduces minimal overhead, yet developers should choose lightweight decorators for high-performance paths while avoiding repeated string or table copies. The app exception should wrap the original exception when performing catch-and-log operations to preserve caller-dependent behavior.

From a testing perspective, the decorator pattern is a gift. The base exporter receives a small table for unit-testing that allows you to verify the JSON output. Each decorator should be tested independently with a fake inner exporter that generates a specific string you can verify against expected logging behavior, masking transformations, and caching results. Each class maintains a single responsibility, so the tests will remain fast and focused.

The decorator pattern implements a design principle that allows your code to be open for extension but closed for modification. When a new cross-cutting need appears—for example, rate limiting for a public API—you write one more decorator and compose it where needed. None of the existing exporters change, and no new subclasses are required to account for every combination. In long-lived ABAP systems, that agility is helpful. The core code stays clean, and you simply add on the needed behavior as an extension.

13.4   Behavioral Patterns

Behavioral patterns concentrate on interaction by showing how objects exchange information, distribute tasks, and execute sophisticated logic as part of a unified operation. The system's dynamic elements receive attention from behavioral patterns, creational patterns handle object creation, and structural patterns manage object arrangement. These patterns help you structure algorithms and distribute work and state management, which results in flexible and maintainable programs that you can easily expand.

Some common behavioral patterns you may encounter include:

- **Chain of responsibility**
  The chain of responsibility pattern passes a request along a chain of handlers until one of the handlers processes it. In ABAP, this could be applied to validation logic with multiple classes of check conditions (such as authorizations, data integrity, and business rules) until one handler either handles the request or fails it.

- **Command**
  The command pattern encapsulates a request as an object, allowing you to parameterize clients and queue, log, or undo actions. In ABAP, this is useful for workflow

tasks or background jobs, where each command object represents a unit of work that can be executed or rolled back.

- **Interpreter**
  The interpreter pattern defines grammar and interprets sentences in that grammar. While not as common in day-to-day ABAP, it can be used for parsing simple business rules, formulas, or domain-specific query languages stored in Customizing tables.

- **Iterator**
  The iterator pattern provides a way to access elements of a collection sequentially without exposing the underlying representation. In ABAP, custom iterator classes can walk over internal tables, selection results, or even BAPI return sets, giving consumers a clean interface to consume data row by row.

- **Mediator**
  The mediator pattern centralizes complex communication between objects, ensuring they don't need to reference each other directly. ABAP apps that coordinate multiple UI controls or modular services can benefit from a mediator that routes messages through a single hub.

- **Memento**
  The memento pattern captures and restores an object's internal state without exposing its internals. In ABAP, this can be useful when you're implementing the undo functionality for user transactions or when you're saving and restoring a complex object's state during batch processing.

- **Observer**
  The observer pattern defines a one-to-many dependency so that when one object changes state, all dependents are notified. ABAP supports this pattern naturally through class events and event handlers, making it a go-to for scenarios where multiple consumers react to a single source of truth.

- **State**
  The state pattern allows an object to alter its behavior when its internal state changes. In ABAP, you might model a sales order lifecycle (i.e., created, released, blocked, and billed) as a set of state objects, each controlling which actions are possible.

- **Strategy**
  The strategy pattern defines a family of algorithms and lets you swap them at runtime. This is a natural fit in ABAP when you want to choose between multiple pricing, tax, or calculation strategies, depending on the customization or context.

- **Template method**
  The template method pattern defines the skeleton of an algorithm in a base class, allowing subclasses to redefine certain steps. SAP frameworks use this extensively, and ABAP developers can apply it in scenarios like invoice processing where the general process is fixed but specific steps vary by document type.

- **Visitor**

  The visitor pattern lets you define new operations on an object structure without changing the objects themselves. Though less common in ABAP, it can be useful in hierarchical structures (such as BOMs or organizational charts), where you want to apply different operations—such as validation, reporting, and transformation—without cluttering the objects with every possible behavior.

We'll explain how you can use behavior patterns in more detail in the following sections, including a closer look at the strategy pattern.

13.4.1 When to Use Behavioral Patterns in ABAP

Behavioral patterns help developers separate the functional aspects of their code from the implementation details. They enable you to create reusable behavior blocks that enable request delegation between multiple handlers, event broadcasting to subscribers, and algorithm replacement. These patterns match the natural features of ABAP language through its implementation of class events, exception handling, and interface-based programming.

ABAP systems that operate for decades with complex business logic benefit from behavioral patterns because these patterns transform complicated IF statement tangles into adaptable designs. SAP's frameworks implement various patterns through workflow engines (command and chain of responsibility), UI event handling (observer and mediator), and app frameworks that provide business process templates (template method). Recognizing these patterns will help you improve your ABAP skills, and you'll be able to better understand and analyze standard SAP code.

13.4.2 Strategy Pattern

The strategy pattern enables clients to select algorithms from a common interface family at runtime. You avoid distributed IF … ELSEIF … branches for calculation selection in the system by moving this decision to the construction or configuration phase, and the rest of the system is free to use the interface. The selection of tax, pricing, and/or rounding rules matches perfectly with Customizing because users make selections based on company code, country, document type, or feature toggle. The strategy pattern maintains explicit choices that remain testable and replaceable.

The calculation of sales item taxes requires careful attention. The requirements differ across countries and sometimes across product groups. The system requires each algorithm to operate independently, but the caller needs to interact with tax strategy without knowing which algorithm runs because the caller only needs to request tax computation for a specific item.

Let's look at an example strategy pattern. First, you'll define the strategy interface—one small, intention-revealing contract (see Listing 13.13).

```abap
INTERFACE zif_tax_strategy PUBLIC.
  METHODS compute_tax
    IMPORTING is_item  TYPE zty_so_item " your domain type
    RETURNING VALUE(rv_tax) TYPE decfloat34
    RAISING   cx_static_check.
ENDINTERFACE.
```

Listing 13.13 Interface Used for a Strategy Pattern

Next, provide a few concrete strategies, as shown in Listing 13.14. Keep each focused on its rule set. For example, we created a US tax strategy that only holds logic related to US-based taxes. We also created a class for an EU tax strategy that only holds logic related to European tax strategies.

```abap
CLASS zcl_tax_strategy_us DEFINITION PUBLIC CREATE PUBLIC.
  PUBLIC SECTION.
    INTERFACES zif_tax_strategy.
ENDCLASS.

CLASS zcl_tax_strategy_us IMPLEMENTATION.
  METHOD zif_tax_strategy~compute_tax.
    " Simplified: destination-based, flat rate example
    rv_tax = is_item-net_amount * CONV decfloat34( '0.060' ).
  ENDMETHOD.
ENDCLASS.

CLASS zcl_tax_strategy_eu DEFINITION PUBLIC CREATE PUBLIC.
  PUBLIC SECTION.
    INTERFACES zif_tax_strategy.
ENDCLASS.

CLASS zcl_tax_strategy_eu IMPLEMENTATION.
  METHOD zif_tax_strategy~compute_tax.
    " Simplified: VAT based on material group
    DATA(lv_rate) = COND decfloat34(
      WHEN is_item-matkl = 'FOOD' THEN '0.04'
      WHEN is_item-matkl = 'BOOK' THEN '0.07'
      ELSE '0.20' ).
    rv_tax = is_item-net_amount * lv_rate.
  ENDMETHOD.
ENDCLASS.
```

Listing 13.14 Implementation Classes of the Different Tax Strategies

A small *context* object uses the strategy pattern. It doesn't know (or care) which strategy it got. Our example in Listing 13.15 shows the implementation of a tax calculator. The purpose of the class is to just calculate the taxes for an item. It shouldn't have to know which country it's calculating taxes for or the different types of strategies there are to calculate taxes. It just needs to know that it has an item and it needs to calculate tax—that should be its sole purpose.

```
CLASS zcl_tax_calculator DEFINITION PUBLIC CREATE PUBLIC.
  PUBLIC SECTION.
    METHODS constructor IMPORTING io_strategy TYPE REF TO zif_tax_strategy.
    METHODS tax_for_item
      IMPORTING is_item  TYPE zty_so_item
      RETURNING VALUE(rv_tax) TYPE decfloat34
      RAISING   cx_static_check.
  PRIVATE SECTION.
    DATA mo_strategy TYPE REF TO zif_tax_strategy.
ENDCLASS.

CLASS zcl_tax_calculator IMPLEMENTATION.
  METHOD constructor.
    mo_strategy = io_strategy.
  ENDMETHOD.

  METHOD tax_for_item.
    rv_tax = mo_strategy->compute_tax( is_item = is_item ).
  ENDMETHOD.
ENDCLASS.
```

Listing 13.15 Implementation of the Tax Calculator Context Object

Finally, wire it up. A strategy can be determined via Customizing, feature flags, or a factory class. Our example in Listing 13.16 shows the strategy being determined by a factory class, where we pass in the country code and that determines which tax strategy is instantiated to be used within the tax calculator.

```
CLASS zcl_tax_strategy_factory DEFINITION PUBLIC CREATE PUBLIC.
  PUBLIC SECTION.
    CLASS-METHODS for_context
      IMPORTING iv_country TYPE land1
      RETURNING VALUE(ro_strategy) TYPE REF TO zif_tax_strategy.
ENDCLASS.

CLASS zcl_tax_strategy_factory IMPLEMENTATION.
  METHOD for_context.
    CASE iv_country.
```

```
      WHEN 'US'.  ro_strategy = NEW zcl_tax_strategy_us( ).
      WHEN 'DE'
      OR   'FR'
      OR   'ES'.  ro_strategy = NEW zcl_tax_strategy_eu( ).
      WHEN OTHERS.
        RAISE EXCEPTION TYPE cx_static_check
          EXPORTING textid = cx_static_check=>default_textid. "handle gracefully
    ENDCASE.
  ENDMETHOD.
ENDCLASS.
```

Listing 13.16 Implementation of the Factory Class That Determines the Tax Strategy

Listing 13.17 shows the usage of this strategy with the tax calculator.

```
DATA(lo_strategy)   = zcl_tax_strategy_factory=>for_context( iv_country = 'US'
).
DATA(lo_calculator) = NEW zcl_tax_calculator( lo_strategy ).
DATA(lv_tax)        = lo_calculator->tax_for_item( is_item = ls_item ).
```

Listing 13.17 Implementation of the Strategy with the Tax Calculator

The solution remains maintainable because each new rule introduction allows developers to create new classes that extend the factory class without impacting existing strategies. ABAP Unit tests can create fake strategies for scenario simulation, while you can test each strategy rule independently. Domain code becomes more readable by including a straightforward method named compute_tax instead of implementing various country and product type branches.

Keep in mind the following when using strategy patterns:

- The interface should not expand too much. The system interface needs to maintain a compact design and should only have one or more related methods that use the same domain types. To address different input needs, strategies receive an optional context structure (zty_tax_context). You should always document its specific field usage.

- Avoid using configuration information within strategies in the main logic. The configuration retrieval should happen once with a constructor or small configuration object, while the logic should be pure and free of side effects. This keeps the process of unit testing straightforward.

- Use interface references as your standard approach in all programming contexts. The zif_tax_strategy interface should be the only reference point for callers, rather than concrete classes.

- You can minimize the performance impact of selecting strategies by either caching the selected instance in each context or maintaining a small registry.

The strategy pattern exists throughout ABAP apps, including tax calculations, pricing logic, rounding functions, number assignment, currency conversion, discount calculation, and serialization through shared interface encoding methods. Any situation requiring a CASE statement to select an algorithm should implement the strategy pattern.

Behavioral patterns establish standardized interaction rules between objects. These patterns define how objects communicate and work together, while creational patterns determine object creation methods and structural patterns determine object assembly methods. ABAP developers can use these patterns to create isolated algorithms and delegated responsibilities and runtime condition changes. The strategy pattern provides a successful implementation that enables runtime logic switching without requiring any modifications to existing calling code. The observer pattern, along with template method and command pattern, share similar design principles because they provide extensible solutions to manage complex logic in a clean manner.

13.5 Summary

In this chapter, we laid the foundation of object-oriented analysis and design, highlighting best practices such as programming to interfaces, applying the single responsibility principle, favoring composition over inheritance, and using dependency injection. These practices provide the groundwork for writing ABAP code that is modular, testable, and adaptable as your requirements evolve. Building on that foundation, we reviewed how design patterns help solve recurring design problems. Patterns provide a shared vocabulary for discussing object-oriented code and a set of reusable approaches for keeping your programs clear and flexible. They appear naturally in object-oriented ABAP development and become even more valuable when applied deliberately.

We then explored the three main categories of design patterns in detail. You learned how creational patterns help you control how objects are instantiated, as seen in the singleton and factory method examples. You saw how structural patterns help you compose objects effectively, as illustrated with the adapter and decorator pattern examples. Finally, you learned how behavioral patterns address the way objects collaborate to carry out logic, demonstrated by the strategy pattern example.

Together, these ideas show how object-oriented best practices and design patterns complement each other. Best practices keep code clean and focused, while design patterns provide reusable structures for solving recurring problems. By applying both in your ABAP development, you can reduce duplication, improve testability, and make your systems more adaptable to future change.

Appendix A
Installing the Eclipse IDE

Installing the ABAP development tools for Eclipse is not too difficult once you under-stand the various steps involved. In this appendix, we'll walk through these steps and show you how to install the ABAP development tools on your local machine.

A.1   Installing the Java SDK

The Eclipse environment runs on top of the *Java Runtime Environment* (JRE), a virtual machine which can run Java code on top of most any OS environment: Microsoft Win-dows, Mac OS, Linux, and so forth. Though you can technically run Eclipse using the standalone JRE, we recommend that you install the complete Java SDK as this contains features that can come in handy for some of the other SAP-related plug-ins you might wish to install alongside your Eclipse installation. The Java SDK can be downloaded from the Oracle Technology Network online at *http://www.oracle.com/technetwork/java/javase/downloads/index.html*. From here, you can click on the Download button highlighted in Figure A.1.

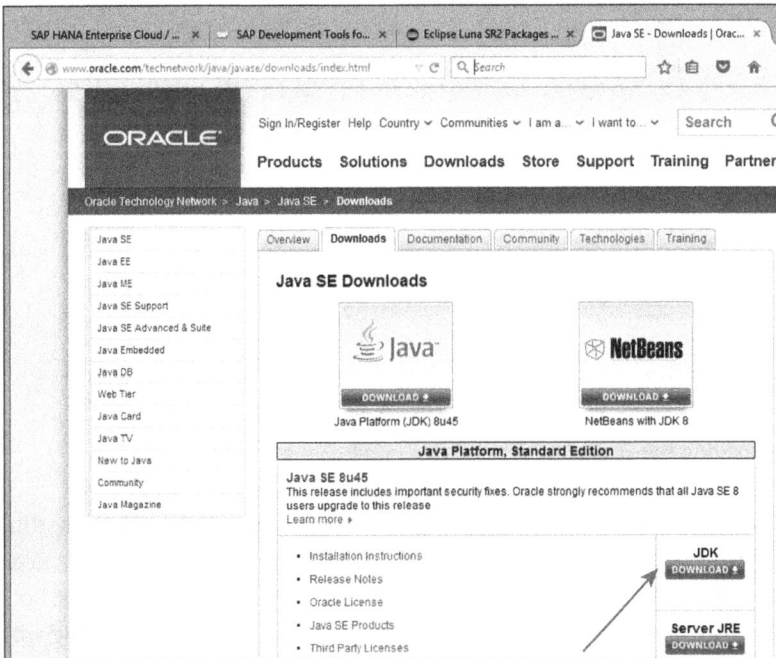

Figure A.1 Downloading and Installing the Java SE SDK

From here, you'll be routed to a Downloads page where you can download the setup executable for your particular OS/environment. The installation process takes moments and is straightforward to complete.

A.2   Installing Eclipse

Once the Java SDK is in place, you can proceed with the installation of the Eclipse IDE itself. To find the appropriate installation package, browse to *https://tools.hana.ondemand.com/#abap* and find the recommended distribution which is compatible with the latest innovation of the ABAP development tools. Figure A.2 shows how you can locate the recommended version to install.

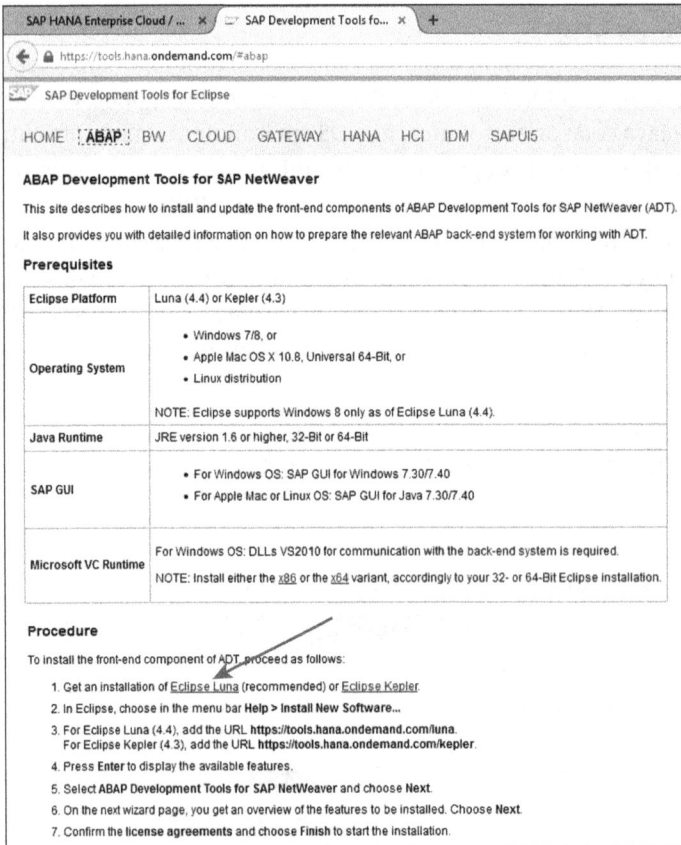

Figure A.2 Downloading the Eclipse IDE (Part 1)

Once you find the appropriate installation for your environment, follow the hyperlink to the Eclipse downloads page as shown in Figure A.3. From here, you can download either the Eclipse IDE for Java Developers or the Eclipse IDE for Java EE Developers. Both installations work just fine for ABAP-related development, so you can't go wrong with either installation. The Java EE edition just contains more plug-ins for Java EE develop-ment and therefore can also be used to develop custom applications.

Figure A.3 Downloading the Eclipse IDE (Part 2)

After you download Eclipse, you can install it by simply unpacking the ZIP archive into a directory on your local machine. From here, you can launch Eclipse the first time by run-ning the Eclipse executable from the root directory of the program installation folder.

> **Tip**
>
> For a variety of reasons, we recommend that you avoid unpacking the Eclipse installa-tion folder too deep within your folder structure. Since some of the Java-related arti-facts have long folder/file names, some plug-ins have problems if the Eclipse installa-tion is buried underneath too many folders.

The first time Eclipse runs, it will prompt you to select a *workspace* folder where proj-ects and related metadata are stored (see Figure A.4). Here, we suggest that you pick a directory path that you'll remember (and keep backed up). While you can of course change this selection after the fact, it's much easier to get it right up front.

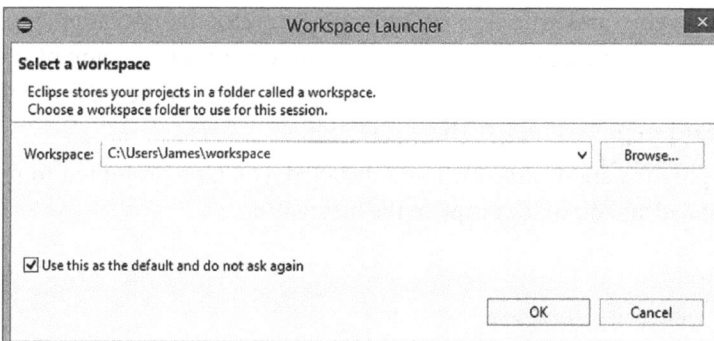

Figure A.4 Determining the Eclipse Workspace Folder

A.3   Installing the ABAP Development Tools

With the Eclipse installation in place, all that's left is to install the ABAP development tools plug-ins. This can be achieved within the Eclipse IDE itself by performing the following steps:

1. First, select the **Help • Install New Software...** menu option as shown in Figure A.5.

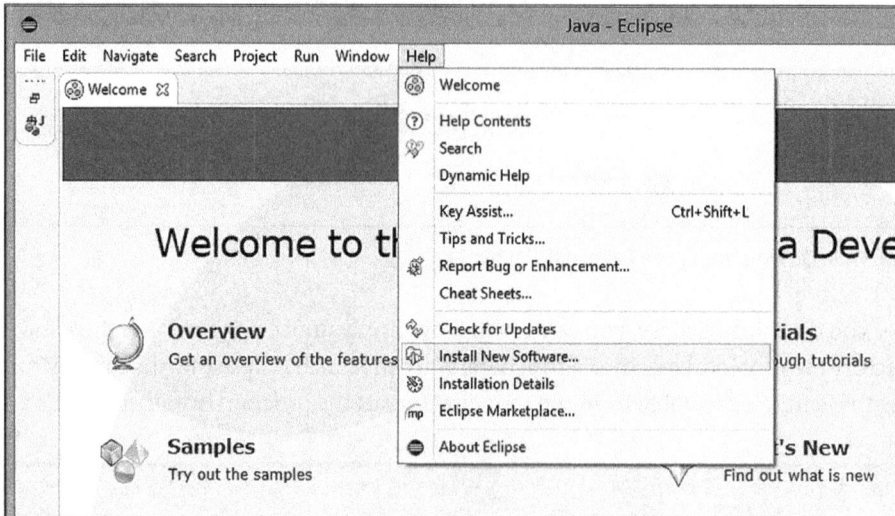

Figure A.5 Installing the ABAP Development Tools (Part 1)

2. This will open up the **Install** dialog box shown in Figure A.6. Here, we must configure a *software site* which points to the ABAP development tools plug-ins. You can find the relevant software site for your particular Eclipse distribution via the SAP Development Tools for Eclipse home page available online at *https://tools.hana.onde-mand.com/#abap* (see Figure A.2). Once you identify this URL, plug it into the **Location** field, give the installation repository a name, and click the **OK** button.

3. At this point, Eclipse will connect to the software site and find a list of provided installation items. As you can see in Figure A.7, the ABAP development tools plug-ins are included in this list, among other things. Here, you can select as many of the items as you like (they're compatible with one another) and then proceed through the rest of the wizard steps to complete the installation.

4. Finally, after the plug-ins are downloaded and installed, you'll be prompted to restart Eclipse. Go ahead and do so to complete the installation.

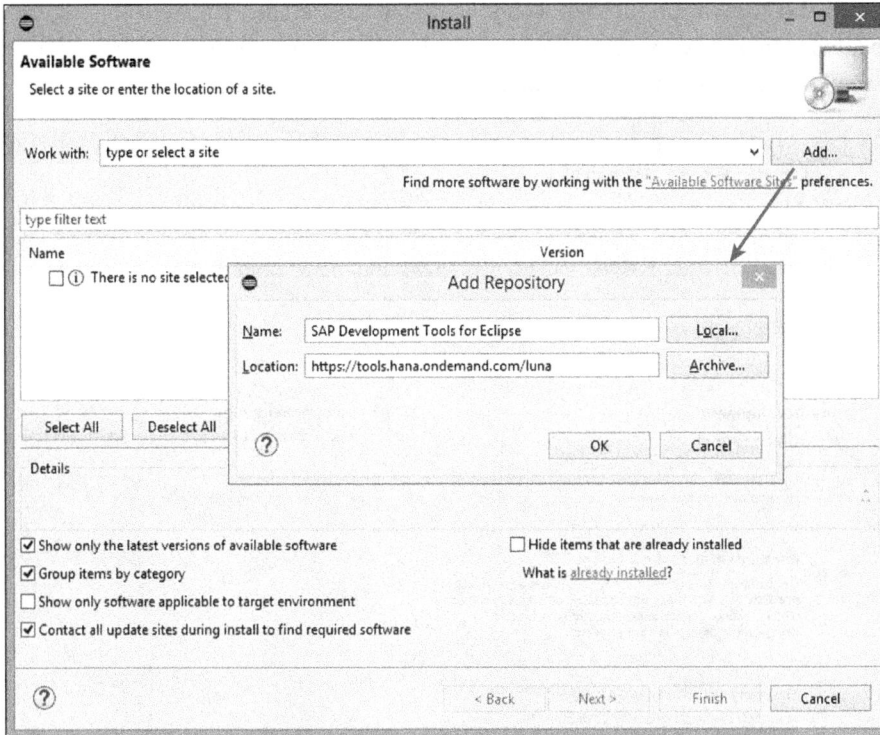

Figure A.6 Installing the ABAP Development Tools (Part 2)

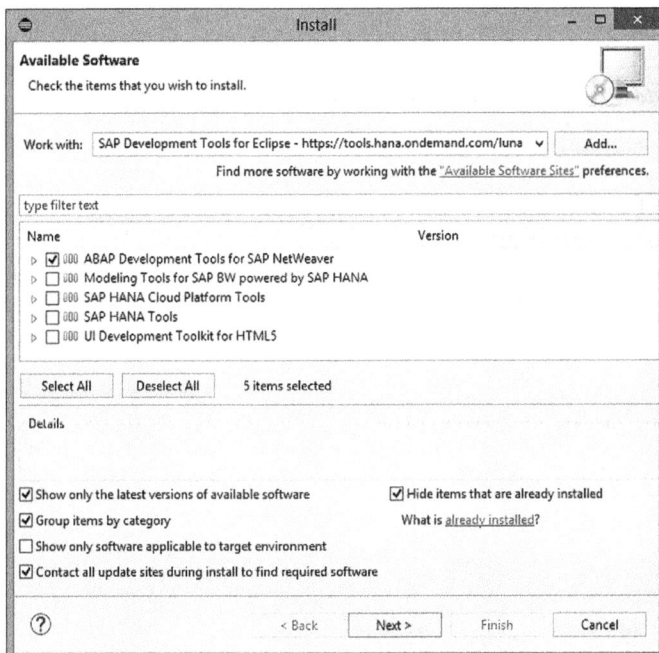

Figure A.7 Installing the ABAP Development Tools (Part 3)

Assuming the installation runs smoothly, Eclipse will restart and the welcome page will be updated to include various ABAP-related links (see Figure A.8). You can click on these links to view tutorials, find supporting documentation, and more.

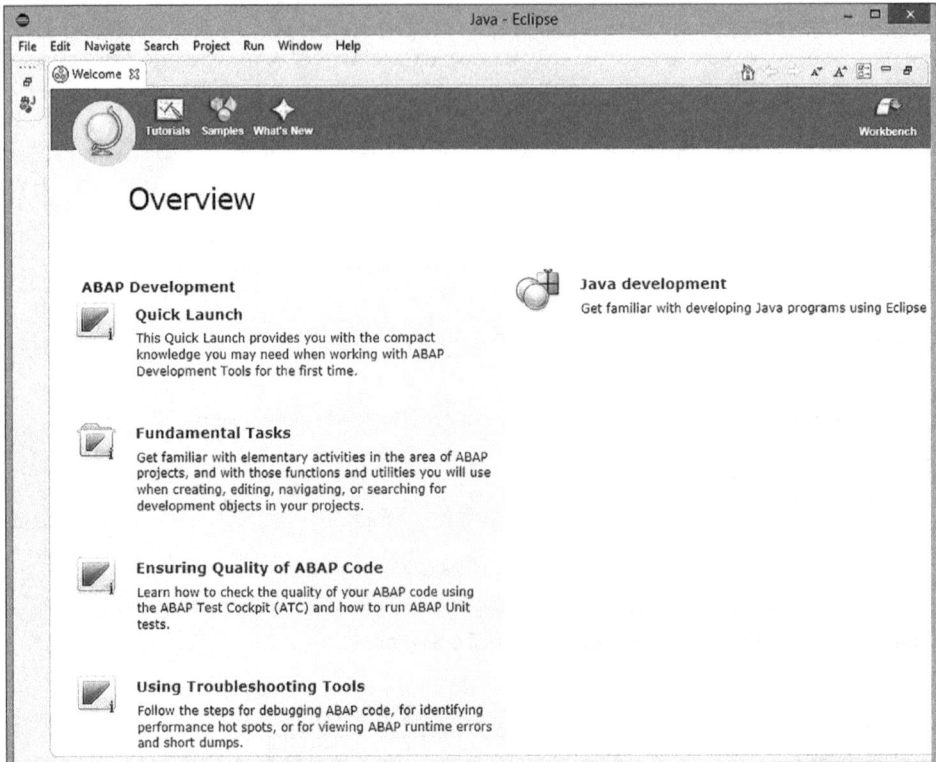

Figure A.8 Verifying the ABAP Development Tools Installation

A.4   Where to Go to Find Help

If the installation process fails for some reason, you can find links to forums and supporting documentation online at *https://tools.hana.ondemand.com/#abap*.

Appendix B
The Authors

Jeffrey Boggess is an enterprise integration specialist at Bowdark Consulting focused on building seamless integrations between SAP, Azure, and Dataverse to solve complex business challenges and provide dependable solutions for clients.

Colby Hemond is a senior technical consultant at Bowdark Consulting, where he is dedicated to bringing innovative ideas and solutions to clients and supporting them in their digital transformation.

James Wood is the founder and CEO of Bowdark Consulting, Inc., a consulting firm specializing in technology and custom development in the SAP landscape.

R. Joseph Rupert is a senior data architect at Bowdark Consulting, Inc. with expertise in migrating complex SAP landscapes to cloud platforms, enabling scalable, secure, and high-performance data ecosystems.

Index

- An everyday reference for ABAP programmers of all levels

- Learn language elements, syntax, concepts, and more

- Explore modularization, modifications, and enhancements

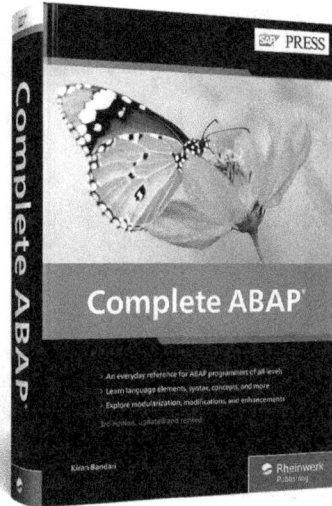

Kiran Bandari

Complete ABAP

Get everything you need to code with ABAP, all in one place! Are you a beginner looking for a refresher on the basics? You'll get an overview of SAP architecture and learn syntax. Already an experienced programmer and looking to improve your ABAP skills? Dive right into modifications and code enhancements. Understand the programming environment and build reports, interfaces, and applications with this complete reference to coding with ABAP!

912 pages, 3rd edition, pub. 09/2022
E-Book: $84.99 | **Print:** $89.95 | **Bundle:** $99.99

www.sap-press.com/5567

Rheinwerk
Publishing

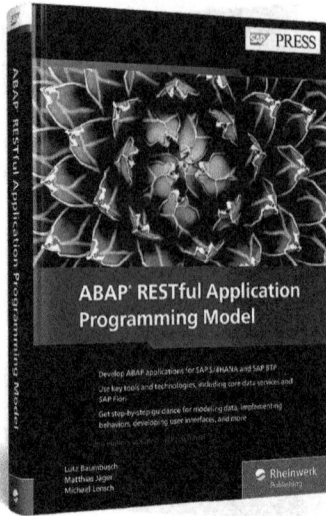

- Develop ABAP applications for SAP S/4HANA and SAP BTP
- Use key tools and technologies, including core data services and SAP Fiori
- Get step-by-step guidance for modeling data, implementing behaviors, developing user interfaces, and more

Baumbusch, Jäger, Lensch

ABAP RESTful Application Programming Model

The ABAP RESTful application programming model (RAP) is the cornerstone of modern development for SAP—get on the cutting edge with this guide! Develop applications from the ground up, from data modeling with CDS to interface generation with SAP Fiori elements. Walk through concrete use cases, including managed and unmanaged scenarios, and then adapt your applications to the SAP BTP, ABAP environment. You're on your way to working with RAP!

576 pages, 2nd edition, pub. 07/2025
E-Book: $84.99 | **Print:** $89.95 | **Bundle:** $99.99

www.sap-press.com/6161